Governing Climate Change in Southeast Asia

This volume showcases the diversity of the politics and practices of climate change governance across Southeast Asia.

Through a series of country-level case studies and regional perspectives, the authors in this volume explore the complexities and contested nature of climate governance in what can be considered as one of the most dynamic and multi-faceted regions of the world. They reflect upon the tensions between authoritarian and democratic climate change governance, the multiple roles of civil society and non-state interventions, and the conflicts between state planning and market-driven climate change governance. Shedding light on climate change mitigation and adaptation efforts in Southeast Asia, this book presents the various formal and informal institutions of climate change governance, their relevant actors, procedures, and policies. Empirical findings from a diverse set of environments are merged into a cross-country comparison that allows for elaborating on similar patterns whilst at the same time highlighting the distinct features of climate change governance in Southeast Asia.

Drawing on case studies from all Southeast Asian countries, namely Brunei Darussalam, Cambodia, Indonesia, Laos, Malaysia, Myanmar, the Philippines, Singapore, Thailand, Timor-Leste, and Viet Nam, this book will be of great interest to students, scholars, and practitioners dealing with climate change and environmental governance.

Jens Marquardt is a Research Associate at the Institute of Political Science at the Technical University of Darmstadt, Germany.

Laurence L. Delina is Assistant Professor in the Division of Environment and Sustainability at the Hong Kong University of Science and Technology.

Mattijs Smits is Assistant Professor at the Environmental Policy Group at Wageningen University, The Netherlands.

Routledge Advances in Climate Change Research

For more information about this series, please visit: www.routledge.com/
Routledge-Advances-in-Climate-Change-Research/book-series/RACCR

"Southeast Asia punches above its weight in terms of energy consumption, carbon emissions, and even geopolitical contests over natural resources. Translating the Paris Agreement into climate action in such a rapidly developing region like Southeast Asia requires context-specific knowledge about politics and institutions. This wonderful book tackles this challenge head on, offering critically important insights into how regional planners, developers and policymakers must balance their development priorities with international commitments. Must reading for anyone who truly cares about the future of our climate."

Benjamin K. Sovacool, *Professor of Energy Policy, University of Sussex, United Kingdom / Distinguished Professor of Business and Social Sciences, Aarhus University, Denmark*

"This exciting volume surveys climate governance and practice in Southeast Asia, demonstrating the importance of examining climate governance in specific contexts, with reference to the material and social histories that shape those responses. The book looks beyond economic success stories in countries such as Singapore or Malaysia, to look instead at how climate governance is deployed in relation to increasing inequalities, human rights violations, and continuous environmental degradation. The book is also accessible and written in an engaging way, and it will be an excellent reference work for researchers working in climate governance and will provide abundant empirical material for undergraduate and postgraduate courses."

Vanesa Castan Broto, *Professor of Climate Urbanism, University of Sheffield, United Kingdom*

"Climate change is a global challenge, but with distinctive local signatures. It 'happens' to people, in places, and is differentially governed locally, nationally and internationally. This important book takes the multi-level character of climate change seriously, acknowledges that governance is contested, and reveals what shape this global challenge is taking across and within the countries of Southeast Asia. All too often, such detail is lost as attention wanders to the global."

Jonathan D. Rigg, *Professor of Human Geography, University of Bristol, United Kingdom*

"*Governing Climate Change in Southeast Asia: Critical Perspectives* is an exciting and tremendously important book, as the regional experts, who have authored the chapters, have all adopted nuanced approaches to examining the challenges and complex politics associated with climate change in a highly diverse and rapidly changing region."

Ian G. Baird, *Professor of Geography, University of Wisconsin-Madison, United States of America*

"Southeast Asia has a higher concentration of nations with extreme exposure to the effects of climate-related events than any other world region. This nuanced and thought-provoking volume shows that governance responses to climate change challenges are nevertheless highly diverse and remain hotly contested, between as well as within countries."

Philip Hirsch, *Emeritus Professor of Human Geography,*
The University of Sydney, Australia

"Compiled by a team of excellent scholars, this is an incredibly timely book on the post-Paris governance issues affecting Southeast Asia. A topical read for academics and practitioners interested in this region, as the detailed empirical overviews for each ASEAN nation yields cogent insights on the tricky socio-political challenges involved in adapting to, and mitigating for, a 1.5°C world."

Winston Chow, *Associate Professor of Science, Technology and Society, Sin-*
gapore Management University, Singapore

"As millions of dollars of international aid continue to pour into Southeast Asia to support climate change mitigation and adaptation, national political will and climate governance within each country remain decisive factors that will make or break the region's efforts in countering the climate crisis. In-depth perspectives in this book can help activists and international aid agencies strategize on how to engage more effectively with different layers of institutions in each Southeast Asian country and beyond."

Sopitsuda Tongsopit, *Solar Expert and Consultant for USAID Clean Power*
Asia, 2016–2021 / Research Data Specialist for California's Office of
Sustainability, Department of General Services,
United States of America

Governing Climate Change in Southeast Asia

Critical Perspectives

Edited by
Jens Marquardt, Laurence L. Delina
and Mattijs Smits

Routledge
Taylor & Francis Group
LONDON AND NEW YORK

earthscan
from Routledge

First published 2022
by Routledge
2 Park Square, Milton Park, Abingdon, Oxon OX14 4RN

and by Routledge
605 Third Avenue, New York, NY 10158

Routledge is an imprint of the Taylor & Francis Group, an informa business

British Library Cataloguing-in-Publication Data
A catalogue record for this book is available from the British Library

Library of Congress Cataloging-in-Publication Data
Names: Marquardt, Jens, editor. | Delina, Laurence L., editor. | Smits,
Mattijs, editor.
Title: Governing climate change in Southeast Asia : critical perspectives /
edited by Jens Marquardt, Laurence L. Delina and Mattijs Smits.
Description: Abingdon, Oxon ; New York, NY : Routledge, 2022. | Includes
bibliographical references and index.
Identifiers: LCCN 2021030020 (print) | LCCN 2021030021 (ebook) |
ISBN 9780367342555 (hardback) | ISBN 9781032154725 (paperback) |
ISBN 9780429324680 (ebook)
Subjects: LCSH: Climatic changes--Government policy--Southeast Asia. |
Climate change mitigation--Government policy--Southeast Asia. | Climatic
changes--Political aspects--Southeast Asia.
Classification: LCC QC903.2.S645 G68 2022 (print) | LCC QC903.2.S645
(ebook) | DDC 363.738/74560959--dc23
LC record available at https://lccn.loc.gov/2021030020
LC ebook record available at https://lccn.loc.gov/2021030021

ISBN: 978-0-367-34255-5 (hbk)
ISBN: 978-1-032-15472-5 (pbk)
ISBN: 978-0-429-32468-0 (ebk)

DOI: 10.4324/9780429324680

Typeset in Times New Roman
by Taylor & Francis Books

Contents

Illustrations

Figures

Tables

Preface

When delegates from 195 countries adopted the Paris Agreement in 2015, they delivered a powerful signal in the fight against climate change. But what became known as a breakthrough in international climate governance only marked the beginning of an even more cumbersome, demanding, contested, and conflictual journey towards implementing meaningful climate action. Since then, governments around the world have formulated their Nationally Determined Contributions. Yet, these individual commitments are not at all sufficient to prevent the world from dangerous anthropogenic climate change and meet the ambitious 1.5°C target agreed upon in Paris. Instead, the Intergovernmental Panel on Climate Change (IPCC) warns us in 2021 that we could already reach this critical tipping point in 2030.

The growing concern over an insufficient response to the climate crisis due to rising tensions between a global climate change governance architecture and complex domestic contexts triggered the idea to write this book about climate change governance in a highly diverse and vibrant region like Southeast Asia. How do countries in the region translate their commitments into national and subnational action? What challenges and power struggles are they confronted with? Which actors, norms, and institutions matter? These questions often remain unanswered when we reduce climate change commitments to emissions reduction targets and techno-optimistic visions of the future. Such a narrow perspective fails to give justice to the utterly different and complex social, political, economic, and environmental contexts in which global aspirations to tackle the climate crisis meet a diverse set of domestic narratives and institutions. These gaps in knowledge are especially true for Southeast Asia, although the region's relevance to tackle climate change is likely to increase over the following decades due to expected growing greenhouse gas emissions. Besides, Southeast Asia's wide range of political and societal systems, from liberal democracies to authoritarian one-party systems, makes it a fascinating region to study in regards to unpacking the politics of climate change.

With this book, we aim to provide in-depth perspectives into the various climate change politics across Southeast Asia. We also hope to raise awareness for the need to contextualize and situate climate change commitments in their conflictual, power-laden domestic contexts. We are thankful to the magnificent group of authors contributing to this book for taking up this challenge. They provide unique insights into all Southeast Asian countries and hopefully spark debates about climate change governance across the region and beyond. While writing, the authors had to deal with the unprecedented impact of the Covid-19 pandemic and regional upheavals like the 2021 military coup in Myanmar. In light of these circumstances, we are *a fortiori* delighted to present this collection, and we are grateful for the commitment and patience each author has shown throughout the process. We wish you an insightful reading experience on the subject of climate change governance in such a fascinating, diverse, and ever-evolving world region.

Jens Marquardt, Laurence L. Delina, Mattijs Smits

Contributors

Alexander Cullen is a Human Geographer and lecturer at the University of Cambridge. His research examines resource conflicts in Southeast Asia and the role of nature in practices of state-making.

Laurence L. Delina is Assistant Professor in the Division of Environment and Sustainability at the Hong Kong University of Science and Technology. His research looks at accelerating just and sustainable transitions, especially of sociotechnical energy systems, in developing Southeast Asia. He is the author of *Strategies for Rapid Climate Mitigation: Wartime Mobilisation as a Model for Action?* and *Accelerating Sustainable Energy Transition(s) in Developing Countries: The Challenges of Climate Change and Sustainable Development* (both from Routledge), as well as of *Climate Actions: Transformative Mechanisms for Social Mobilisation* and *Emancipatory Climate Actions: Strategies from Histories* (both from Palgrave Macmillan).

Monica Di Gregorio is an Associate Professor in Environmental Politics and Governance at the Sustainability Research Institute at the University of Leeds. She is an environment social scientist working on the Global South with regional expertise on Indonesia and Brazil. She has undertaken research on climate change politics, environmental social movements, policy networks, and land use institutions and policies in the Global South. Her work on climate change policies focuses on land-based mitigation and synergies between climate change mitigation, adaptation, and sustainable development in the land use sector. She has a broader research interest in natural resources governance in the tropics and has published work on climate policy networks, multi-level and transnational climate governance, property rights, and collective action institutions in natural resource management, and coalition work in the environmental justice movement. She is a Research Associate at the Priestley International Centre for Climate and the Centre for Climate Change Economics and Policy and has been a Senior Associate of the Center for International Forestry Research.

Tim Frewer is an independent researcher based in Cambodia who focuses on critical development issues and agrarian change.

Natasha Hamilton-Hart is Professor in the Department of Management and International Business and Director of the New Zealand Asia Institute at the University of Auckland. Prior to joining the University of Auckland, she was Associate Professor at the National University of Singapore. Her research lies in the fields of international and comparative political economy, with a focus on Southeast Asia.

Miles Kenney-Lazar is Assistant Professor in the Department of Geography, National University of Singapore. His work examines the changing political ecologies of land and property in the Mekong region, especially how the capitalization and commodification of land produces unequal agrarian and environmental geographies with significant livelihood ramifications. He has written about the possibilities for resistance by the rural poor to the dispossession of their lands and their capacity to influence governance processes. Empirically, his research has focused on land contestation related to the expansion of Chinese, Vietnamese, and Burmese agro-industrial plantations and special economic zones in Laos and Myanmar.

Antonio G.M. La Viña is a teacher, thinker, lawyer, and human rights and climate justice advocate. In addition to his long-time advocacy of human rights and good governance, he is known for his environment and climate change expertise. He teaches Environmental Law and Policy in the Environmental Science Department of the Ateneo de Manila University and has served many times as lead climate negotiator for the Philippines. He was the lead negotiator and spokesperson of the Philippine delegation during the 2015 Paris climate negotiations. In addition to contributing to this compilation, he has authored or edited numerous journal articles, books, and anthologies, mostly on topics regarding the environment, law, and governance. He was formerly an undersecretary of the Department of Environment and Natural Resources, and is currently the director of the Energy Collaboratory of the Manila Observatory.

Jens Marquardt is a Research Associate at the Institute of Political Science at the Technical University of Darmstadt. He has previously worked on non-state climate action at the Department of Political Science at Stockholm University after conducting research on the relation between climate science and politics at Harvard University's Program on Science, Technology and Society. His research interests include environmental governance, power relations, and issue of contestation around climate change politics. Jens is the author of the book *How Power Shapes Energy Transitions in Southeast Asia* (Routledge) and is currently committed to a joint research project on how to institutionalize climate change mitigation in the Global South.

Zeeda Fatimah Mohamad is an Associate Professor at the Department of Science and Technology Studies, Faculty of Science, University of Malaya. She has a MSc degree in Environmental Management and Policy from

Lund University and a PhD in Science and Technology Policy from Sussex University. Her research interest is primarily to understand the relationship between the development of science, technology, and innovation and environmental protection, particularly within the context of sustainable development and associated challenges to late-industrializing countries. She has also applied her interdisciplinary perspective into the emerging field of Sustainability Science.

Koos Neefjes holds a master's degree in Land & Water Use from Wageningen University. He has worked in about 30 countries over nearly 35 years – the last 20 of these years being spent in Viet Nam. Koos is currently consulting on climate change policies, as well as practical climate change adaptation and greenhouse gas emissions reduction challenges. He is advising the Government of Viet Nam, aid agencies and businesses, and is investing in a dual land-use agriculture-solar PV system to draw lessons for scaling this up and out. He was a policy advisor on climate change with UNDP in Viet Nam and on sustainable development for Oxfam Great Britain.

Romeo Pacudan is an Associate Professor in Energy and Environmental Policy and Management at the Institute of Policy Studies, Universiti Brunei Darussalam. He was the Interim Chief Executive Officer and Chief Researcher of the Brunei National Energy Research Institute, a policy think tank focusing on national and regional energy and climate change issues. He has more than 20 years of work experience in the energy sector and has been engaged in several energy research studies and projects as energy economist, and policy, regulatory and planning expert by various governments, private companies, and international multilateral institutions such as the Asian Development Bank and the World Bank. He has a doctorate in Applied Economics from the Université Grenoble-Alpes (France) and a master's of engineering in Energy Technology from the Asian Institute of Technology (Thailand).

Oliver Pye teaches Southeast Asian Studies at Bonn University. He has taught several courses on 'The Politics of Climate Change in Southeast Asia' and has organized study trips to the Climate Summits in Copenhagen (2009) and Paris (2015). His research foci include political ecology, forestry, development, and social movements. He is the author of *Khor Jor Kor. Forest Politics in Thailand* (2005, White Lotus), editor of *The Palm Oil Controversy in Southeast Asia. A Transnational Perspective* (2012, ISEAS, with Jayati Bhattacharya) and *A Political Ecology of Agrofuels* (2015, Routledge, with Kristina Dietz, Bettina Engels and Achim Brunnengräber).

Jameela Joy M. Reyes is a lawyer whose practice is mostly in the fields of environmental and human rights law. She obtained her Juris Doctor degree from the University of the Philippines and her undergraduate degrees in Psychology and Political Science from the Ateneo de Manila University.

She currently serves as a legal consultant in a number of environmental and human rights organizations, and is the Senior Communicator and Legal and Policy Research Associate of the Manila Observatory.

Adam Simpson is Senior Lecturer, International Studies, in Justice & Society at the University of South Australia. He has held a six-month Visiting Research Fellowship at the Centre for Southeast Asian Studies, Kyoto University, and Visiting Scholar positions at SOAS, University of London, Queen Mary University of London, and Keele University. His research adopts a critical perspective and is focused on the politics of the environment and development in Southeast Asia, particularly Myanmar and Thailand. He has published in international journals including *Environmental Politics, Society & Natural Resources, Third World Quarterly,* and *Pacific Review.* He is the author of *Energy, Governance and Security in Thailand and Myanmar (Burma): A Critical Approach to Environmental Politics in the South* (Routledge 2014; updated paperback edition, NIAS Press 2017) and is lead editor of *Routledge Handbook of Contemporary Myanmar* (2018) and *Myanmar: Politics, Economy and Society* (Routledge 2021).

Mattijs Smits is Assistant Professor at the Environmental Policy Group of Wageningen University and Research. He researches and teaches in the fields of (renewable) energy policy and politics, environment, sustainability, (rural) development, and climate finance. During his academic and professional career, he spent extended periods living and working as researcher and consultant in Southeast Asia, notably in Laos, Thailand, and Vietnam. He holds degrees from four different universities on three continents: a BSc and MSc from the University of Utrecht and Wageningen University, and PhD degrees from The University of Sydney and Chiang Mai University. He is the author of the book *Southeast Asian Energy Transitions: Between Modernity and Sustainability* (Ashgate).

Ashley South has 20 years' experience as an independent author, researcher, and consultant. He has a PhD from the Australian National University, an MSc from SOAS (University of London), and is a Research Fellow at Chiang Mai University. His main research interests include: ethnic conflict and peace processes in Burma/Myanmar and Mindanao; forced migration (refugees and internally displaced people); politics of language and education; climate change, varieties of adaptation, and resilience. Most of Ashley's publications are available at www.AshleySouth.co.uk.

Irina Safitri Zen is an Assistant Professor at the Urban and Regional Planning Department, Kulliyyah of Architecture and Environmental Design, International Islamic University Malaysia. She is also deputy director of Sejahtera Center for Sustainability and Humanity, IIUM. Irina has received a first degree in Ecology, a master's degree in Environmental Management, and a PhD in Environmental Management and Policy from the National University of Malaysia. Her research interests focus on sustainability, environmental management, climate change, and resource management for a better understanding of human relationships with the environmental system. She has worked in various fields of research in the inter- and trans-disciplinarily realm of sustainability science for the past ten years.

Part 1

Introduction

1 Governing climate change in Southeast Asia

An introduction

Jens Marquardt, Laurence L. Delina and Mattijs Smits

Southeast Asia is arguably one of the most diverse and rapidly changing regions in the world. Ranging from Cambodia, Laos, Malaysia, Myanmar, Thailand, and Viet Nam on continental Southeast Asia (Mainland Southeast Asia) to the maritime Southeast Asian countries of Brunei Darussalam, Indonesia, the Philippines, Singapore, and Timor-Leste, the region covers a broad range of cultures, ethnicities, and religions. From constitutional monarchies over socialist one-party systems to parliamentarian democracies, many different political systems reflect the region's heterogeneity (Croissant & Lorenz 2018). Simultaneously, cooperation, interdependencies, and exchange have increased over the past decades, bringing the countries of Southeast Asia closer together. All Southeast Asian countries, except for Timor-Leste, are members of the Association of Southeast Asian Nations (ASEAN), founded in 1967. ASEAN aims to foster a regional identity and facilitates a collective debate on regional concerns such as economic development, transboundary air pollution, and climate change.

As one of the world's most vulnerable regions to climate change, Southeast Asia is confronted with increasingly frequent and more devastating extreme weather events like typhoons and droughts (Ha, Fernando & Mahmood 2016; Yusuf & Francisco 2009). While it is impossible to predict the social and economic damages caused by climate change, the Asian Development Bank (ADB) reckons that Southeast Asia will be among the world's hardest-hit regions in the world (Raitzer et al. 2015). As a consequence, climate adaptation measures and efforts to increase resilience have gained widespread public attention. But with the region's rising emissions, changing consumption patterns, and economic growth, climate change mitigation and cleaner production have also entered the forefront of debate. Heavy smoke from forest fires on islands like Borneo or water scarcity in the Mekong region create tensions and conflicts over natural resources and environmental degradation while demonstrating how the causes and effects of climate change are intertwined with social, political, and economic complexities. Over a decade ago, the ADB (2009) argued in a much-noticed report that while Southeast Asia could benefit a lot from low-carbon development, the region could experience substantial losses from the devastating effects of climate change. More than ten

DOI: 10.4324/9780429324680-2

years after the ADB report was published, we are taking stock of the region's response to climate change and the different modes by which climate change is governed.

According to the Paris Agreement under the United Nations Framework Convention on Climate Change (UNFCCC), all nations – including countries from the Global South[1] – should join forces in the global fight against the climate crisis (UNFCCC 2015). Besides, the post-Paris climate change governance regime establishes a system of 'hybrid multilateralism' that rests upon strong support and commitments by non-state actors (Hale 2016; Kuyper, Linnér & Schroeder 2018). Parties to the UNFCCC must present, update, and improve their Nationally Determined Contributions (NDCs) in which they delineate their voluntary commitments to mitigate climate change, outline their adaptation measures, and project future emissions trajectories. Yet, scholars like Jen Iris Allan (2019) argue that these plans are insufficient to achieve the emissions reductions required to prevent the world from dangerous anthropogenic climate change and "limit the temperature increase to 1.5°C above pre-industrial levels" (UNFCCC 2015), as outlined in the Paris Agreement. Simultaneously, urging countries in regions like Southeast Asia to question their emissions-intensive growth models and request more ambitious climate change mitigation measures are not free from contestation either. Given the historical emissions of consumption-intensive, high-income countries and the conflicting priorities related to climate change measures such as economic development, poverty alleviation, or industrialization, the countries covered in this volume are also confronted with various social, political, economic, and environmental challenges in governing climate change.

The international community's pressure on the Global South to outline their mitigation efforts provokes issues of justice, equity, and the overall effectiveness of a global regime that hinges on the commitments of all parties (Agarwal, Narain & Sharma 2017; Okereke & Coventry 2016; Puaschunder 2020). In this volume, we contribute to a more contextualized and nuanced understanding of how climate change is heterogeneously governed in Southeast Asia. We also acknowledge the region's diversity in our analyses. While countries like Laos or Cambodia represent the region's least developed countries with minimal per capita emissions, other states like Singapore or Brunei Darussalam have developed into emissions-intensive societies. Yet, in contexts like the UNFCCC, these countries actively represent the interests of the Global South.

The authors in this volume point at the numerous issues of contestation and conflicts arising from the manifold governance arrangements and institutional frameworks for climate action in Southeast Asia. They collectively stress the need for a more nuanced understanding of climate change governance to shed light on the world's struggle to tackle one of the most pivotal issues of our time. Doing so, they focus on the social and political contexts in which climate change policies are formulated and implemented and where climate change measures are confronted with resistance and opposition. Climate change governance, as broadly referred to in this volume, describes all modes of governing climate change by state governments and various sub- and non-state actors in society. This concept also relates to "the processes of interaction and decision-making among the actors involved in a collective problem that lead to the creation, reinforcement, or reproduction of social norms and institutions" (Hufty 2011, p.405). A better understanding of these complex governance contexts is a prerequisite for the success of the global commitment to tackle climate change.

Climate change politics and the Global South

For decades, the principle of 'common but differentiated responsibilities' guided international climate change negotiations. Countries that have historically contributed very little to climate change (mainly those from the Global South) were expected to have less ambitious climate change commitments due to their need for economic development (Brunnée & Streck 2013). At the same time, the climate crisis can be insufficiently tackled without urgent and almost immediate action to reduce or limit the growth of emissions in the Global South. The Paris Agreement – together with the United Nations' Sustainable Development Goals (SDGs) – has blurred the divide between developed and developing nations regarding who is responsible for global environmental challenges. This has become more important since, from 2010, the annual collective emissions in Global South countries have been exceeding emissions in the OECD world (Peters et al. 2012) and because most of the additional emissions in the future are also expected to come from Global South countries (IEA 2015).

Social science studies point at the adverse social and economic impacts of climate change in the Global South, which are further aggravated by widespread poverty and lack of adequate public infrastructures in these places (Araos et al. 2017; Beer 2014; Petzold et al. 2020). With many Global South countries characterized by different development levels, highest population densities, and lowest per capita income levels and their political, social, and economic instabilities, many future climate impacts, including severe weather events (Sen Roy 2018), could be magnified. The Intergovernmental Panel on Climate Change (IPCC 2014) reports that the Global South faces the most severe climate impacts, including a general reduction in potential crop yields in most tropical and sub-tropical regions, decreased water availability for populations in many water-scarce areas, and widespread increase in the risk of flooding in many human settlements. Overall, the relative percentage of damage from climate extremes will be substantially more significant in the Global South than in the Global North (Eckstein et al. 2020; Ravindranath & Sathaye 2002).

Global South regions are extremely vulnerable to already occurring and impending climate impacts, with some areas more vulnerable than others. Projected to most likely increase over the next decades are sea-level rise in low-lying islands in the Pacific, extended periods of droughts in Northern and

Sub-Saharan Africa due to decline in precipitation, and extreme weather events like heatwaves, floods, or typhoons in Southeast Asia. Besides, most Global South countries are inadequately prepared for these impacts in the near future. High population densities in areas that are also more climate-vulnerable, limited access to adaptation resources, and poor infrastructures challenge these countries' coping strategies. Climate impacts are also experienced indirectly in these countries in terms of the spread of infectious diseases in areas like the Ethiopian highlands, or failure in crop yields in parts of Asia due to untimely rainfall (Sen Roy 2018).

In this context, it is crucial to explore the various, implicit or explicit, attempts to govern climate change in the Global South. Climate change governance-related research, particularly in emerging economies like Brazil, China, and India, has already taken off, given their rising economic relevance and political power (Never 2012). Understanding how climate change policy is implemented in an extraordinary heterogeneous group of countries in the Global South, however, is still at an early stage. In-depth studies of how climate is governed in these countries remain exceptions. Some of these limited examples include the case of Mexico, where Jose Maria Valenzuela (2014) demonstrated how a divided authority at the vertical level and political fragmentation within the Mexican state had challenged effective climate change governance in that country. In their three-province comparative analysis in China, Daphne Ngar-yin Mah and Peter Hills (2014) critically examined how central-local government relations in that country may facilitate or impede climate change policy learning. Eduardo Viola and Matías Franchini (2013) discussed policies and programs established to evaluate Argentina's climate change governance performance, where most of the announced ambitious climate targets failed to be implemented. The Argentine situation looks similar to South Africa, where the national government struggled to enforce its comprehensive climate change policies due to competing interests (Masters 2013). In one of the few edited volumes about climate change governance in developing countries, David Held and colleagues (2013, p.4) suggested that "the locus of climate change policymaking seems to be shifting" towards domestic climate plans, policies, and targets away from the global regime context. Their collection discussed how "underlying interests, ideas and institutions" shape climate change policymaking and identified "uneven domestic capacity and divergent interests among different institutions" (Held, Roger & Nag 2013, p.9) as crucial barriers to successful climate change policy implementation.

Tackling climate change is intrinsically linked to questions of prosperity and development in the Global South. Both issues represent long-term challenges and "call for responses ranging from the local to the global level, and demand actions by multiple stakeholders, including governments, the private sector, and civil society" (Tanner & Horn-Phathanothai 2014, p.49). Effective climate mitigation requires countries in the Global South to contribute to global emissions reductions; however, they would also need assistance from the Global North in transitioning towards low-emissions development pathways for both instrumental and ethical reasons. Notably, mitigation action can also contribute in fundamental ways to the prosperity and development of Global South countries. This suggests that development and climate change challenges should not be tackled in isolation; instead, many development and climate outcomes can be causally linked and co-determined by development choices and pathways. In accordance, Thomas Tanner and Leo Horn-Phathanothai (2014, p.84) also call for "integrative responses […] that are sensitive to development needs and development efforts that are informed by climate-awareness." This volume brings together insights from 11 Southeast Asian countries on how the demands for prosperity, development, and climate are met – or not – in this rapidly changing world region.

Southeast Asia in an era of a climate crisis

Southeast Asia is one of the world's most vibrant and fastest-growing developing regions. The World Bank (2018) would often present Southeast Asian countries like Thailand or Indonesia as positive examples of a growth-oriented development model that should be emulated. Economic progress, however, is mainly driven by fossil fuel combustion, leading to air pollution, adverse health impacts, and massive increases in greenhouse gas (GHG) emissions. Figure 1.1 shows that GHG emissions from all 11 Southeast Asian countries grew from slightly over 415,000 tons of CO_2 equivalent (tCO_2e) in 1990 to more than 1.6 million tCO_2e in 2019.

This picture of the region's annual per capita emissions reveals the unequal distribution of emissions across these countries. The biggest per capita emitting

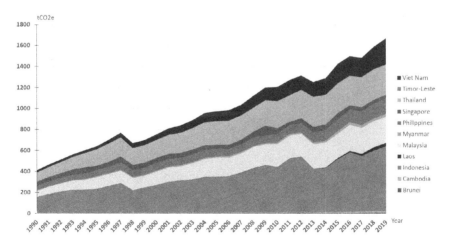

Figure 1.1 Collective GHG emissions in Southeast Asia in tons of CO_2 equivalent, excl. LULUCF (1990–2019)[2]
Source: Illustration based on data from https://ourworldindata.org

countries include Brunei Darussalam (20.99 tCO_2e), Malaysia (7.83 tCO_2e), and Singapore (6.71 tCO_2e). In comparison, Cambodia (0.97 tCO_2e), Myanmar (0.49 tCO_2e), and Timor-Leste (0.43 tCO_2e) emit significantly less than one tCO_2e per capita, reflecting the countries' different states of carbon consumption, and relatedly their economic development.

Given the region's rapid economic development, demographic growth, and steadily increasing demand for natural resources, Southeast Asia has faced severe environmental problems for decades (Hirsch 2016; Rigg 2003). Climate change, the ADB predicts, could cost the region some 11 percent of its GDP by 2100 (Raitzer et al. 2015). Heavily populated low-lying areas, where most of the region's urban centers are located, make millions of Southeast Asians vulnerable to climate impacts such as heavy flooding. Temperature rise also threatens the region's agriculture sector (Prakash 2018). The Global Climate Risk Index 2020 lists four Southeast Asian nations (i.e., Myanmar, the Philippines, Viet Nam, Thailand) among the top 10 countries most affected by climate-related extremes between 1999 and 2018 (Eckstein et al. 2020). Not surprisingly, many Southeast Asian countries emphasize the need for climate adaptation measures and resilience as national priorities instead of focusing on mitigation through emissions reduction. The Philippine government, for instance, has, for many years, expressed healthy skepticism towards mitigation efforts and, instead, been focusing on adaptation measures like early warning systems or protection of typhoon-prone areas.

All 11 Southeast Asian countries have ratified the Paris Agreement and submitted their NDCs to the UNFCCC registry. These submissions reflect the region's diversity in terms of GHG emissions profiles and development pathways. Reviewing these NDCs in 2021 reveals that Brunei, Laos, Myanmar, and Timor-Leste have refused to set quantitative emissions reduction targets; nonetheless, they outline actions related to their national forest sector or their

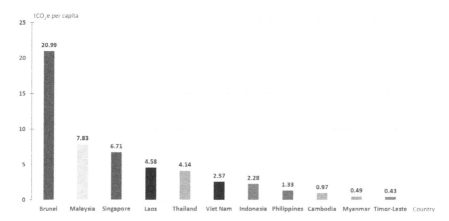

Figure 1.2 Annual per capita GHG emissions in tCO_2e, excl. LULUCF (2019)
Source: Illustration based on data from https://ourworldindata.org

national renewable energy development strategies. Cambodia makes its climate mitigation efforts conditional on the availability of support from the international community. Indonesia has pledged to reduce 29 percent of its emissions compared to the business-as-usual scenario by 2030, which will be increased to 41 percent with international assistance. Similarly, Viet Nam aims to decrease their emissions by 9 percent by 2030 and up to 27 percent with international support, according to the country's updated NDC. Thailand intends to reduce its emissions by 20 percent from the projected business-as-usual level by 2030, which could increase up to 25 percent depending on technology transfer, financial resources, and capacity building. The Philippines pledged to cut its carbon emissions by 70 percent until 2030 depending on international assistance. Malaysia describes a green growth strategy and aims to reduce its GDP's emissions intensity by 45 percent by 2030 relative to 2005. Singapore aims to reduce its emissions intensity by 36 percent from 2005 levels, stabilize its emissions, and reach its peak around 2030.

According to the IPCC (2014), the Global South will face the most significant burden on climate impacts, particularly extreme weather events. In Cambodia, Myanmar, and Thailand, climate adaptation measures are a national priority with substantial effects on human and economic development. Based on their specific vulnerability, countries propose a variety of adaptation priorities. For example, Cambodia intends to improve the adaptive capacity of communities, while Brunei and Myanmar highlight the need for reforestation efforts. Laos describes flooding as a significant economic threat, which requires a more resilient economy. Myanmar describes resilience in the agriculture sector by developing early warning systems and forest preservation measures as the country's priority adaptation sectors. Timor-Leste prioritizes what it calls adaptation areas, which include food security, water resources, and health.

The contributions of this volume

Tackling climate change requires collective global efforts towards emissions reduction and decarbonization. Supporting developing countries in mitigating climate-related emissions or adapting to the severe effects of climate change today can prevent future lock-ins on a fossil fuel-based economic system and help establish more climate-resilient societies. Although promoting sustainability transitions and climate change policies have become dominant paradigms for both developed and developing countries, these terms, concepts, and policies remain fuzzy, highly contested, and mostly market-oriented. Taking a closer look at how Southeast Asian countries respond to climate change from within, this volume surfaces new ideas, alternative concepts, and multiple perspectives on governing climate change.

Although national and local climate efforts may take the Paris Agreement as a collective point of reference, existing governance arrangements in the region are diverse and complex. Across Southeast Asia, climate change policies are formally in place, albeit they are fragmented in terms of arrangements, locations, and actors. Both internal and external actors attempt to strengthen these efforts by centralizing climate change governance, but evidence of their effects on climate change mitigation and adaptation is still limited. While national governments struggle to provide sufficient regulatory frameworks for climate action across their jurisdictions, non-state actors and networks have also been fostering diverse activities such as initiatives for climate-resilient development (Gallagher 2018), urban climate planning (Daniere & Garschagen 2019), or ASEAN-wide coordination efforts (Elliott 2012). Grounded on the need to understand these multi-layered socio-political environments where climate action flourishes or fails and to shed light on these multiple environments, this volume contributes to the literature by providing a more nuanced understanding of climate change governance in the Global South, particularly in Southeast Asia.

This volume provides an overview of the politics of climate change in contemporary Southeast Asia. Acknowledging that climate change politics differ significantly across all 11 Southeast Asian countries, the chapters in this volume offer a wide-ranging yet easy-to-read overview of the contents, structures, and processes of climate change governance across the region. The authors either provide situated and contextualized accounts of climate change governance or share the challenges and opportunities from a practitioner's perspective. This fluidity in contributions makes these chapters particularly relevant for academics, decision-makers, and policy-oriented researchers alike. In demonstrating the complexity and contested nature of climate change governance (e.g., concerning the role of formal and informal social, cultural, and political factors), the chapters describe various notions of globalized versus context-specific, apolitical versus political, and technocratic versus bureaucratic governance frameworks fostered by politicians, development practitioners, or environmental governance scholars. The chapters point at the tensions between a globalized vision of climate change governance (e.g., as articulated through the Paris Agreement) and more localized and context-specific political cultures, norms, and visions in the respective counties. While all chapters elaborate on different themes, tensions, and potentials for interventions, they also allow – to a certain degree – for a comparative reflection of common themes and striking differences, which we provide in the concluding chapter for this volume.

Finally, we collectively contribute to the variety of conceptual debates related to development and climate change governance. Moving beyond universalized macro-level development paradigms, such as catch-up development or sustainability as frameworks for analysis (Ashoff & Klingebiel 2014), this volume focuses on national contexts and domestic circumstances such as institutional frameworks, historically grown central-local relations, and the influence of big industries on climate change policymaking. We thus acknowledge that the issues surrounding climate action are intrinsically linked to questions of development and are dependent upon the unique socio-political contexts of these countries (Croissant & Lorenz 2018; Schreurs 2010). The chapters also

engage beyond a global climate change governance discourse primarily dominated by universal environmental governance concepts based on a Western
understanding of liberal democracies (Hermwille 2018). They do so by highlighting and surfacing some of the understudied, often marginalized, discourses
and concepts that are indigenously Southeast Asian (such as a post-colonial critique) and calling for more contextualization in global environmental governance
(e.g., Biermann et al. 2016). To prevent any conceptual lock-in and avoid the
constraints of working around a particular analytical framework, some of the
chapters follow a rather explorative and inductive approach towards their analyses. Grounded in empirical work, policy analysis, and conceptual reflections,
these chapters speak to various theoretical debates related to climate change
policy implementation and the different types of knowledge production, participation, and coordination.

This volume speaks to practitioners, students, and researchers alike interested in climate change governance in the Global South, climate change policymaking, and complex environmental governance frameworks. We aim to
spark debates around how climate change politics can be triggered and sustained where environmental degradation and economic pressures are at their
highest. Practitioners promoting climate mitigation and adaptation efforts in
the Global South will benefit from the chapters' various empirical insights
and analytical approaches that capture climate change measures within complex governance arrangements. This variety ranges from institutional perspectives and historical timelines to broader frameworks like political ecology.

The chapters ahead

This volume provides a concise overview of climate change governance and
practices across Southeast Asia, highlighting the conflicts and tensions between
a globalized climate change governance regime and a set of highly country-
specific, situated, and context-dependent governance practices across a diverse
and rapidly changing region. This series of country-level case studies explores
the complexity and contested nature of climate change governance in one of
the world's most rapidly developing regions (Litsareva 2017). While countries
like Singapore, Malaysia, or Thailand, for instance, are often described as
economic success stories, the region is also, as a whole, characterized by societal challenges such as inequality, human rights violations, and environmental
degradation (Greenough & Tsing 2003). Following centuries of colonial rule
(which all countries in Southeast Asia, except for Thailand, have been subjected to), the countries cooperate with each other through regional institutions,
such as the ASEAN, which provide them platforms for intergovernmental
cooperation and regional integration.

This volume also sheds light on the region's various climate change mitigation and adaptation efforts. Instead of giving a comprehensive overview of
climate action in the entire region, the authors provide their individual and
context-specific snapshots of climate change governance in these countries.

They investigate the formal and informal institutions of climate change governance, as well as their relevant actors, procedures, practices, and policies.

The findings from this diverse set of country-specific contributions and a regional overview are summarized in a concluding chapter in which we elaborate on the similarities, differences, and patterns across the various country cases presented here. We point at some of the distinct features of climate change governance in Southeast Asia, focusing on the role of the state, non-state actors, and the market. This chapter also critically reflects upon some traditional environmental governance concepts embedded or revised, or neglected in and by Southeast Asian countries. Bringing together the insights from 11 Southeast Asian countries, this volume finally sheds light on the diversity of national and regional responses to accelerating climate change in one of the world's most dynamic yet climate-vulnerable regions. We hope that this collection provides new insights into this fascinating world region and contributes to broader debates about the relationships between environmental protection, social justice, and economic development in the Global South in the face of the climate crisis.

In varying degrees, the authors present the emergence of climate-related institutions, discuss the role of state- and non-state actors at different levels, shed light on historically grounded continuities, and provide some rich empirical case studies. Based on these country-specific snapshots, the authors draw their conclusions and formulate recommendations for moving forward in climate change politics. The chapters are diverse, focus on multiple levels and actors, and cover a broad range of climate-related topics. At the same time, a strong emphasis on the practices of climate change politics and a thorough understanding of the country-specific context holds them together.

- In chapter 2, **Romeo Pacudan** discusses climate change governance in *Brunei Darussalam*. This Southeast Asia's smallest country in terms of land area is equipped with very high conventional energy resources relative to its population size. Pacudan sheds light on the country's challenge to balance economic diversification with climate change mitigation. He concludes that coordination at the ministerial level is useful in developing a comprehensive climate change policy. Still, practical implementation, Pacudan suggests, depends on the ability of the sectoral governing system to innovate, introduce, and enforce climate change policies.
- In chapter 3, **Tim Frewer** sheds light on climate change governance in *Cambodia* by examining the effects of climate programming in terms of institutional effects, geographic flows of climate funds, and effects on governing relations. He does so by investigating a loose web of relations between donors, government departments, non-governmental organizations, and program recipients.
- In chapter 4, **Monica Di Gregorio** tackles the challenge of climate change policy integration in *Indonesia's* land-use sector. She concludes that sectoral ministries' willingness to recognize the importance of climate change

objectives and synergies between mitigation and adaptation will be crucial to moving toward a more effective climate change policy integration.

- In chapter 5, **Miles Kenney-Lazar** examines novel forms of climate action in *Laos*, where the country's government, together with international donors, has pursued a range of reforestation and sustainable energy production policies and projects for climate mitigation. Moving beyond rules-based approaches to environmental governance, this chapter analyses the power-laden interactions between key actors working on climate change in Laos to advance a relational understanding of climate change politics and governance in Southeast Asia.

- In chapter 6, **Irina Safitri Zen** and **Zeeda Fatimah Mohamad** explore how the ongoing socio-political turmoil in *Malaysia* has resulted in an uncertain environment to consistently implement climate change policies and an overarching governance strategy. They discuss how non-partisan efforts and cooperation by non-state actors in the private sector and academia have played an essential bridging role in sustaining climate change policy development in Malaysia's complex climate change governance arrangements.

- In chapter 7, **Adam Simpson** and **Ashley South** shed light on climate change governance in *Myanmar*. They argue that governing climate change is afflicted by the same political constraints that characterize all areas of policymaking. Their analysis focuses on the cultural and institutional legacies of half a century of authoritarian governance and maldevelopment, which are constantly reinforced by the military's ongoing and constitutionally enshrined dominance within critical areas of the country's governance structures.

- In chapter 8, **Antonio Gabriel La Viña** and **Jameela Joy M. Reyes** present the *Philippines*' complex climate change governance arrangement. Although the country has established a robust institutional framework over the years to tackle climate change, the authors identify significant challenges and delays when it comes to translating commitments into action within the Philippines' complex multi-level governance arrangements.

- In chapter 9, **Natasha Hamilton-Hart** provides insights into climate change governance in *Singapore*. She argues that Singapore's approach to climate change policies has minimized its international negotiations' responsibilities. Its mitigation efforts prioritize efficiency gains and measures consistent with a pro-economic growth agenda while deflecting attention from areas where there are trade-offs between mitigation and growth.

- In chapter 10, **Adam Simpson** and **Mattijs Smits** draw attention to the tension between top-down climate change governance and bottom-up activism in *Thailand*. Adopting a political ecology approach, this chapter analyses the role of environmental activists in influencing Thailand's climate and energy governance. The authors also assess the risks and likely outcomes of developments in the future.

- In chapter 11, **Alexander Cullen** examines climate change governance in *Timor-Leste* as it is negotiated across competing spaces of epistemological and ontological differences from the national to the local level. Utilizing a case study of a southern coastal village, he shows how failures to consider complex customary epistemologies and residual socio-relations to the land can produce conflict.
- In chapter 12, **Koos Neefjes** investigates how the history of environmental disasters, as well as low emissions, has shaped *Viet Nam's* climate change governance. He shows how climate change policies and politics are about contested environments, including access to land, clean water, and air, and competition for external investment capital.
- In chapter 13, **Oliver Pye** takes a *regional perspective* on climate change politics in Southeast Asia. He argues that although the region can leap-frog developments in the Global North on carbon-neutral production, current power structures, and vested interests, particularly those related to fossil fuels, prevent a sustainable transformation. While the climate justice movement presents alternative development scenarios, Pye highlights the challenges to shift the power imbalance.

We conclude this volume in chapter 14 by identifying common themes and patterns across the preceding chapters and providing some theoretical reflections based on these diverse contributions. We do this by placing the region's climate change governance systems vis-à-vis various concepts, including, but not limited to, multi-level environmental governance (Wälti 2010), polycentric governance (Ostrom 2009), innovative climate change governance approaches (Turnheim, Kivimaa & Berkhout 2018), and critical interventions such as political ecology (Robbins 2004).

Notes

1   We are using the term *Global South* here to categorize a group of countries which have been historically disadvantaged and underprivileged through various modes of power and force. Yet, the 'Global South' or 'developing countries' remain highly contested concepts. They neither refer to a homogenous group of countries, nor should they be understood as neutral terms. Andrea Hollington et al. (2015) provide an overview on a number of definitions and perceptions of the Global South. Although countries of the Global South cannot be easily defined, they are understood here as countries located predominately in the geographical South, but more importantly, they share a history shaped by "colonialism, a struggle to address widespread poverty, and a predominant policy concern with the process of 'development' over the past few decades" (Dubash & Morgan 2012, p.263). With regard to climate change, there is not a clear-cut division between the Global North and the Global South, but rather a diverse spectrum of countries in terms of emissions, commitments, and vulnerability.

2   These data are shown here to give a comparable overview on the region. Yet, country-specific data used in the following chapters might be more specific and up-to date, yet inconsistent with the data provided here. For example, countries do not publish their inventories every year and report their data to the UNFCCC which

leads to interpolations in the data shown here. Emissions from land use, land use change and forestry (LULUCF) are not included. These are particularly high in countries like Indonesia with high rates of deforestation.

References

ADB 2009, *The Economics of Climate Change in Southeast Asia: A Regional Review*, Manila.

Agarwal, A., Narain, S. & Sharma, A. 2017, 'The Global Commons and Environmental Justice—Climate Change', in J. Byrne, L. Glover & C. Martinez (eds), *Environmental Justice: Discourses in International Political Economy*, Routledge, London, pp. 171–201.

Allan, J.I. 2019, 'Dangerous Incrementalism of the Paris Agreement', *Global Environmental Politics*, vol. 19, no. 1, pp. 4–11.

Araos, M., Ford, J., Berrang-Ford, L., Biesbroek, R. & Moser, S. 2017, 'Climate Change Adaptation Planning for Global South Megacities: The case of Dhaka', *Journal of Environmental Policy and Planning*, vol. 19, no. 6, pp. 682–696.

Ashoff, G. & Klingebiel, S. 2014, 'Transformation eines Politikfeldes: Entwicklungspolitik in der Systemkrise und vor den Herausforderungen einer komplexeren Systemumwelt', in C. Jakobeit, F. Müller, E. Sondermann, I. Wehr, & A. Ziai (eds), *Entwicklungstheorien: Weltgesellschaftliche Transformationen, entwicklungspolitische Herausforderungen, theoretische Innovationen*, Nomos, Baden-Baden, pp. 166–199.

Beer, C.T. 2014, 'Climate Justice, the Global South, and Policy Preferences of Kenyan Environmental NGOs', *The Global South*, vol. 8, no. 2, pp. 84–100.

Biermann, F., Bai, X., Bondre, N., Broadgate, W., Arthur Chen, C.T., Dube, O.P., Erisman, J.W., Glaser, M., van der Hel, S., Lemos, M.C., Seitzinger, S. & Seto, K.C. 2016, 'Down to Earth: Contextualizing the Anthropocene', *Global Environmental Change*, vol. 39, pp. 341–350.

Brunnée, J. & Streck, C. 2013, 'The UNFCCC as a Negotiation Forum: Towards Common but More Differentiated Responsibilities', *Climate Policy*, vol. 13, no. 5, pp. 589–607.

Croissant, A. & Lorenz, P. 2018, 'Government and Political Regimes in Southeast Asia: An Introduction', in A. Croissant & P. Lorenz (eds), *Comparative Politics of Southeast Asia*, Springer, Cham, pp. 1–14.

Daniere, A.G. & Garschagen, M. (eds) 2019, *Urban Climate Resilience in Southeast Asia*, Springer, Wiesbaden.

Dubash, N.K. & Morgan, B. 2012, 'Understanding the Rise of the Regulatory State of the South', *Regulation and Governance*, vol. 6, no. 3, pp. 261–281.

Eckstein, D., Künzel, V., Schäfer, L. & Winges, M. 2020, *Global Climate Risk Index 2020: Who Suffers Most from Extreme Weather Events?*, Germanwatch, Bonn.

Elliott, L. 2012, 'ASEAN and Environmental Governance: Strategies of Regionalism in Southeast Asia', *Global Environmental Politics*, vol. 12, no. 3, pp. 38–57.

Gallagher, E. 2018, *Building Climate Resilience in Southeast Asia: A Framework for Private Sector Action*, BSR.

Greenough, P. & Tsing, A. L. 2003, *Nature in the Global South: Environmental Projects in South and Southeast Asia*, Duke University Press, Durham & London.

Ha, H., Fernando, R. L. S. & Mahmood, A. (eds) 2016, *Strategic Disaster Risk Management in Asia*, Springer India, New Delhi.

Hale, T. 2016, '"All Hands on Deck": The Paris Agreement and Nonstate Climate Action', *Global Environmental Politics*, vol. 16, no. 3, pp. 12–22.

Held, D., Roger, C. & Nag, E.-M. (eds) 2013, *Climate Governance in the Developing World*, Polity Press, Cambridge.

Hermwille, L. 2018, 'Making Initiatives Resonate: How Can Non-State Initiatives Advance National Contributions Under the UNFCCC?', *International Environmental Agreements: Politics, Law and Economics*, vol. 18, no. 3, pp. 447–466.

Hirsch, P. (ed.) 2016, *Routledge Handbook of the Environment in Southeast Asia*, Routledge, London.

Hollington, A., Salverda, T., Schwarz, T. & Tappe, O. 2015, *Concepts of the Global South, Voices from around the world*, Global South Studies Center Cologne, Cologne.

Hufty, M. 2011, 'Investigating Policy Processes: The Governance Analytical Framework (GAF)', in U. Wiesmann & H. Hurni (eds), *Research for Sustainable Development: Foundations, Experiences, and Perspectives*, pp. 403–424.

IEA 2015, *World Energy Outlook 2015*, International Energy Agency, Washington D. C.

IPCC 2014, *Climate Change 2014: Synthesis Report*, Geneva.

Kuyper, J. W., Linnér, B. O. & Schroeder, H. 2018, 'Non-State Actors in Hybrid Global Climate Governance: Justice, Legitimacy, and Effectiveness in a Post-Paris Era', *Wiley Interdisciplinary Reviews: Climate Change*, vol. 9, no. 1, pp. 1–18.

Litsareva, E. 2017, 'Success Factors of Asia-Pacific Fast-Developing Regions' Technological Innovation Development and Economic Growth', *International Journal of Innovation Studies*, vol. 1, no. 1, pp. 72–88.

Masters, L. 2013, 'Reaching the Crossroads: The Development of Climate Governance in South Africa', in D. Held, C. Roger & E.-M. Nag (eds), *Climate Governance in the Developing World*, Polity Press, Cambridge, pp. 258–276.

Never, B. 2012, *Macht in der globalen Klima-Governance*, German Institute of Global and Area Studies, Hamburg.

Ngar-yin Mah, D. & Hills, P. R. 2014, 'Policy Learning and Central-Local Relations: A Case Study of the Pricing Policies for Wind Energy in China (from 1994 to 2009)', *Environmental Policy and Governance*, vol. 24, no. 3, pp. 216–232.

Okereke, C. & Coventry, P. 2016, 'Climate Justice and the International Regime: Before, During, and After Paris', *Wiley Interdisciplinary Reviews: Climate Change*, vol. 7, no. 6, pp. 834–851.

Ostrom, E. 2009, 'Beyond Markets and States: Polycentric Governance of Complex Economic Systems', *American Economic Review*, vol. 100, no. 3, pp. 641–672.

Peters, G., Marland, G., Le Quéré, C., Boden, T., Canadell, J.G. & Raupach, M. R. 2012, 'Rapid Growth in CO2 Emissions After the 2008–2009 Global Financial Crisis', *Nature Climate Change*, vol. 2, no. 1, pp. 2–4.

Petzold, J., Andrews, N., Ford, J. D., Hedemann, C. & Postigo, J. C. 2020, 'Indigenous Knowledge on Climate Change Adaptation: A Global Evidence Map of Academic Literature', *Environmental Research Letters*, vol. 15, no. 11.

Prakash, A. 2018, 'Boiling Point', *Finance and Development*, viewed 30 September 2020 at, <https://www.imf.org/external/pubs/ft/fandd/2018/09/southeast-asia-clima te-change-and-greenhouse-gas-emissions-prakash.htm>.

Puaschunder, J. 2020, *Governance & Climate Justice: Global South & Developing Nations*, Palgrave Macmillan, Cham.

Raitzer, D.A., Tavoni, M., Orecchia, C., Srinivasan, A., Bosello, F., Marangoni, G. & Samson, J.N.G. 2015, *Southeast Asia and the Economics of Global Climate Stabilization*, Asian Development Bank, Manila.

Ravindranath, N.H. & Sathaye, J.A. 2002, *Climate Change and Developing Countries*, Springer, London.

Rigg, J. 2003, *The Human Landscape of Modernization and Development*, Routledge, London.

Robbins, P. 2004, *Political Ecology. A Critical Introduction*, Blackwell Publishing, Oxford.

Schreurs, M.A. 2010, 'Multi-Level Governance the ASEAN Way', in H. Enderlein, S. Wälti & M. Zürn (eds), *Handbook on Multi-level Governance*, Edward Elgar, Cheltenham, pp. 308–320.

Sen Roy, S. 2018, 'Climate Change in the Global South: Trends and Spatial Patterns', in S. Sen Roy (ed.), *Linking Gender to Climate Change Impacts in the Global South*, Springer, Cham, pp. 1–25.

Tanner, T. & Horn-Phathanothai, L. 2014, *Climate Change and Development*, Routledge, London.

Turnheim, B., Kivimaa, P. & Berkhout, F. (eds) 2018, *Innovating Climate Governance: Moving Beyond Experiments*, Cambridge University Press, Cambridge.

UNFCCC 2015, *Paris Agreement, 21st Conference of the Parties*, United Nations Framework Convention on Climate Change, Paris.

Valenzuela, J.M. 2014, 'Climate Change Agenda at Subnational Level in Mexico: Policy Coordination or Policy Competition?', *Environmental Policy and Governance*, vol. 24, no. 3, pp. 188–203.

Viola, E. & Franchini, M. 2013, 'Discounting the Future: The Politics of Climate Change in Argentina', in D. Held, C. Roger & E.-M. Nag (eds), *Climate Governance in the Developing World*, Polity Press, Cambridge, pp. 113–133.

Wälti, S. 2010, 'Multi-Level Environmental Governance', in H. Enderlein, S. Wälti & M. Zürn (eds), *Handbook on Multi-Level Governance*, Edward Elgar, Cheltenham, pp. 411–422.

World Bank 2018, *Riding the Wave: An East Asian Miracle for the 21st Century*, World Bank, Washington D.C.

Yusuf, A. & Francisco, H. 2009, 'Climate Change Vulnerability Mapping for Southeast Asia', *East*, December, pp. 1–19.

Part 2
Country perspectives

2 Whole-of-nation approach to climate change governance in Brunei Darussalam

Romeo Pacudan

Introduction

Brunei Darussalam's climate change policy launched in July 2020 states that the government adopts a whole-of-nation approach in addressing adverse changing climate patterns as the country proceeds to move towards a low-carbon and a climate-resilient economy (BCCS 2020, p.6). This all-encompassing climate change governance approach aligns with the government's current aim to involve all stakeholders, public and private, and the whole society to work together to achieve the economic development envisaged in the Wawasan 2035 (National Vision 2035) (Prime Minister's Office 2007). The government recognizes that every stakeholder (public, private, non-government organizations) has a specific role to play. Their active engagement is crucial in achieving the long-term government vision.

The whole-of-nation approach appears to be a natural progression from the whole-of-government approach. The previous cabinet ministers used the concept of a whole-of-government approach in governance while the current ones align themselves to the whole-of-nation concept. For example, the Digital Government Strategy 2015–2020 stipulated the whole-of-government approach in innovation and service provision (Prime Minister's Office 2015). The same progression can also be observed in Singapore as the public service goes beyond the whole-of-government approach to the whole-of-nation approach (Yahya 2018; Low 2016, p.14). The whole-of-government approach (also known as inter-sectoral collaboration) refers to government agencies "working across boundaries to achieve shared goals and an integrated government response. In contrast, the whole-of-nation approach widens this aperture to include all organizations to reach common goals" (Doyle 2019, p.106). The "whole-of-government focuses on internal stakeholder engagement while the whole-of-nation looks at external non-government stakeholder engagement" (World Health Organization 2015, p.98).

The Brunei National Climate Change Policy is a result of the whole-of-nation approach to policy formulation. The policy document contains the objectives, strategies, tools, and performance indicators for the monitoring framework. Still, the record does not present how the whole-of-nation

DOI: 10.4324/9780429324680-4

approach was carried out to achieve these policy outcomes. The critical questions raised in this chapter are the following: i) how the cooperation among stakeholders was established during climate change policy formulation, and ii) as the country moves into policy implementation, what are the emerging challenges and their implications for climate change governance?

This chapter is organized as follows: presentation of the government's rationale for climate change intervention, discussion of the national challenges in the transition to a low carbon and climate-resilient economy, presentation of the whole-of-nation approach in policy formulation, presentation of the policy strategies, and discussion of potential challenges and the governance implications in policy implementation.

The rationale for climate change intervention

Brunei Darussalam (Brunei) is situated on the northwest coast of the island of Borneo, with flat coastal plains and an equatorial climate. Brunei is vulnerable to extreme weather events and rising sea levels. Since the 1970s, Brunei's annual mean temperature has been observed to increase by 0.25°C per decade, annual rainfall intensity has increased by 100mm per decade, and around 40 percent of wildlife biodiversity has been lost due to forest degradation (BCCS 2020, p.6, 7). Coral bleaching has been observed in some shoals near Brunei-Muara District, and climate changes have also been expected to lengthen the transmission seasons of vector-borne diseases such as dengue, malaria, and Zika.

In response to climate change threats, Brunei commits to international treaties and cooperation arrangements that address global climate change issues. The country acceded to the United Nations Framework Convention on Climate Change in 2007 and the Kyoto Protocol in 2009. Brunei Darussalam ratified the Doha Amendment to the Kyoto Protocol in 2014 and the Paris Agreement in 2016. Brunei's climate change policy has been formulated to be the basis for developing a response framework for the requirement imposed on the Parties to the Paris Agreement. This international agreement aims to strengthen the global response to climate change threats and limit the global temperature rise to below 2°C above pre-industrial levels. Signatories to the Paris Agreement are required to submit a Nationally Determined Contribution which specifies each country's efforts to reduce greenhouse gas emissions and to adapt to impacts of climate change (UNFCCC 2015). A global stocktake will be undertaken every five years to monitor the progress in achieving the collective target.

Challenges in the transition to low carbon economy

Addressing climate change in Brunei requires a prudent approach to balance the interaction between critical economic and environmental parameters. The country faces various financial and technical challenges in achieving a balanced mix of measures that satisfies multiple developmental concerns. The country's climate policy and strategies were formulated against the backdrop of long-standing national circumstances and economic policies that encourage higher energy consumption and consequently higher carbon emissions. The government's main challenge is to design policy measures that sustain these developmental trends but decouple economic development with environmental degradation and emissions. In the following list, we briefly outline priorities for five policy areas.

Oil and gas-based economy: With high conventional energy resources relative to its population size, the oil and gas sector plays a dominant role in Brunei's economy. The oil and gas industry contributed around 57 percent of the gross domestic product in 2019 (Department of Statistics 2020, p.8). Oil and gas industries are also envisaged to remain critical pillars of Brunei's economy. Oil production is projected to increase to 650,000 barrels of oil equivalent per day in 2035 from the baseline of 408,000 barrels of oil equivalent per day in 2010, and the energy sector is expected to increase its contribution to the gross domestic product from B$10 billion in 2010 to B$42 billion in 2035 (EDPMO 2014, p.12). Oil and gas production are dominant sources of fugitive greenhouse gas emissions. Though the country's fugitive emissions declined by 65 percent during the period 2010 to 2018 due to rejuvenation projects in both onshore and offshore facilities, its contribution in 2018 remained high at around 18.1 percent of the total emissions (BCCS 2020, p.22). Besides, greenhouse gas emissions from the production, processing, and transport of oil and natural gas are significant sources of greenhouse gas emissions in the country. With a projected increase in production, both fugitive emissions and emissions associated with the production, processing, and transport of oil and natural gas will also increase.

Energy-intensive industries: The plummeting international oil and gas prices and declining energy reserves indicate an urgent need to diversify the country's economic base. Aware of the risks related to high dependency on energy exports and a growing population with a high unemployment rate, the government promotes economic diversification in its economic development plan. The Wawasan 2035 (National Vision 2035) aims to shape Brunei into a dynamic and sustainable economy, transform it into a well-educated society with a place in the top-ten list of countries with the world's highest living standards, as well as to have the highest income per capita worldwide (Prime Minister's Office 2007). Among the key pillars to achieve these goals that aim to diversify the economy is the promotion of foreign direct investment and private sector investment in the downstream oil and non-oil industries through joint venture arrangements with large international companies. With the abundance of hydrocarbon resources and low population densities, the emerging trend under the economic diversification strategy is increased investments in energy-intensive downstream industries. Essential investments in energy-intensive manufacturing industries include methanol production, fertilizer manufacturing, and refinery and aromatics production facilities (Oxford Business Group 2016). These industries are associated with high carbon and air pollutant emissions per unit of gross domestic product.

Subsidized energy prices: Brunei pursues an energy pricing policy that sets energy prices, in the case of petroleum products, below the international market reference prices or, in the case of electricity services, below the long-run marginal cost (Pacudan & Hamdan 2019, p.333). Low energy prices in the country have resulted in a rapid increase in energy consumption beyond what can be explained by factors such as rising income levels and population growth. Energy consumption per capita in Brunei is one of the highest in the Asia-Pacific region. Besides, heavily subsidized energy prices sway investment decisions towards energy-intensive projects and encourage wasteful consumption of electricity. However, lower energy prices render investments in energy efficiency and renewable energy technologies less competitive with existing energy supply technologies (Pacudan 2018a, p.368; Pacudan 2018b, p.284).

Dependency on fossil fuels: Brunei is one of the countries in the world with a very high income per capita. This affluence combined with low prices for energy services (which is typical to most energy-exporting countries) creates a society with a very high demand for energy services. Among the 21 countries under the Asia-Pacific Economic Cooperation, Brunei Darussalam ranked seventh with 2.2 tons of oil equivalent per capita (APERC 2020, 20). The top six countries are Canada (4.8 tons of oil equivalent per capita), the United States of America (4.2 tons of oil equivalent per capita), Australia (3.1 tons of oil equivalent per capita), Russia (2.9 tons of oil equivalent per capita), New Zealand (2.7 tons of oil equivalent per capita), and Korea (2.5 tons of oil equivalent per capita). Due to the abundance of conventional energy resources in the country, power generation is almost 100 percent fossil fuel-based. Brunei's installed power generation capacity amounted to 902.3 megawatts in 2019. Natural gas-fired power plants accounted for 98.5 percent of the total installed capacity, followed by diesel power plants with 1.32 percent and solar PV with 0.14 percent (Ministry of Energy 2020a, p.4). Similarly, domestic passenger transport is dominated by road transport. There were 446,393 registered vehicles in 2019, which translates to almost one car per inhabitant (Ministry of Energy 2020b, 3). The demand for fossil fuel-based energy supply would continue to increase with the expected economic and population growth prospects.

High forest cover and pristine forest reserves: Aside from energy resources, the country also possesses a diverse and complex forest ecosystem. Since 2010, the government has managed to cease forest land conversions and maintain around 72 percent of the land area with forest cover. On the other hand, 41 percent of the land area has been designated as forest reserves (Ministry of Primary Resources and Tourism 2020, p.6). These resources are important carbon sinks, and their preservation is an important strategy to mitigate climate change. Balancing conservation and sustainable management while optimizing resource benefits would remain a key challenge in managing Brunei Darussalam's natural resources.

Greenhouse gas emissions reduction potential

Greenhouse gas emissions inventory

Between 2016 and 2017, climate change-related policies and international negotiations were coordinated and carried out by the then Energy and Industry Department at the Prime Minister's Office. During this period, Brunei submitted its first and second national communications to the United Nations Framework Convention on Climate Change in 2016 and 2017, respectively. The first national communication covered greenhouse gas inventory for 2010 while the second estimated greenhouse gas emissions for the period 2010 until 2014.

In the process of preparing the Brunei Climate Change Policy, the Brunei Climate Change Secretariat updated the greenhouse gas inventory. It estimated the country's greenhouse gas emissions for 2015 and 2018. Compared with other ASEAN countries, Brunei's total gross emissions are relatively small. The gross emissions declined from 11.6 million tons of CO_2eq in 2015 to 10.1 million tons of CO_2eq in 2018. This was primarily driven by reducing emissions from the upstream oil and gas industries due to measures implemented to reduce natural gas venting and flaring.

While the country's gross emissions reached 10.1 million tons of CO_2eq in 2018, net emissions amounted to -1.3 million tons of CO_2eq, indicating that Brunei is a net sink for emissions (BCCS 2020, p.16). The gap between the gross and net emissions represents the forestry sector's contribution as a carbon sink. The energy sector contributed the highest share, accounting for 91.2 percent of the total gross emissions. Other sectors' contributions were marginal: waste sector contributed 2.2 percent, agriculture and other land use 1.4 percent, and residential sector 0.9 percent. On the other hand, with 72 percent of the country's land area covered with forest, the forestry sector is estimated to absorb around 11.4 million tons of CO_2eq from the atmosphere (BCCS 2020, p.17).

Emissions reduction scenario

Together with key stakeholders, the Secretariat facilitated the preparation of the new Climate Change Policy, which outlines Brunei's emissions reduction targets for 2035 and stipulates strategies to achieve these targets. The Secretariat is also responsible for preparing the country's Nationally Determined Contribution. Brunei submitted its Intended Nationally Determined Contribution before the 21st Conference of Parties (COP21) in December 2015. Based on sectoral targets during that period, this Intended Nationally Determined Contribution sets out three goals to reduce emissions by 2035: reduce 63 percent of energy consumption relative to a business-as-usual scenario, aspire to 10 percent of the energy mix from the exploitation of renewable energy, and reduce 40 percent of CO_2 emissions from morning peak-hour vehicle use compared to the business-as-usual scenario (MOD 2015). The Intended Nationally Determined Contribution recognized the importance of the energy sector in the country's total emissions. Thus, the goals set out for emissions reductions were focused on energy industries. Following the 21st Conference of Parties, Parties to the Paris Agreement were allowed to convert their Intended Nationally Determined Contributions into Nationally Determined Contributions. The Brunei government, however, opted to submit a new document outlining their Nationally Determined Contributions.

The country's climate policy will pave Brunei's low carbon resilient pathways for a sustainable nation. This new policy, instead of specifying an aggregate emissions reduction target, indicates an emissions reduction scenario. Under the business-as-usual scenario, gross emissions of greenhouse gases were projected to increase by 4.9 percent annually from 2015 until 2035. The gross business-as-usual greenhouse gas emissions was estimated to reach 30.2 million tons of CO_2eq by 2035, as shown in Figure 2.1.

The emissions reduction scenario appears to be ambitious as it aims to reduce greenhouse gas emissions by more than 50 percent in 2035 from the business-as-usual emissions level. These reductions could be achieved through the implementation of various policy strategies (outlined in the following section) such as those that i) reduce industrial emissions, ii) increase forest cover, iii) increase deployment of electric vehicles and iv) renewable energies, v) improve energy efficiency in the power sector, vi) introduce carbon pricing

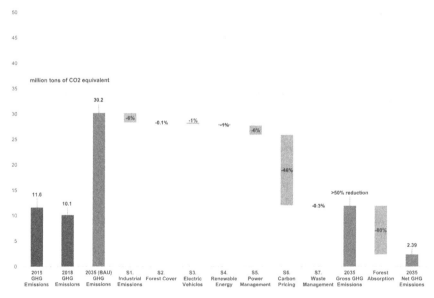

Figure 2.1 Brunei's greenhouse gas emissions trajectory and emissions reduction scenarios

Source: Compiled by the author

and vii) waste management. Among these measures, carbon pricing was envisaged to contribute the most significant reduction, amounting to around 46 percent of the business-as-usual greenhouse gas emissions.

In addition to these, existing forests were also estimated to absorb CO_2 emissions from the atmosphere, equivalent to around 50 percent of the gross greenhouse gas emissions. This would leave the net greenhouse gas emissions in 2035 to amount to only about 8 percent of the gross business-as-usual greenhouse gas emissions.

Collaborations in climate change governance

Institutional coordination

The complex nature of climate change mitigation and adaptation calls for a much broader response to climate change governance. Brunei established the first intersectoral collaboration flatform spearheaded by the National Climate Change Council in 2010. It was co-chaired by the Minister of Energy and Minister of Development with members ranking from Deputy Ministers and Permanent Secretaries from the Prime Minister's Office and five other ministries (see Figure 2.2) with its Secretariat housed at the Ministry of Development.

With a new set of cabinet ministers installed in 2018, and through the initiative of the Minister of Energy, the Council was re-organized, with both the Ministers of Energy and Development remaining as co-chairs while the number of its members were reduced to the Minister of Primary Resources and Tourism, Minister of Transport and Infocommunications, and Deputy Minister of Energy. One of the main lessons learned from the past structure is that convening a meeting with a bigger number of high-level government officials would be logistically tricky. Thus, the members of the new Council were limited to ministers of key sectors that will be the main target for climate change mitigation and adaptation. The Council's main mandate is to support

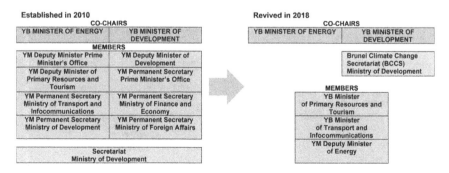

Figure 2.2 Organizational structure of Brunei's National Climate Change Council
Source: Compiled by the author

global and national efforts to combat climate change and honor the country's commitments to climate change treaties, including the Paris Agreement. The Council has the authority to formulate national policies to address and adapt to climate change. The institutional arrangement of the Council, which convened its first meeting in October 2018, is also shown in Figure 2.2.

In the process, the Council established its Secretariat, the Brunei Climate Change Secretariat, which at the same time serves as Brunei's national focal point to the UNFCCC. The Secretariat was initially created with selected officers from the Ministry of Energy. The Secretariat was recently moved to the Ministry of Development since it represents the government in international environmental and climate change cooperation and international treaties. Besides, the Secretariat not only serves the Council but also the Executive Committee and the three working groups (discussed in the following subsection). The Secretariat supervises the formulation of the country's long-term emissions reduction target with assistance from the Brunei National Energy Research Institute. It also takes the lead in the formulation of the Brunei National Climate Policy and the preparation of strategy road maps; oversees the development of a white paper that sets out detailed strategies and action plans for achieving the proposed long-term emissions reduction target; sets up a monitoring, reporting, and verification system; and assesses climate change impacts.

An inclusive stakeholder engagement structure

Along with the Council's revival, the Executive Committee on Climate Change and three technical working groups were also established, as shown in Figure 2.3 and Figure 2.4. Members of the Committee and the groups were appointed to include representatives from key ministries, the private sector, and non-governmental organizations. It must be noted that at the outset, the

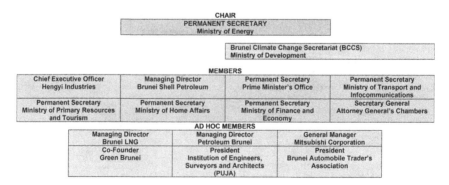

CHAIR
PERMANENT SECRETARY Ministry of Energy

Brunei Climate Change Secretariat (BCCS) Ministry of Development

MEMBERS			
Chief Executive Officer Hengyi Industries	Managing Director Brunei Shell Petroleum	Permanent Secretary Prime Minister's Office	Permanent Secretary Ministry of Transport and Infocommunications
Permanent Secretary Ministry of Primary Resources and Tourism	Permanent Secretary Ministry of Home Affairs	Permanent Secretary Ministry of Finance and Economy	Secretary General Attorney General's Chambers

AD HOC MEMBERS		
Managing Director Brunei LNG	Managing Director Petroleum Brunei	General Manager Mitsubishi Corporation
Co-Founder Green Brunei	President Institution of Engineers, Surveyors and Architects (PUJA)	President Brunei Automobile Trader's Association

Figure 2.3 Organizational structure of Brunei's Executive Committee on Climate Change

Source: Compiled by the author

Figure 2.4 Three technical working groups for climate change issues
Source: Compiled by the author

Committee considers that all stakeholders need to be engaged right at the very beginning from policy formulation.

The Committee's responsibility is to ensure that a national accord on climate change policy will be consistent with the social, economic, and fiscal policies, strategies and activities, and support the development of climate monitoring infrastructure, data collection, analyses, and dissemination. Also, the Committee is tasked to provide guidance to climate mitigation or adaptation proposals that will have an economic impact on Brunei.

Three technical working groups were also formed to discuss pertinent climate change issues. Each working group member must ensure that climate activities are aligned with the policies of their respective agencies, in addition to securing performance and achievement of all targets (Figure 2.4). These working groups will provide recommendations and implement climate policies and strategies in respective sectors. The Climate Change Mitigation Working Group examines alternative possibilities to reduce Brunei Darussalam's emissions and determines the potential groundwork required for long-term climate change mitigation. The Adaptation and Resilience Working Group evaluates Brunei Darussalam's vulnerability and its resilience to the effects of climate change. It facilitates adaptation efforts to improve the country's resistance to climate change effects. The Support Framework Working Group examines Brunei's technical and funding requirements for the effective operation of climate action, facilitates efforts to decrease GHG emissions, and supports climate action in all sectors.

Drafting of climate change policy and operational documents

The Committee established an ad-hoc committee, the Climate Policy Drafting Committee (CPDC), with representatives appointed from key stakeholders (public, private, and non-governmental organizations) to prepare the climate change policy. A two-day workshop was held to draft the policy, and during this event, all stakeholders presented their views and alternative solutions and participated in constructive discussions. The main outcome of the workshop was a draft climate change policy. This draft policy document underwent further reviews and consultation iterations before it was finalized. The final draft was also presented to youth representatives for additional inputs. The final form was reviewed and endorsed by the Committee and the Council before submitting it to the highest authority. The policy document was approved and launched in July 2020.

A similar process was undertaken by the Committee in preparing for strategy road maps. Relevant stakeholders (public, private, and non-governmental organizations) were identified and invited to participate in preparing strategic road maps. A two-day operational document task force workshop was organized to draft a strategy road map for each strategy. Relevant units of each Ministry assigned for policy implementation took the lead in outlining strategy road maps. These road maps are, in fact, strategic plans of government units that were given to implement climate change policy strategies.

Low carbon intervention measures

The policy outlined ten intervention strategies to transition Brunei Darussalam into a low carbon economy: six strategies are emission reduction strategies (reducing industrial emissions, increasing deployment of electric vehicles and renewable energies, improving energy efficiency under power sector management, introducing carbon pricing, and managing wastes), one strategy is to increase carbon sink (increasing forest cover), one strategy is on adaptation (climate resilience and adaptation), and two strategies are supporting strategies (carbon inventory and awareness and education). Potential contributions of these strategies in terms of reducing the impact of global climate change were presented in Figure 2.1. The planned interventions and their performance indicators are summarized below.

Industrial emissions (strategy 1): Despite the decreasing emissions from oil and gas industries from 2010 to 2018 due to rejuvenation projects, industrial emissions remain relatively high, accounting for around 18 percent of the gross greenhouse gas emissions in 2018. The government aims to reduce overall emissions in the industrial sector (upstream and downstream oil and gas industries) through zero routine flarings, as defined by World Bank Standards and the As Low as Reasonably Practicable (ALARP) level. The initiative of zero routine flarings was introduced by the World Bank in 2015, aiming to reduce the wasteful release of associated gas from the extraction of crude oil.

Increasing forest cover (strategy 2): At present, around 72 percent of the total land area is covered by forest. The government aims to increase the forest reserve from 41 percent to 55 percent of the total land area. Under this strategy, the government aims to plant 500,000 new trees by 2035 through reforestation and afforestation.

Electric vehicles (strategy 3): The country is highly dependent on private transportation. As of 2018, there were over 426,000 registered vehicles in the country. The number of vehicles was estimated to increase by 2 percent per year from 2018 until 2035. The government aims to increase the penetration of electric vehicles, accounting for 60 percent of total vehicle sales in 2035.

Renewable energies (strategy 4): Electricity generation from renewable energies accounts for around 0.14 percent of total generation at present. Among renewable energy resources, solar energy is the most abundant in the country and could be exploited using mature and commercially viable technologies. This strategy seeks to increase the total share of renewable energy to at least 30 percent of the power generation mix's total installed capacity.

Power management (strategy 5): Over 99 percent of the power generation in the country is generated from fossil fuel-based power plants, mainly natural gas. The country's total installed power plant capacity in 2018 amounted to 889 megawatts. This strategy aims to reduce greenhouse gas emissions by at least 10 percent through improving demand- and supply-side energy efficiency and conservation.

Carbon pricing (strategy 6): Carbon pricing will be implemented in a later stage once a proper monitoring, verification, and reporting system is established. Carbon pricing will internalize external costs associated with carbon emissions. This strategy aims to introduce carbon pricing to industrial facilities and power utilities.

Waste management (strategy 7): Methane gas emissions from municipal solid waste will be reduced through waste minimization and the adoption of waste-to-resource technologies. The plan aims to reduce household waste to be disposed of in landfill sites to 1 kilogram per person per day.

Climate resilience and adaptation (strategy 8): Brunei Darussalam is vulnerable to flooding, forest fires, strong winds, and landslides. This strategy seeks to strengthen the country's resilience against climate change risks and increase its capacity to adapt to climate impacts.

Carbon inventory (strategy 9): This strategy requires all facilities that emit and absorb greenhouse gases to report their greenhouse gas data. The mandatory reporting promotes transparency and robustness in the national carbon emissions and sink data, which provide a better understanding of the country's greenhouse gas emissions.

Awareness and education (strategy 10): This strategy seeks to promote awareness and educate stakeholders on the impacts and measures to mitigate and adapt to climate change. Specific actions include incorporating climate change mitigation and adaptation concepts in the school curriculum and co-curriculum programs; and communicating, promoting, and socializing the national climate change policy to all stakeholders.

Implementation challenges and critical governance implications

Each strategy presented above was assigned to relevant Ministries, including the Secretariat for implementation or for undertaking further studies. The Ministry of Energy was tasked to implement strategies 1 (industrial emissions), 4 (renewable energies), and 5 (power management); the Ministry of Primary Resources and Tourism for strategy 2 (increasing forest cover); Ministry of Development for strategy 7 (waste management); Ministry of Transport and Infocommunications for strategy 3 (electric vehicles); and the Secretariat was tasked to coordinate and undertake preparatory studies for strategy 6 (carbon pricing), 8 (climate resilience and adaptation), 9 (carbon inventory), and 10 (awareness and education). For each strategy, the climate change policy document also identified tools and specified a monitoring framework that identifies performance indicators. In addition, the Secretariat, together with all stakeholders, prepared implementation road maps. Despite this there remain challenges in the implementation of these policies. Some of the key challenges are discussed in this section, namely sectoral collaboration, financing, regulatory frameworks, carbon pricing, and governance implications.

Sectoral collaboration: The strategies outlined in the policy encompass various sectors and governance structures. The Secretariat acts mainly as a focal agency and coordinator concerning the implementation of core mitigation policies. The Ministry of Energy will carry out intervention measures targeting the energy sector. Those related to transportation, forestry, and waste management will be under the Ministry of Transport and Infocommunications, Ministry of Primary Resources and Tourism, and Ministry of Development, respectively. In addition, various government agencies will be involved in climate change adaptation. Each Ministry however, has its specific mandate and programmed activities, and issues related to climate change may not be their main priority. Effective coordination may be needed to ensure that climate change issues are aligned with each Ministry's central agenda and would be given equal priority in terms of budget allocation and implementation.

Financing: Brunei's public sector is a dominant economic sector in the country. In 2019, the government's consumption and capital formation accounted for more than 27 percent of the country's gross domestic product (Department of Statistics 2020, p.10). Climate change intervention (and investment) measures in sectors whose services are mainly provided by the public sector (transportation/energy infrastructures and forestry) would potentially increase the financial requirement needed from the government. Required infrastructure investments include charging stations for electric vehicles, renewable power plants, energy efficiency technologies for public buildings, adaptation measures, reforestation/afforestation measures, and others. The government, however, could tap the private sector in meeting some of these financial requirements. In the case of the forestry sector, some private companies, through their corporate social responsibility departments, have committed to supporting government reforestation programs. For example, Brunei Shell Petroleum has taken the lead in rehabilitating a peat swamp forest (Peat Swamp Rehabilitation Project 2020) in Belait District, while Hengyi Industries have committed to reforesting around 200 hectares of land in Brunei-Muara District (Hengyi Industries 2020). Planting trees could easily attract private sector support. Still, for other infrastructure projects, the introduction of regulatory frameworks may be required, which would allow private sector participation in providing various services and ensuring them sufficient returns of their investments (Pacudan 2016, p.495). Public-private partnerships may take different forms in different sectors. For power generation, this could be in the form of independent power producers, performance contracting in energy efficiency improvements, and perhaps the private provision of vehicle charging stations. However, these would require some form of sectoral reforms, and each sector's reform process may not be easy and would take time to implement.

Regulatory frameworks: Regulatory frameworks mentioned above are those that promote public-private partnerships, while frameworks referred to in this section are those that promote energy efficiency or mandate industries and individuals to adopt practices that reduce carbon and other environmental emissions. Reducing industrial emissions (strategy 1) requires the establishment of emission targets as well as technology standards consistent with the World Bank's zero routine flarings in oil and gas extraction. The deployment of electric vehicles (strategy 3) requires new or revision of existing vehicle regulations. Various frameworks need to be introduced, such as net metering, green certificates, building integration of RE, etc, to increase private sector deployment of renewable energy technologies (strategy 4). To improve electricity production efficiency on the supply-side, a regulation on power plant efficiency needs to be established, and on demand-side efficiency, appliance standards and labelling frameworks need to be introduced (strategy 5). Carbon pricing regulations will be required to provide the legal basis in internalizing carbon costs (strategy 6). These regulations will be formulated and implemented by each concerned government Ministry. Introducing a new law could take time in Brunei. Still, the main challenge will be on aligning and prioritizing the required climate change-related regulations with other regulatory requirements and the priorities of each Ministry.

Carbon pricing: As presented earlier, carbon pricing's potential contribution in reducing carbon emissions is relatively high at around 46 percent of the business-as-usual gross emissions in 2035. The policy indicated that the carbon pricing policy will be implemented no earlier than 2025 and that further studies will be carried out between now and that period. Based on global experiences, the government could either introduce a carbon taxation or an emissions trading scheme (also known as a cap-and-trade scheme). Under carbon tax policies, the price of carbon is raised through a tax or fee on greenhouse gas emissions or fossil fuels' carbon content (The World Bank 2020, p.17). Under a cap-and-trade scheme, the total allowable volume of emissions in a particular period from a specified set of sources is defined, and economic actors are allowed to trade their emission rights (ICAP 2020, p.37; The World Bank 2020, p.7). A carbon tax provides stable carbon prices while a cap-and-trade scheme ensures that the target emissions reduction is achieved. Carbon taxes and cap-and-trade schemes have both advantages and disadvantages, but the selection of an appropriate carbon pricing mechanism for Brunei would depend on critical economic, institutional, and political factors. Whatever the government's choice is, it remains to be seen whether the selected pricing mechanism could deliver around a 46 percent reduction of carbon emissions in 2035.

Governance implications: As discussed in the institutional coordination section Brunei's climate change governance in policy formulation involved a strong intersectoral collaboration at the highest level (Council) and stakeholder engagement at the operational level, particularly during the policy drafting stage where all stakeholders (public, private, and non-government organizations) contributed to identifying and selecting appropriate strategies and measures that address climate change issues. The combination of institutional coordination and stakeholder engagement in policy formulation aims to secure public support and establish the policy's legitimacy, empowers non-governmental organizations, and ensures the sustainability of policy interventions.

Before these policy strategies could be implemented, mitigation strategies (reducing industrial emissions, increasing deployment of electric vehicles and renewable energies, improving energy efficiency under power sector management, introducing carbon pricing, and managing waste) require additional regulatory frameworks formulated and implemented at the sectoral level. The institutional coordination at the Council level needs to be further strengthened by internal collaboration at the departmental level, especially in the formulation of the sectoral policies and regulatory frameworks that support climate change mitigation. Similarly, the Secretariat, who is responsible for carbon pricing, climate resilience, adaption, and support strategies (carbon inventory, and awareness and education) needs to strengthen its collaboration with other government agencies. Collaboration at the departmental level would be necessary, particularly in preparing for regulatory frameworks related to its assigned tasks such as carbon pricing and in anchoring some of its programs and activities related to education, awareness raising, and capacity building. In formulating different policies and regulatory frameworks, each assigned government agency needs to embrace the whole-of-nation approach concept by engaging, in addition to other government agencies, the private sector, and non-governmental organizations.

Each stakeholder will then play different roles in the actual implementation of climate change policies and regulatory frameworks. Sectoral agencies will be responsible for the enforcement. At the same time, the private sector would be involved in infrastructure and technology investments, either as part of public-private partnerships or as compliance with the climate change mitigation regulations. Non-governmental organizations will be expected to support the implementation of government policies, including education and awareness programs. The Secretariat will be responsible for reporting progress to various bodies of the UNFCCC. It will be working closely with key government agencies to monitor and verify compliance with the climate change-related policies and regulations.

Conclusion

Brunei, being vulnerable to extreme weather events and sea-level rise, responded to global climate change threats by committing to international treaties such as the Paris Agreement, which aims to stabilize the global temperature rise to below 2°C above the pre-industrial levels by the end of this century. As part of this commitment, the country prepared a climate change policy that outlines its voluntary actions to reduce greenhouse gas emissions. The planned climate change interventions between now and 2035 were carefully crafted to balance the country's developmental needs and global environmental protection. The balancing of economic and ecological objectives was made possible by adopting a whole-of-nation approach to policy formulation. The government established a climate change governing Council which is represented by key ministries that would be instrumental in climate change mitigation and adaptation, and created an executive committee on climate change and technical working groups which represented the government and non-government sectors.

While Brunei's whole-of-nation model appears to be influential in for-mulating a national climate change policy, it faces several challenges in implementing these policies. The main challenge for the whole-of-nation approach to climate change policy implementation is to demonstrate: i) effective coordination among ministries ensuring alignment and budget prioritization of climate change policies with those of core policies and pro-grams of key ministries; ii) timely introduction of regulatory frameworks promoting public-private partnerships and accessing private sector financing; iii) establishment of policy frameworks fostering investments of energy-effi-cient technologies and adopting practices that reduce carbon and other environmental emissions; and iv) introduction of a carbon pricing framework that could effectively and significantly reduce greenhouse gas emissions.

Addressing these challenges requires a more complex approach to governance in policy implementation. The whole-of-nation approach to policy formulation needs to evolve with more defined internal collaboration arrangements at each Ministry's operational level. As key strategies were assigned to various minis-tries, each Ministry needs to formulate supporting policies and regulations fur-ther and prioritize the climate change agenda in its specific sectoral plan and in allocating financial resources. While the implementation and enforcement of these measures rest with the assigned Ministries, the Secretariat needs to strengthen collaboration with these agencies in monitoring compliance with cli-mate change policies and regulations. At the external level, cooperation and compliance would be required from the private sector, while government pro-grams are expected from non-governmental organizations.

References

APERC 2020, *APEC Energy Statistics 2017*, Tokyo, Energy Statistics and Training Office, Asia Pacific Energy Research Centre.

BCCS 2020, *Brunei Darussalam Climate Change Policy*, Brunei, Ministry of Development, Brunei Darussalam.

Department of Statistics 2020, *Gross Domestic Product Second Quarter 2020*, Brunei, Ministry of Finance and Economy.

Doyle, B. 2019, 'Lessons on Collaboration from Recent Conflicts: The Whole-of-Nation and Whole-of-Government Approaches in Action', *InterAgency Journal*, vol. 10, pp. 105–122.

EDPMO 2014, *Brunei Darussalam Energy White Paper*, Brunei, Energy Department, the Prime Minister's Office.

Hengyi Industries 2020, '500 Trees Planted in Hengyi 3rd Reforestation Project', viewed 20 December 2020, <https://www.hengyi-industries.com/media/press-releases/500-trees-planted-in-hengyi-3rd-reforestation-project/>.

ICAP 2020, *Emissions Trading Worldwide Status Report 2020*, Berlin, International Carbon Action Partnership.

Low, J. 2016, 'Singapore's Whole-of-Government Approach in Crisis Management', *Ethos*, 16, pp. 14–22.

Ministry of Energy 2020a, 'Strategy 4: Renewable Energy' (slide presented during the Operational Document Task Force Workshop, 2–3 September 2020, Brunei Darussalam).

Ministry of Energy 2020b, 'Strategy 3: Electric Vehicles' (slide presented during the Operational Document Task Force Workshop, 26 August 2020, Brunei Darussalam).

Ministry of Primary Resources and Tourism 2020, 'Strategy 2: Forest Cover' (slide presented during the Operational Document Task Force Workshop, 7–8 September 2020, Brunei Darussalam).

MOD 2015, *Brunei Darussalam's Intended Nationally Determined Contribution* [Online], viewed 20 December 2020 at, <https://www4.unfccc.int/sites/submissions/INDC/Published%20Documents/Brunei/1/Brunei%20Darussalam%20INDC_FINAL_30%20November%202015.pdf/>

Oxford Business Group 2016, *The Report: Brunei Darussalam 2016* [Online], viewed 20 December 2020 at, <https://oxfordbusinessgroup.com/brunei-darussalam-2016/>.

Pacudan, R. 2016, 'Implications of Applying Solar Industry Best Practice Resource Estimation on Project Financing', *Energy Policy*, vol. 95, pp. 489–497.

Pacudan, R. 2018a, 'Feed-in Tariff vs Incentivized Self-consumption: Options for Residential Solar PV Policy in Brunei Darussalam', *Renewable Energy*, vol. 122, pp. 362–374.

Pacudan R. 2018b, 'The Economics of Net Metering Policy in the Philippines', *International Energy Journal* 18, 283–296.

Pacudan, R. & Hamdan, M. 2019, 'Electricity Tariff Reforms, Welfare Impacts, and Energy Poverty Implications', *Energy Policy*, vol. 132, pp. 332–343.

Peat Swamp Rehabilitation Project 2020, 'Badas Tree Planting' [Online], viewed 20 December 2020 at, <https://badastreeplanting.org/blog/>.

Prime Minister's Office 2007, 'Wawasan 2035' [Online], viewed 20 December 2020 at, <https://www.gov.bn/SitePages/Wawasan%20Brunei%202035.aspx/>.

Prime Minister's Office 2015, *Digital Government Strategy 2015–2020*, Prime Minister's Office, Brunei.

UNFCCC 2015, 'Paris Agreement' [Online], viewed 20 December 2020 at, <https://unfccc.int/process-and-meetings/the-paris-agreement/the-paris-agreement/>.

World Bank 2020, *State and Trends of Carbon Pricing 2020*, The World Bank Group, Washington D.C.

World Health Organization 2015, 'Health in All Policies Training Manual' [Online], viewed 20 December 2020 at, <https://www.who.int/social_determinants/publica tions/health-policies-manual/en/>.

Yahya, Y. 2018, 'Public Service to Go from "Whole-Of-Government" To "Whole-of-Nation"', *The Straits Times*, 9 May 2018, viewed 15 January 2021 at, <https://www. straitstimes.com/politics/public-service-to-go-from-whole-of-government-to-whole-of-nation/>.

3 The rise and fall of a climate change assemblage in Cambodia

Tim Frewer

Climate change in Cambodia as an assemblage

Within much of the literature on climate change adaptation in the Global South, the (democratic) state is seen to play a central role (e.g., Agrawala & Van Aalst 2008). This view includes a focus on building state capacities and new institutions to conduct adaptation and mitigation activities (Hallegatte et al. 2015), as well as facilitating private finance into 'green' industries (Hallegatte et al. 2016). Such a view is often informed by liberal democratic assumptions, where it is held that state-conducted climate interventions have been pushed by and remain accountable to an informed and progressive civil society (e.g., World Bank 2019). This view also assumes that the degree to which a state builds climate change institution and conducts climate change projects is an index of how democratic and accountable its institutions are. In the context of Cambodia, such sentiments and positions are unable to appreciate the way in which an authoritarian state can simultaneously conduct donor-driven climate change projects while maintaining authoritarian tendencies. Nor are they capable of explaining why states, companies, and international development actors with a poor track-record on climate change (e.g. the Australian and US governments, the World Bank) are providing funding for climate change interventions in Cambodia.

In order to try and avoid some of these problems, this chapter examines climate change interventions in Cambodia as part of an assemblage. The advantage of viewing climate change interventions in Cambodia as part of an assemblage is that through describing assemblages, one must describe the relations that constitute the assemblage and enable its day-to-day existence (DeLanda, 2016). For the purposes of this chapter, I will briefly describe three features of assemblages, taken from DeLanda (2016) that help to better understand the climate assemblage in Cambodia. Firstly, the component parts – policies, organizations, flows of finance – of an assemblage have an independent existence outside of the assemblage. The component parts are not entirely defined by the assemblage and can easily be plugged into other assemblages. This means that the actors who are involved in the assemblage have interests and agendas that go beyond climate change interventions. This

DOI: 10.4324/9780429324680-5

helps to make sense of seemingly contradictory positions – that people in villages both resist and participate in climate change projects, that state actors both conserve and exploit forests, and that NGOs dogmatically pursue their institutional agendas while also having staff committed to philanthropy. Secondly, the assemblage has emergent properties meaning that the whole is greater than the sum of its parts. This characteristic defines an assemblage in opposition to a simple collection of things. Through a myriad of projects, documents, and meetings, the climate assemblage emerges as a material reality (a bureaucracy and network of NGOs and experts) that can pursue funding from international donors and shape policies and discourse on climate change. Thirdly, the assemblage is composed of material and expressive components. The expressive components are particularly important in creating and maintaining the relations that make the assemblage a reality and give it emergent properties. This includes discourses such as 'climate change adaptation' and 'resilience' that provide a language and rationality to climate change interventions and policies. Although the myriad of programs and activities conducted in Cambodia under the banner of climate change could be considered as a series of fragmented and differentiated programs, I argue that they are linked together by material and expressive components and should therefore be seen as an assemblage. For instance, while there have been hundreds of different climate change projects over the last decade, these have all been funded by less than ten donors. These donors themselves tend be well coordinated in terms of the agendas and rationalities that govern the various programs they fund. Also, there is a relatively small group of Cambodia-based government employees, consultants, technical assistants, and NGO personnel who act as brokers within these projects and who are well connected.

The first thing to highlight about what I am terming a 'climate assemblage' in Cambodia is that the assemblage is not a unified whole that is internal to the Cambodian state. Many of the reports, project documents, and literature that emerge from the assemblage frame climate change interventions as coming under the sovereign authority of the Cambodian state and being confined to its territory. Yet the climate assemblage in Cambodia is composed of several autonomous parts that geographically go beyond the Cambodian state, including the various multilateral donor institutions which have provided finance to Cambodia, bilateral donors, private companies from various countries, technical assistants, various experts from outside Cambodia, and international NGOs. As such, climate interventions in Cambodia are composed of elements that have their origins outside of Cambodia.

Because climate assemblages go beyond state borders, I insist that any serious attempt to understand climate interventions in Cambodia (or other countries in Southeast Asia that are recipients of climate finance) must start with a recognition that logics, rationalities, and finance that give coherence to climate interventions in these places have been forged in Europe, the US, Australia, New Zealand, and Japan,[1] often as part of broader political and economic projects. My use of a specifically 'Cambodian' climate assemblage is thus simply used as a heuristic to provide focus on what is a much larger assemblage. In doing this, I avoid framing climate change interventions in Cambodia either as coming from the Cambodian state, or as merely a top-down, donor-driven agenda implemented in a peripheral space in conflict with local realities (Craig & Porter 2006; Goldman 2006; Young 2002).

Building a climate change assemblage in Cambodia

Making sense of the Cambodian climate assemblage first requires an appreciation of the sudden influx of development finance, NGOs, and expertise from the early 2000s. From the late 1990s, the Cambodian economy and governance system undertook several major changes where the country was opened to international capital and integrated into a free-market system while building a bureaucracy nominally based on a liberal democratic model (Hughes 2003). Importantly, the Cambodian government opened up to Western financial and development institutions at a time when international development had become more focused on environmental issues and more focused on institutional development and the governance of vulnerable groups (Craig & Porter 2006). This was in response to three developments: i) the new global institutional arrangements and agreements which promised more funds for biodiversity loss and climate change (Goldman 2006; Young 2002), ii) the failures of international development to halt forest and biodiversity loss, and iii) the failure to include marginalized groups in development (Rich, 2013). In this sense, Cambodia became a laboratory for an emerging set of development ideas, concepts, and practices that had been forged out of the failures of prior decades of development. The plethora of development programs focusing on climate change funded by the World Bank, USAID, and UNDP and UNEP of the 1990s and 2000s can thus be seen as experiments in new phase of development orientated toward climate change adaptation and mitigation.

Before 2000 Cambodia had virtually no policies or institutions focused on climate change. These emerged through donor-funded projects and climate policies and institutions under the tutelage of international donors and experts. In 1996 Cambodia ratified the UN Convention on Climate Change. With very little knowledge of the international negotiations, nor with bureaucrats who had the technical capacity to engage in the international process in a meaningful way, climate change remained a largely unknown and peripheral concern. Yet Cambodia still ratified the UN Framework Convention on Climate Change (UNFCCC) as a non-Annex 1 party in 1995. In 1999 the Global Environment Fund (GEF) and United Nations Environment Program (UNEP) provided $100,000 for a 'climate change enabling activity' which sought to sow the seeds of a climate bureaucracy in Phnom Penh while enabling key bureaucrats to participate in international climate negotiations. The government received a further $230,000 in 2002 from GEF and UNITAR to prepare the county's first national communication (outlining its own carbon emissions and needs in terms of adaptation). In 2002 it signed the Kyoto Protocol, which was followed by UNEP, EU, and the Japanese International Cooperation Agency (JICA) providing grants for capacity development on the Clean Development Mechanism (CDM). With money from a range of donors, Cambodia also began CDM projects, including a multi-million-dollar nationwide biodigester program (which never achieved CDM accreditation).

In 2003, two USD$325,480 GEF grants (with an additional $100,000 top up) were provided to prepare a National Biodiversity Strategy and Action Plan and a formal communication to the UNFCCC (where 10 percent of each grant was automatically taken as administration fees by the GEF), and more than 50 percent of the remaining money was given to foreign consultants from UNDP, FAO, and private companies (GEF 2020). The rest of the money was provided to senior Ministry of Environment (MoE) bureaucrats for capacity development within their offices (ibid.). Both reports made links between Cambodia's violent history, status as a poor developing country, and its vulnerability to climate change. They also emphasized land use change and forestry as a future potential source of greenhouse gas emissions (GHGs). Consequently, Cambodia became imagined not as a net sink of greenhouse gas emissions but as a net polluting country.[2] Both projects also emphasized the urgent need for capacity development and further inflows of aid to improve institutional capacity.

The next major important donor intervention was another GEF-funded program (through the Least Developed Countries Fund (LCDF)) – the National Adaptation Program of Action to Climate Change (NAPA) in 2006. The NAPA report essentially acted as a shopping list for future donor-funded climate change projects (under the finance from the LDCF), where it identified 39 such projects (worth $300 million). Cambodia successfully attracted $8,782,500 worth of funding to implement five of these projects. Although this represented only a small fraction of the requested money, Cambodia has received the second highest amount of LDCF finance of any country in the world and overall has received 5 percent of all LDCF funds (for comparison, this represents the combined amount given to multiple African countries).[3] The GEF took in close to one million in agency fees[4] and both UNDP and UNEP as administering agencies took between 10–25 percent of the remainder (which does not include salaries paid to consultants or evaluators). The program was an important precedent in terms of bringing major climate finance to the upper echelons of government ministries and gave key bureaucrats and government departments, under the tutelage of UNDP and UNEP, important experience in carrying out climate change projects.

Following this, carbon aid came in greater quantities from a range of donors, and in response, the Cambodian government amped up its own carbon bureaucracy and carbon legislation. By 2003 the MoE had established its own Climate Change Office which became a fully-fledged department in 2008. Based on a review of Between 2000 and 2013, there have been a total of

275 climate change projects[5] across all ministries. Just in 2011, total climate expenditure was $170 million. For comparison, in the preceding decade (1990–2000) there were 98 projects which broadly dealt with climate change and disasters representing $98 million (Ministry of Environment, Royal Government of Cambodia 2006). The MoE and FA have experienced significant bureaucratic expansion as conservation and climate change finance and expertise have increasingly flown into them. By 2016 they had 30 and 41 different offices respectively. In 2016 the MoE, hosting over 600 employees, even had to be relocated to a new $8 million-dollar, seven-story facility to cater for expansion. By 2013 climate change finance represented more than one-third of total resources managed by the MoE (worth USD$4.5 million). In an MoE strategic plan, it was stated that the ministry would try and secure 'several hundred million' of climate finance in the upcoming years (Ministry of Environment, Royal Government of Cambodia 2006).

While not all aims and goals of the aforementioned projects were met, this did not necessarily impede further climate finance. A paradigmatic example concerns the Clean Development Mechanism. In 2002 the Netherlands Ministry of Foreign Affairs granted the MoE $250,000 through UNEP for a project titled Capacity Development for the Clean Development Mechanism (and topped up with another $50,000 in 2003). The project worked with senior elements of the MoE to provide technical assistance, capacity building, and training on identifying and implementing CDM projects. The MoE quickly established itself as the focal point for CDM in Cambodia (with the Minister as the direct focal person) and created a CDM national authority. The expectation was that the CDM mechanism was going to unleash major inflows of finance that would be directed toward the MoE. The MoE, while having limited experience in this field, began enthusiastically promoting CDM (see McDonald-Gibson 2003). After spending $300,000 on technical assistance and capacity development (not to mention assistance from other projects of which capacity development on CDM was a secondary aspect), Cambodia is left with ten registered CDM projects – all of which are small-scale.[6]

One of the most important donor-sponsored programs in terms of institutional development has been the Cambodia Climate Change Alliance (CCCA). This program is part of a much larger fund – the Global Climate Change Alliance (GCCA) – which is administered and implemented by the EU.[7] The CCCA was created as a multi-donor initiative, supported by UNDP, the EU, the Swedish International Development Cooperation Agency (SIDA), and the Danish International Development Agency (DANIDA) to more systematically "strengthen the national institutional framework for climate change" (Global Climate Change Alliance 2012). The CCCA worked in two distinct phases – phase one, worth $10.8 million, ran from 2010–2014 and worked closely with the (now defunct) National Council on Climate Change to produce, amongst other things, the Cambodia Climate Change Strategic Plan. During this phase the CCCA began to build new government networks across the bureaucracy to "mainstream climate change" by pushing for ministry-specific climate change plans.

In the first phase, the CCCA provided grants worth $2.6 million for 11 NGO and government-run climate change projects (that tended to involve capacity development or technical assistance at the provincial or village level). In the second phase (2014–2019), worth $13 million, the CCCA provided another round of grants to NGOs, and developed a 'climate change knowledge platform' (in the form of a website), as well as assisting with the completion of another MoE report – the Cambodia Climate Change Action Plan. The latter aspect of the program was important in establishing links to NGOs and focused heavily on concepts such as 'climate resilience' and 'climate adaptation'.

The largest package of climate finance to come to Cambodia however, has come through the Strategic Program for Climate Resilience (SPCR). Funded through the World Bank-administered Climate Investment Funds (CIF), SPCRs were rolled out in 13 countries with the aim to "pilot and demonstrate ways in which climate risk and resilience may be integrated into core development planning" (Asian Development Bank 2012, p.1). Due to a major controversy over a World Bank-funded land titling project, it was decided that Climate Investment Funds would be channelled and administered by the Asian Development Bank which had a long-established (and less controversial) history in Cambodia. In 2011 a SPCR was created and endorsed by the government and the ADB, which outlined four overall project areas comprising nine distinct projects worth a total of US$531 million ($120 million worth of grants and concessional loans from the CIFs and $421 in co-financing). Collectively, these projects represent the largest quantity of climate finance to be dedicated to Cambodia to date. Once again, the MoE established itself as a key government actor in organizing and facilitating the flow of this finance and implementation of these projects.

The overall logic of the SPCR program was to use grants from the CIFs to fund 'soft interventions' (capacity building, technical assistance, institutional development) and use concessional loans from the ADB and elsewhere to fund 'hard interventions' (physical infrastructure). Most of the projects were in fact already a part of the ADB's extant portfolio in Cambodia – where CIFs were used to add on additional climate components to larger projects. Each of the projects had 'soft' and 'hard' components which corresponded to the allocation of grants and loans. For instance, the rural roads improvement project included not only the physical building of roads (the 'hard component') but also a 'soft' capacity development and training component (for village level participants and local level authorities) funded through a $2 million grant to the Ministry of Rural Development.[8] The only entirely grant-funded component was the 'technical assistance component' which was initially funded through a $7 million loan but which later received an additional $4 million. Seemingly overlapping with the activities of the Cambodia Climate Change Alliance, the technical assistance component aimed to "mainstream climate resilience into development planning" (Asian Development Bank 2012, p.7) and resulted in yet another strategic report authored by the MoE – the Cambodia Climate Change Response Strategic Plan.

After Cambodia joined the UN Reduction of Emissions through Deforestation and Degradation (REDD+) program in 2010, and the Forest Carbon Partnership Facility (FCPF) in 2009, forestry issues also began to increasingly come under the banner of climate change. Forestry in Cambodia has a much longer history as a controversial issue[9] between donors and the government (Cock 2017), so REDD+ provided seemingly new opportunities for donors to address forestry issues. REDD+ activities have occurred in two overlapping phases – the first was the piecemeal implementation of privately run commercial forestry carbon projects that utilized the emerging concept of REDD + developed since the Bali 2007 Conference of Parties. All these projects involved major amounts of aid money and have involved significant investments (time, energy, and finance) from NGOs, the Forestry Administration (FA), and MoE. With the establishment of a REDD+ secretariat, a REDD+ taskforce, four technical REDD+ working groups, and the development of a National REDD+ strategy, donor money flowed into both voluntary carbon mitigation projects and national level REDD+ preparedness.

The second phase of REDD+ involved an attempt to 'scale up' these different commercial projects to the national level, in line with a new emerging agreement on REDD+ at the CoPs. Here, countries would be provided with finance based on quantified and verifiable results that showed forests were being protected (and carbon conserved) above a baseline. This involved an enormous amount of expertise to establish an overall monitoring system that could accurately detail changes in biomass across the entire national territory, as well as a methodologically sound way of establishing a national baseline and the constant nationwide monitoring of any derivations from this baseline. At the same time, global negotiations also required a set of comprehensive safeguards, standards and monitoring instruments, evaluation and consultative processes, not to mention a financial mechanism capable of delivering significant flows of finance to different agencies and bodies. As part of this, the FA and MoE received US$6.7 million to pursue 'REDD+ readiness' which included the above activities. In addition to the FCPF, the FA also received funding for REDD+ readiness of US$10 million from JICA through the Cam-REDD Program, $4.2 million from the UNREDD Program, and smaller amounts from UNDP (US$500,500 worth of small grants for REDD+) in addition to separate grants from bilateral donors such as the US, EU, Norway, and INGOs. For instance, the EU provided $1.8 million for a project titled 'Sustainable Forest Management and Rural Livelihood Enhancement through Community Forestry and REDD Initiatives'.

Outside of these major climate change programs, there have also been smaller projects run by NGOs that receive internal funding, funding from smaller donors, or bilateral funding separate to the large packages mentioned above. At the time of writing, there were also over 80 NGOs involved in the climate change assemblage (including one NGO that was created solely to focus on climate change – the Cambodia Climate Change Network). Four different universities were involved with the assemblage (each being involved in several donor-funded climate change projects). Several donors have, for instance, given funds for small-scale disaster risk reduction and climate adaptation projects. International NGOs such as Oxfam and Plan International have received funds from parent organizations, which in turn come from bilateral donors, to implement various adaptation projects in Cambodia. Australia, Norway, and the US have also given funds to these NGOs and others to implement small-scale projects.

The emergent climate change assemblage in Cambodia

The climate assemblage typically establishes relations between component parts through the conduct of programs and projects. These development projects generally follow a specific form of managerialism, starting with an aim and agenda (outlined by the donor) as well as a series of quantifiable and measurable outcomes (to be assessed by a third party) (Roberts, Jones & Fröhling 2005). They have a specific time frame and specific geographic location that they target. These projects and their structure bring the various component parts – government officials, NGO workers, consultants and experts – into the assemblage. Virtually all of the projects highlighted above follow this format, only with variations in aims, outcomes, time frame, geographic location, and measurable outcomes. All such projects work either directly with the government or through NGOs that are registered with the government (and receive permission from relevant government actors).

At this point some basic things can be observed regarding these projects. Firstly, projects are created by donors and not by the people who are the target beneficiaries. They are funded by donors and carried out by implementing agencies according to the aims and outcomes that have been created by the donor. Even projects that have extensive local level training sessions or participation components (such as the Cambodia Climate Change Alliance), tend to provide funding to smaller NGOs only when these NGOs frame climate change in terms of concepts, discourses, and rationalities that donors have already decided upon. This immediately dispels any notion of climate change projects organically arising from 'civil society' or of such projects operating autonomously or in opposition to the state, as all projects are either dependent upon government permission and cooperation, or run through NGOs that are registered with and monitored by the government (Frewer 2013).

Following on from this, it is also important to note the way in which climate change projects tend to demonstrate success through reports and meetings, i.e. that development success is performative (Lewis & Mosse 2006). The project, as a unique form of intervention, is structurally dependent upon a series of documents that can verify whether program activities occurred and whether relevant actors participated. This is done through meetings, mid-term evaluations, progress reports, assessments, topical reports, and media interventions. In the case of climate change projects in Cambodia, there tends to be great emphasis on recording project activities and disseminating project information through websites. It is also very common for larger projects to have multiple meetings (typically at expensive hotels in Phnom Penh) involving significant numbers of government and NGO staff. Participation in this context often means attending meetings and so large lists of participants within such projects often belie the fact that small numbers of people (typically consultants or technical advisers) are doing the majority of analytical work that goes into producing reports. Such reports often fail to give a sense of project challenges and problems, and instead focus on abstract benchmarks and outcomes. For instance, in the case of the first REDD+ project, the large output of project documents gave the sense of overall project success, focusing on the project's pioneering efforts in terms of gender and conservation. In reality, the project had entirely collapsed on the ground and been abandoned by project proponents, much to the distress of beneficiaries who been left without support (Frewer 2017).

For all the projects analysed here, involvement of village level beneficiaries was typically measured through the number of participants at trainings or meetings. Although 'participation' of beneficiaries in climate change interventions is often a key discursive focus of many projects, logistically it is challenging to meaningfully incorporate significant numbers of village level participants into project structures (Käkönen et al. 2014; Work et al. 2019). Therefore, the interests and agendas of state and NGO workers in the climate assemblage tend to be more closely aligned together than either is to the target beneficiaries.

It is here that it is important to consider some of the discursive or expressive components of the climate assemblage. One key term that all climate change projects in Cambodia rest upon is vulnerability. All projects find their reason of existence in addressing a particular target group of people who are vulnerable to climate change. Many key documents produced over the last decade have carved out various target groups of Cambodians who are vulnerable to climate change. For instance, the Cambodia Climate Change Action Plan and the Understanding Public Perceptions of Climate Change reports, both produced under support from the CCCA as well as the first communication under the UNFCCC, have all emphasized the vulnerability of poor rural Cambodians to climate change. Some reports have even tried to quantify this vulnerability as discrete risks. For example, the Maplecroft Climate Vulnerability Index (Maplecroft 2019) has consistently ranked Cambodia as one of the most vulnerable countries in the world while insurance and reinsurance companies such as Munich Re and Standard & Poor have also ranked Cambodia as extremely vulnerable to climate risks (Ball 2015). Climate interventions in Cambodia are structurally dependent on government and NGO actors, under the guidance of consultants and technical experts, who address the vulnerability of rural, climate-vulnerable Cambodians. On one side are the vulnerable as a target of development interventions and on the other are those who will work on this vulnerability by increasing resilience.

This discursive orientation of focusing on rural, vulnerable populations has important material implications. Geographically, the climate assemblage is centred on Phnom Penh where the relevant government ministries and NGO offices are located and many of the day-to-day activities of projects (meetings, workshops, document writing) happen. Thus, individuals, whether in the government, NGOs, or consultant positions, tend to come into contact with one another within particular spaces in Phnom Penh. Most of the individuals who are a part of the climate assemblage, and especially those who act as important conduits through their brokering activities, will spend most of their time in Phnom Penh. Yet all projects are discursively centred on the rural. The effect of this is that those within the assemblage (both government and NGO) tend to be literally closer together than either is to the vulnerable people in whose name they work.

There are some other important discursive elements of these projects that need to be elucidated. One is the focus on 'capacity development' and resilience which all climate change projects share. As mentioned previously, even seemingly 'hard' project interventions involve substantial capacity development components. This forms part of a trend within the wider international development assemblage where top-down state interventions involving the distribution of public goods and services have been heavily critiqued (Chandler 2014). The prevailing logic of capacity development and resilience instead focuses on channelling immaterial benefits to key actors who will be able to promote and enhance resilience amongst target populations (Evans & Reid 2014). Thus, rather than directly providing goods and services to the vulnerable, the aim of capacity development and resilience building is to enhance the conditions surrounding the vulnerable so they themselves can become resilient (Grove 2014). Seemingly paradoxically, this in turn tends to rest upon large amounts of labor being directed at specific state actors so they can increase the resilience of target populations (cf. Graeber 2015). For instance, in the Mainstreaming Climate Resilience into Development Planning project (Asian Development Bank 2012), 170 person-months of international consultants, and 720 person-months of national consultants were contracted representing US$11 million in contracts. Even 'harder' components such as the Climate Resilient Rice Commercialization Sector Development Program involved over 635 person-months of consultancy (including 14 months for community mobilization specialist, 8 for marketing specialist, 49 for policy development specialists, 39 for climate change specialists and 32 for gender specialists) where $600,000 was spent on capacity development workshops. Hence the bureaucratic expansion of government committees and departments is predicated on their increased ability to enhance the resilience of the vulnerable. The Ministry of Agriculture, Forests and Fisheries as well as the

Ministry of Environment have been involved in numerous projects which aim to enhance the resilience of small-scale farmers by teaching them how to cultivate drought-resistant crops (i.e. in small gardens). Yet both departments have shown a very limited ability to tackle widespread problems which plague rural farmers including: low agricultural commodity prices, loss of land due to economic land concessions, tenure insecurity, and extremely limited rural infrastructure and state services. (Lang 2018; Milne & Mahanty 2019; Milne et al. 2019)

The examples related to vulnerability show that climate change interventions in Cambodia do not merely favor the market over the state as the privileged site in which to address the negative impacts of climate change. They also tend to follow a deeper logic which attempts to institute a particular relation between the state and the governed where the state facilitates the resilience of the governed. The vulnerable are to be made resilient through state-expert led interventions that are targeted at enhancing their capacities rather than directly improving their material conditions. This explains why so much expenditure is focused on experts and bureaucracy in Phnom Penh while the material conditions of the rural poor that place millions in a position of desperation seem immune to climate change interventions.

In some cases, 'climate adaptation' is conflated with the spread of markets. For instance, within the SPCR (CIF 2011) the 'Climate Proofing Infrastructure' component is geographically focused on what the ADB has identified as an economic corridor; a group of areas sketched out by the ADB which are identified as being able to facilitate market growth across the country (CIF 2011, p.54). Similarly, the primary logic of the Rural Roads Improvement project is 'improved access to markets and jobs' (CIF 2011, p.36) and one of the main indicators used to evaluate outcomes of the project was 'percentage of rural people in Kampong Cham province with year-round access to markets' (CIF 2011, p.23). The Promoting Climate-Resilient Agriculture component also had a heavy emphasis on developing business plans for smallholders, and expanding inputs needed for 'high value crop production' including fertilizers, irrigation systems, and even insurance (CIF 2011, p.57). Thus, much of the logic of the SPCR interventions revolve around the extension of physical and social infrastructure which market expansion is dependent upon and should thus be seen as an attempt by donors to re-brand pre-conceived programs and funding under the banner of climate change adaptation.

In projects run by NGOs, the logic of resilience is often more biopolitical, targeting the biological existence of vulnerable populations and aiming to reduce their biological and ecological impact on vulnerable environments. A good example of this logic at work is UNDP and SIDA's Community Based Adaptation Program. In an article produced by UNDP titled 'Villages Take Entrepreneurial Step in Solving Water Woes', a village is described where 'entrepreneurs' took an NGO-provided loan to expand a local water reservoir to establish a user pays system (UNDP Cambodia 2015). The project was

even extended by villagers to 'water uses' outside of the village. In this framework, rural villagers are celebrated for their ability to embrace the market
as a solution to their problems. Similarly, in a video produced by UNDP on
Building Resilient Communities, a poor rural family is presented who received
agricultural training and a gas stove (UNDP Cambodia 2015). The video
weaves together the family's vulnerability to climate change with the benefits
of the program such as providing "a clean source of renewable energy that
can reduce dependence on charcoal and the cutting of forests". Beneficiaries
of the project also received training on revolving funds, on the expansion of
home gardens, and the use of solar panels. This is the resilient subject par
excellence – who is off the grid and uses market-provisioned solar power
rather than the state grid, who uses village raised revolving funds rather than
low-interest state-backed loans, who purchases a gas stove to reduce their own
impact on the biosphere, who voluntarily reduces their dependence on forests,
who embraces the market to meet their basic needs, and who can survive at a
low material standard of living.

In other cases, climate change projects are more orientated toward the
creation of profit. For instance, the four voluntary REDD+ projects that have
been conducted in Cambodia are all orientated toward the generation of
profit through the creation of novel carbon credits. They involve coalitions of
government bureaucracies, experts, NGOs, and carbon brokering companies
who produce commodities from REDD+ programs which are to be sold to
Western companies. So too Carbon Nexus, based in Phnom Penh, created by
eight NGOs to act as a platform to "bring together businesses, local communities, and development entrepreneurs to improve lives and deliver sustainable climate benefits" offers "superior, direct access to carbon credit
projects that support long-term livelihood, health, and environmental benefits
for communities in developing countries" (Carbon Nexus 2016). Carbon
Nexus not only provides support to both NGOs and businesses running
small-scale carbon projects but also markets carbon products and allows
companies to directly offset their emissions by providing carbon footprint
analytics and carbon credits generated in Cambodia. In both these cases,
these projects are more than just using the market to achieve environmental
goals; they are about creating novel commodities and promoting the markets
which underpin them.

The politics of the climate change assemblage

The assumption that the Cambodia bureaucracy can deliver basic public
goods and services, as separate from the hierarchical class politics that dominate the state, poses major logistical problems and difficulties for donor-
driven climate change projects (Pak et al. 2007). In some instances, such as in
the MoE, there are key individuals who have formed close relations with
donors and technical assistants and have been able to conduct individual
projects with at least some level of autonomy from the demands of the

hierarchical Cambodian People's Party (Milne et al. 2015). In other instances, such as within the FA, individuals who are high within the CPP hierarchy have played a key brokering role between donors and the CPP where climate change projects have had the direct support of Hun Sen.

Another important political aspect of the climate assemblage concerns finance. Finance is what keeps the climate assemblage together by forging bonds between donors, government officials, NGOs, experts, and consultants through distinct projects and programs. Very little of the finance provided as a part of climate change interventions reaches beneficiaries (in the form of income) (Mattar, Kansuk & Jafry 2019) and in this sense the poor, rural beneficiaries of climate change interventions should be considered a peripheral part of the assemblage. By 2019 the political situation inside and outside Cambodia had changed compared to previous years when major climate financing was flowing to Cambodia. This also coincided with the ending of several major climate projects such as the CCCA and Strategic Program for Climate Change (SPCC). Efforts to expand REDD+ were also strained as early voluntary projects such as the Oddar Meanchey Project had entirely failed (Lang 2018), and a UN mechanism that would see the flow of finance into national REDD+ participating countries had failed to materialize (Clouse 2020). Much of the climate bureaucracy, including committees and taskforces that had been built by climate finance, had started to disintegrate by this time due to the completion of major projects. This was part of a pattern whereby Western-financed programs that aimed to 'mainstream' certain agendas within the Cambodian bureaucracy had failed to attract sustained funding or see intuitional changes (e.g. in relation to good governance, deforestation, and the Clean Development Mechanism) (Cock 2017; Hughes 2009). Hence, it is likely that many government and NGO officials view climate change programs in terms of short-term benefits where they can performatively commit to the long-term goals of programs without any meaningful commitment to changing state institutions.

Conclusion

This chapter has examined climate change interventions as emerging from a broader climate change assemblage. Through examining the relationships, discourses, and rationalities that form the climate assemblage, the politics of climate change assemblages in Cambodia becomes clear. As opposed to approaches which analyse climate change interventions against either the goals and standards of these projects themselves, or analyse them according to abstract ideals derived from liberalism about how such state/civil societies interventions should work, I have kept the focus on the materiality of the assemblage within Cambodia and the particular political problems which arise from it. By using key insights from assemblage theory, I have shown how state actors, donors, and NGOs – although having seemingly differentiated interests and agendas – form relations together around the finance that flows into climate change projects. NGOs, advisers, experts, and donors accommodate state actors who pursue their political projects through the all-powerful CPP. Simultaneously, state actors accommodate NGOs, experts, and donors who intervene in state operations. The focus on capacity development, which is often amorphous, provides flexibility for these actors to pursue their different interests. Performative adherence to climate adaptation and resilience within meetings and reports belies the shorter term (and more material) interests of actors (e.g., salaries, work experience, funding for state departments, and institutional power). While the climate assemblage is discursively fixated on poor, rural Cambodians, its material existence is centred on Phnom Penh. This means that not only is there a large geographic distance between those in the assemblage and those who are supposedly beneficiaries, but there is also a large political divide.

There are important implications here, relevant to climate governance more generally in Southeast Asia. The lion's share of climate finance given to developing countries goes to the expansion of state bureaucracy which is increasingly composed of western, or western trained, technical experts and NGOs. Institutions, consultants, and experts high up the aid chain take a significant portion of finance relative to village level beneficiaries. This reveals one of the key limitations of the type of western funded climate assemblages that I consider in this chapter; that the aims and agendas of those within the climate assemblage do not often resonate with the aims and agendas of those that the assemblage claims to represent and bring benefits to. For instance, climate change projects in Cambodia rarely directly address the way in which capitalist relations have spread across rural Cambodia in the last three decades resulting in a declining size of landholdings for smallholder farmers and an agrarian livelihood crisis in which smallholders cannot generate sufficient income from the sale of commodity crops. They also tend to remain silent on state territorialization, where over the same time period smallholders and forest-dependent communities have been increasingly alienated from rural lands and forests. These two issues are, of course, not specific to Cambodia and arguably climate assemblages across the region suffer the from the same problem. Climate change projects – although discursively and logistically flexible enough to incorporate state and donor interests – are rarely able to meaningfully engage with livelihood challenges of the rural poor.

Notes

1 I specifically focus on these countries as they have been important contributors of climate finance which has flowed through multilateral and bilateral financial institutions (Such as the World Bank, Asia Development, IMF, USAID) as well as development institutions (the various UN agencies and international NGOs).
2 For the first five CoPs (Conference of Parties), Cambodia was considered as having zero emissions. Even though there had been no actual accounting of GHG emissions, Cambodia's small, predominantly rural population, extremely small number of motor vehicles, almost no industry, and minimal energy production, made its

zero emissions status a foregone conclusion. Indeed, as the per capita energy usage in the US across the 1990s averaged 7,800 kg (oil equivalent), in Cambodia it was only 290 kg. In 2000 US per capita electricity consumption was 13,671 kWh. In Cambodia it was a mere 32 kWh (427 times smaller than the US). Yet by 2012 things had started to change as Cambodia's forests were now included within estimates of GHG emissions (owing to the fact that satellite images had made it possible to make estimates of the carbon content of terrestrial biomass). With forest cover change estimates (which were also made possible through satellite imagery), it became possible to not only calculate deforestation rates, but also the GHG emissions associated with deforestation. This saw Cambodia for the first time in its history become classified as a net GHG emitter where by the 2015 Paris CoP, Cambodia was even committing to cutting down its GHG emissions as part of its commitments to the Paris Agreement.

3   Cambodia has been involved in 68 large-scale programs that have received financing under various UNFCCC funds and which form a part of Annex 1 country pledges to provide finance to non-Annex 1 countries under various agreements reached at the different CoPs.

4   The GEF has made $17 million from administering all its LDCF projects (which are financed from various bilateral sources).

5   'Climate change project' here means a project that has 'climate change' in its title and directly addresses issues related to climate change. In comparison, the Overseas Development Institute in its review of climate finance in Cambodia which included 'low relevance' climate change projects in its analysis (projects that only address climate change as a secondary issue) concluded that there have been 750 climate change projects just in the period from 2009–2011. It is problematic to include these 'low relevance' projects because it gives the impression that Cambodia has received a much greater amount of 'new' and 'additional' climate finance than it actually has.

6   92 percent of all Cambodia's generated Certified Emission Reductions come from only three hydropower projects that seemingly fail to adhere to the principle of additionality as they are all financed though Chinese state based low-interest loans, meaning the projects would have gone ahead regardless of carbon finance and were all already planned before official CDM registration (McDonald-Gibson 2003).

7   The GCCA was established in 2007. It operates 51 programs in 38 countries with a fund of US$830 million (as of 2016). As of 2015 the Cambodia program was the largest country program (in terms of finance committed).

8   As another example, out of the $7 million provided to the Climate-Resilient Agriculture in Mondulkiri and Koh Kong project – seemingly a 'hard' project – over $3 million out of $7 million was allocated to capacity development, training, workshops, and consultants.

9   Since the 1970s Cambodia's primary rainforests have dramatically reduced in size from 70 percent of land cover in 1970 to only 3.1 percent in 2007 (although the country still has 40 percent overall forest cover). In the latter 2000s deforestation was particularly high – within the top three countries in the world for most years.

References

Agrawala, S. & Van Aalst, M. 2008, 'Adapting Development Cooperation to Adapt to Climate Change, *Climate Policy*, vol. 8, no. 2, pp. 183–193, doi:10.3763/cpol.2007.0435.

Asian Development Bank. (2012). *Kingdom of Cambodia: Mainstreaming Climate Resilience into Development Planning (Financed by the ADB Strategic Climate Fund)*. Tokyo: Asia Development Bank.

Ball, J. 2015, 'Who Will Pay for Climate Change?' *The New Republic*, 4 November 2015.

Carbon Nexus 2016, 'Introduction', viewed at <https://nexusfordevelopment.org/>.

Chandler, D. 2014, 'Beyond Neoliberalism: Resilience, the New Art of Governing Complexity, *Resilience*, vol. 2, no. 1, pp. 47–63, doi:10.1080/21693293.2013.878544.

CIF 2011. *Strategic Program for Climate Resilience – Cambodia*, Climate Investment Funds, Cape Town.

Clouse, C. 2020, 'In the Battle to Save Forests, a Make-or-Break Moment for REDD +', *MongaBay*, viewed at, <https://news.mongabay.com/2020/07/in-the-battle-to-save-forests-a-make-or-break-moment-for-redd/>.

Cock, A. 2017, *Governing Cambodia's Forests*, NIAS, Copenhagen.

Craig, D. & Porter, D. 2006, *Development Beyond Neoliberalism?: Governance, Poverty Reduction and Political Economy*, Cambridge University Press, Cambridge.

DeLanda, M. 2016, *Assemblage Theory*, Edinburgh University Press, Edinburgh.

Ear, S. 2009, 'The Political Economy of Aid and Regime Legitimacy in Cambodia', in J. Öjendal and M. Lilja (eds) *Beyond Democracy in Cambodia: Political Reconstruction in a Post-conflict Society*, Nias Press, Singapore, pp. 151–188.

Ear, S. 2013, *Aid Dependence in Cambodia: How Foreign Assistance Undermines Democracy*, Columbia University Press, Columbia.

Evans, B. & Reid, J. 2014, *Resilient Life: The Art of Living Dangerously*, John Wiley & Sons, London.

Frewer, T. 2013, 'Doing NGO Work: The Politics of Being "Civil Society" and Promoting "Good Governance" in Cambodia', *Australian Geographer*, vol. 44, no. 1, pp. 97–114.

Frewer, T. 2017, *Climate Assemblages: Governing the Vulnerable in a Neoliberal Era* (PhD), University of Sydney, Sydney.

GEF 2020, 'Enabling Cambodia to Prepare its First National Communication in Response to its Commitments to UNFCCC', viewed at, <https://www.thegef.org/project/enabling-cambodia-prepare-its-first-national-communication-response-its-commitments-unfccc>.

Global Climate Change Alliance 2012, *Paving the Way for Climate Compatible Development: Experiences from the Global Climate Change Alliance*, London, European Union.

Godfrey, M., Sophal, C., Kato, T., Vou Piseth, L., Dorina, P., Saravy, T., & Sovannarith, S. 2002, 'Technical Assistance and Capacity Development in an Aid-dependent Economy: The Experience of Cambodia', *World Development*, vol. 30, no. 3, pp. 355–373.

Goldman, M. 2006, *Imperial Nature: The World Bank and Struggles for Social Justice in the Age of Globalization*, Yale University Press, Yale.

Graeber, D. 2015, *The Utopia of Rules: On Technology, Stupidity, and the Secret Joys of Bureaucracy*, Melville House, New York.

Grove, K. J. 2014, 'Adaptation Machines and the Parasitic Politics of Life in Jamaican Disaster Resilience, *Antipode*, vol. 46, no. 3, pp. 611–628.

Hallegatte, S., Bangalore, M., Bonzanigo, L., Fay, M., Kane, T., Narloch, U. & Vogt-Schilb, A., 2015, *Shock Waves: Managing the Impacts of Climate Change on Poverty*, The World Bank, Washington D.C.

Hughes, C. 2003, *Political Economy of the Cambodian Transition*, Routledge, Oxon.

Hughes, C. 2009, *Dependent Communities: Aid and Politics in Cambodia and East Timor*, Cornell University, Ithaca.

Käkönen, M., Lebel, L., Karhunmaa, K., Dany, V. & Try, T. 2014, 'Rendering Climate Change Governable in the Least-Developed Countries: Policy Narratives and Expert Technologies in Cambodia', *Forum for Development Studies*, vol. 41, no. 3, pp. 351–376.

Chakrya, K., Sassoon, A. M., Seangly, P., Crane, B. & Kaliyann, T. 2016, 'Scenes From a Drought: Across the Country, Water Scarcity is Threatening Health and Livelihoods', *Phnom Penh Post*.

Lang, C. 2018, 'Virgin Atlantic Has Stopped Buying Carbon Credits from the Oddar Meanchey REDD Project', *REDD-Monitor*, viewed at, <https://redd-monitor.org/2018/01/10/virgin-atlantic-has-stopped-buying-carbon-credits-from-the-oddar-mea nchey-redd-project/#:~:text=On%209%20January%202018%2C%20Virgin,emissio ns%20from%20the%20aviation%20sector>.

Lewis, D. & Mosse, D. 2006, *Development Brokers and Translators: The Ethnography of Aid and Agencies*, Kumarian Press, London.

Maplecroft 2019, 'Climate Vulnerability Index', viewed at, <https://www.maplecroft. com/risk-indices/climate-change-vulnerability-index/>.

Mattar, S., Kansuk, S., & Jafry, T. 2019, 'Global Climate Finance is Still Not Reaching Those Who Need It Most', *The Conversation*, viewed at, <https://theconversa tion.com/global-climate-finance-is-still-not-reaching-those-who-need-it-most-115268/>.

McDonald-Gibson, C. 2003, 'CDM: Clean and Green, but Are the Prospects Too Lean?' *The Phnom Penh Post*, Phnom Penh.

Milne, S. & Mahanty, S. 2019, 'Value and Bureaucratic Violence in the Green Economy', *Geoforum*, vol. 98, pp. 133–143

Milne, S., Mahanty, S., To, P., Dressler, W., Kanowski, P., & Thavat, M. 2019, 'Learning From "Actually Existing" REDD+: A Synthesis of Ethnographic Findings', *Conservation & Society*, vol. 17, no. 1, pp. 84–95, doi:10.2307/26554473.

Milne, S., Pak, K., & Sullivan, M. 2015, 'Shackled to Nature? The Post-Conflict State and Its Symbiotic Relationship with Natural Resources', in S. Milne & S. Mahanty (eds), *Conservation and Development in Cambodia: Exploring Frontiers of Change in Nature, State and Society*, Routledge, Oxon, pp. 28–50.

Ministry of Environment, Royal Government of Cambodia 2006, 'National Adaptation Programme of Action to Climate Change', Ministry of Environment, Phnom Penh.

Morgenbesser, L. 2017, 'The Failure of Democratisation by Elections in Cambodia', *Contemporary Politics*, vol. 23, no. 2, pp. 135–155.

Pak, K., Horng, V., Eng, N., Ann, S., Kim, S., Knowles, J., & Craig, D. 2007, *Working Paper 34: Critical Literature Review on Accountability and Neo-Patrimonialism: Theoretical Discussions and the Case of Cambodia*, Cambodia Development Resource Institute, Phnom Penh, Cambodia.

Rich, B. 2013, *Foreclosing the Future: The World Bank and the Politics of Environmental Destruction*, Island Press, Washington D.C.

Roberts, S. M., Jones, J. P.III & Fröhling, O. 2005, 'NGOs and the Globalization of Managerialism: A Research Framework, *World Development*, vol. 33, no. 11, pp. 1845–1864, doi:10.1016/j.worlddev.2005.07.004.

UNDP 2011, *UN Collaborative Programme on Reducing Emissions from Deforestation and Forest Degradation in Developing Countries National Programme Document*, UNDP, Phnom Penh.

UNDP 2013, 'Forest Carbon Partnership Facility REDD+ Readiness Project', viewed at, <https://www.kh.undp.org/content/cambodia/en/home/operations/projects/environm ent_and_energy/forest-carbon-partnership-facility-redd–readiness-project-.html>.

UNDP 2015, 'Building Resilient Communities in Cambodia', viewed at, https://www.adaptation-undp.org/resources/videos/building-resilient-communities-cambodia.

UNDP Cambodia 2015, 'In Cambodia Villages Take Entrepreneurial Step in Solving Water Woes', *UNDP Newsletter*, 30 June 2015.

Work, C., Rong, V., Song, D. & Scheidel, A. 2019, 'Maladaptation and Development as Usual? Investigating Climate Change Mitigation and Adaptation Projects in Cambodia', *Climate Policy*, vol. 19 (sup1), pp. S47–S62, doi:10.1080/14693062.2018.1527677.

World Bank 2019, 'Engaging Civil Society for Better Development Impact', viewed at, <https://www.worldbank.org/en/news/feature/2019/08/20/engaging-civil-society-for-better-development-impact>.

Young, Z. 2002, *A New Green Order? The World Bank and the Politics of the Global Environment Facility*, Pluto Press, New York.

4 The politics of climate policy integration and land use in Indonesia

Monica Di Gregorio

Introduction

Indonesia started to seriously engage in national climate change policy development at the time of 13th Conference of the Parties of the United Nations Framework Convention on Climate Change, which the country hosted in Bali in 2007. President Yudhoyono saw the climate change agenda as a way to enhance Indonesia's and his own role in global diplomacy and to attract support for disaster mitigation and relief (Zhida 2012). In 2009, he announced mitigation plans to reduce greenhouse gas emissions by 26 percent without and 41 percent with international support by 2020 compared to business-as-usual, establishing himself as a climate change leader in the region. In its first Nationally Determined Contributions (NDCs) to the UNFCCC, Indonesia have extended the deadline to 2030, but maintains high ambitions to curb its emissions from the land use sector. At the same time, the Indonesian archipelago has long been vulnerable to several climate change impacts, in particular rising sea levels and extreme weather events. El Niño events in 2003 and 2006 led to droughts that reduced the ability of households to meet their food requirements (Boer & Perdinan 2008; Boer & Subbiah 2005, Rizaldi & Suharnoto 2012) reinforcing the need for action. However, most focus of global climate change processes under the UNFCCC has been on climate change mitigation.

Over two thirds of greenhouse gas emissions in Indonesia derive from the land use sector and are primarily linked to the conversion of forests into large-scale agriculture, predominantly palm oil monoculture plantations. Small-scale agriculture, often part of multifunctional landscape mosaics, is the second largest driver of conversion of forest (van Noordwijk et al. 2006; Global Forest Watch 2020). Indonesia's energy sector contributes about one third of greenhouse gas emissions and its share will increase over time. Both a fossil fuel producer and large exporter of coal, Indonesia faces major lock-ins in the energy sector (Fünfgeld 2019; Edwards 2019). Yet, at present the land use sector remains by far the country's main source of emissions, and pressure on forest remains high. Weaknesses in forest governance and climate change impacts linked to El Niño-related droughts interact periodically and create

DOI: 10.4324/9780429324680-6

feedback loops that further increase carbon emissions, leading to extensive forest fires with devastating environmental and health impacts not just in Indonesia, but also in Singapore and other neighbouring Southeast Asian countries (Koplitz et al. 2016).

Effective integration of climate change mitigation, adaptation, and economic development objectives is particularly important in the land use sector, where uncoordinated efforts will reduce both effectiveness and equity of climate action (Di Gregorio et al. 2017). This is because the land use sector supports a large number of smallholders and provides valuable natural resources for economic development (Nunan 2017). Moreover, unlike other sectors like energy, interaction between climate change mitigation and adaptation is ubiquitous in land use, which makes it particularly important to address trade-offs between policy objectives linked to emissions reductions and those aimed at reducing vulnerability to climate change in order to achieve effective, cost-efficient, and equitable climate action (IPCC 2019; Klein, Schipper & Dessai 2005; Locatelli et al. 2015).

This chapter focuses on investigating the policies, politics, and practice of climate policy integration in the land use sector in Indonesia, with a particular focus on integrating mitigation and adaptation responses and mainstreaming climate change in land use and development policies. It first illustrates the state of climate change policy and the level of climate policy integration in Indonesia. Then, it discusses the political economy of climate policy integration, its main challenges, and possible ways forward.

The state of climate policy integration in Indonesia's land use sector

Climate policy integration in the land use sector requires two important processes to occur. First, policy processes and resulting policy objectives in the climate change mitigation and adaptation subdomains need to be integrated with each other. Such "internal climate policy integration" is necessary to minimize trade-offs and better exploit synergies between climate change responses (Locatelli et al. 2015). Second, it also requires 'external climate policy integration', which refers to mainstreaming either mitigation or adaptation objectives into sectoral policy processes within individual sectors (vertical dimension) as well as across different sectors (horizontal dimension), such as forestry, agriculture, and broader development policy processes (Di Gregorio et al. 2017). The investigation of Indonesia's climate change policy processes sheds light on the opportunities and challenges that the country faces in achieving effective integration of mitigation and the adaptation agendas into development policies.

The establishment of the climate change policy architecture

In Indonesia, the first climate change committee was established in 1992 under the State Ministry for the Environment in preparation for its National

Climate Communications under the UNFCCC. It would take another 15 years and Indonesia's hosting of the 13th Conference of Parties in 2007 for the first National Action Plan Addressing Climate Change to be released. The Plan stresses the need to address the interlinked agrarian, water, and infrastructure crises. It further identifies priority areas for mainstreaming mitigation through a triple-track strategy – a 'pro-poor', 'pro-job', and 'pro-growth' agenda – and the need to mainstream climate change adaptation into national development plans. Yet, it is a very general document lacking details concerning mandates, identification of climate actions, as well as resource allocation (RoI 2007). After Bali, a semi-independent National Council on Climate Change, led by a former businessman and composed of 16 ministerial representatives, was set up to coordinate mainstreaming of climate change in sectoral and national development policies. The Council reported directly to the Executive Office of the President. Its seven working groups had multistakeholder representation, including government officials, academics, NGOs, private sector and community representatives (Purnomo et al. 2013).

The climate change mitigation agenda: 'Going it alone'

Climate change action became a priority for Indonesia's former President Yudhoyono, when, at the 2009 G20 meeting, he announced Indonesia's pledge to reduce emissions by 26 percent from a business-as-usual baseline by 2020 and up to 41 percent with international support, very much to the surprise of relevant ministries. Since the 2007 Bali Roadmap, Indonesia embraced the international agenda on Reducing Emissions from Deforestation and Forest Degradation (REDD+), largely thanks to financing from the government of Norway and further multilateral and bilateral funding. The Letter of Intent (LoI) between Norway and Indonesia led also to the establishment of an independent REDD+ Task Force (KepPres 25/2011) and the subsequent REDD+ Agency (PerPres 62/2013). The Task Force's mandate was to design REDD+ policies, while the REDD+ Agency was responsible for the implementation of REDD+. These semi-independent bodies operated directly under the Office of the President through the President's Delivery Unit for Development Monitoring and Oversight.

Another development linked to the agreement with Norway was the release in 2011 of the Moratorium on the Clearing of Primary Forests (KepPres 10/2011). Together with the National Action Plan on GHG Emission Reductions (KepPres 61/2011) and the Presidential Regulation on GHG Inventories (PerPres 71/2011), these three regulations set out Indonesia's main trajectory on climate change mitigation. That year, Indonesia's forests and peatlands became the main targets for national emission reductions, representing 87 percent of planned reduction. REDD+ would be the main mechanism to achieve these reductions. Agriculture played a minimal role in mitigation plans contributing just 1 percent (Bappenas 2011). The emission reductions plan emphasizes that climate mitigation is an integral part of the Indonesian

National Development Plan, and frames mitigation targets as co-benefits of development activities.

Major changes in the mitigation agenda occurred at the end of 2014 with the election of President Widodo, whose policy priorities did not include climate change. To streamline government, he merged the forestry and the environment ministries. The newly established Ministry of Environment and Forestry gained control of the REDD+ agenda, leading to the dismantling of the semi-independent National Council on Climate Change and the REDD+ Agency. Their functions were incorporated under the new Ministry's Directorate General of Climate Change Oversight (PerPres 16/2015; Widiaryanto 2015). The new Ministry also established a Steering Committee on Climate Change that included government and non-government actors. These changes transformed the constellation of power around climate change within government. Consequently, the earlier vertical form of climate change policy architecture, where all major agencies reported directly to the President, was replaced by a horizontal approach under the lead of the new Ministry and coordination support from the Ministry of National Development Planning.

Further developments occurred after the massive forest fires of 2015, which caused over 100,000 premature deaths in Southeast Asia (Koplitz et al. 2016). Evidence shows that the fires were linked to the expansion of palm oil concessions and that emissions from fires were highest in degraded peatlands (Marlier et al. 2015a; Marlier et al. 2015b). In response, in 2016 the President established the Peatland Restoration Agency (PerPres 1 2016) which aimed to restore 2 million hectares of degraded peatlands by 2020 (Sari, Dohong & Wardhana 2021). At the same time, the President made the temporary Moratorium on the Clearing of Primary Forests permanent (Inpres6 2017). And the following year, a three-year Moratorium on New Oil Palm Concessions on Peatland followed (Inpres8 2018). Overall, mitigation strategies and efforts in the land use sector on the part of the new Ministry of the Environment and Forestry have occurred largely in isolation from climate change adaptation, which denote a lack of internal climate policy integration.

Indonesia's first NDC submitted to the UNFCCC in November 2016 extends the deadline of the 29 percent and 41 percent (with international support) emission reduction targets from 2020 to 2030 (RoI 2016). The NDC indicates that emissions reductions will come primarily from the forest and energy sectors, the vast majority still from forestry (17 percent out of 29 and 23 percent out of 41 respectively). Since 2009, the mitigation targets made Indonesia a climate change leader in Southeast Asia. In 2020, Indonesia qualified for REDD+ compensation payments for USD$56 million for 11.2 tCO$_2$e of emission reductions achieved during 2017 (Todung 2020). The payment has attracted some criticism, as deforestation rates stabilized in 2018 and 2019 and forest fires still threaten permanence. Major reductions in the deforestation rates, including in 2020, are attributed to unusually low oil palm prices (Gaveau et al. 2021). Additionally, major reductions in 2020 are attributed to the combination of low prices and a very wet year. Whether

NDC targets will be met will depends on future trajectories (Wijaya et al. 2017). Finally, in the 2020 NDC revisions process, Indonesia indicated that it would not have the intention to increase its mitigation ambitions, citing COVID-19 impacts on the economy as one of the reasons (Jong 2021).

The climate change adaptation agenda and the search for integration with mitigation

Compared to the mitigation agenda, the climate adaptation agenda progressed much more slowly and remains underdeveloped in terms of scope, breadth, and funding. The coordinating Ministry of National Development Planning is the country's lead adaptation agency. Since 2012, it has hosted the National Coordination Team on Climate Change chaired by the Deputy Minister of Natural Resources and Environment (Kep.38/M.PPN/HK/03/ 2012 2012). Later that year, the Ministry released the Strategy for Mainstreaming Adaptation into National Development Planning, and two years later the National Action Plan on Climate Change Adaptation (Bappenas 2013). The two policies are meant to guide both sectoral and cross-sectoral climate adaptation actions, direct priorities for adaptation, and improve coordination across sectors and government levels. The action plan is directed to "(a) reduce the effects of climate change to a minimum, (b) increase the resilience and/or reduce the level of vulnerability of natural system, livelihood, programs and activities to the impacts of climate change" (Bappenas 2014, p.29).

Priority areas for adaptation are the health and disaster management sectors, where considerable interactions exist with climate impacts in Indonesia. The two objectives that are most relevant for the land use sector are food security, which focuses on the agricultural sector, and maintaining forest, other ecosystems, and biodiversity. Food security is to be achieved through the adjustment and development of climate-resilient farming systems. Key climate-resilient action to reduce climate risk include food diversification and climate-proof irrigation, infrastructure improvements, the use of adaptive technologies, and the development and optimization of land use, maintenance of water and genetic resources, and improved climate information and communication systems. Ecosystem resilience focuses on securing and protecting water resources from extreme weather events, avoiding ecosystem and biodiversity loss, ensuring the sustainability of water supply, and conservation of ecosystems and biodiversity. These objectives include various forest management activities, forest and land rehabilitation, improved governance of conservation areas, and control of forest and land fires, which are also major mitigation actions.

Unlike the mitigation agenda, since its inception in 2014 the Indonesian adaptation agenda makes a substantive effort to link to existing forest-based mitigation actions, highlighting the synergies between adaptation and mitigation. It is also geared largely to complement the national economic development agenda. Adaptation and development objectives are seen largely as complementary, underlying synergies between the two policy aims (Bappenas 2012; Di Gregorio et al. 2015b). However, potential trade-offs between short-term and long-term objectives of both development and adaptation aims risk remaining unexplored, potentially threatening resilience (Kok & de Coninck 2007).

Limited integration of climate change objectives into development plans

Considerable efforts still need to be made to effectively integrate climate change objectives in economic development policies. The 2010 Presidential Regulation on the National Medium-Term Development Plan contains brief references to climate mitigation targets as well as to the need to address adaptation to ensure food security and strengthen disaster management. The Plan only mentions climate change mainstreaming once and in very general terms (Bappenas 2010, p.32). The Masterplan on the Acceleration and Expansion of Indonesian Economic Development 2011–2025 released in 2011, while mentioning Indonesia's emission reduction policy and the plans to develop REDD+, does not elaborate on climate change or on its linkages to development plans. Instead, it highlights economic growth targets, such as the potential to expand palm oil, mining, and forest plantations in Kalimantan, and neither considers nor discusses potential trade-offs with mitigation targets. The only brief mention of a concrete climate impact refers to the effect of droughts on rice production. In addition, it neglects to discuss adaptation measures or the need to mainstream adaptation into development strategies. The same applies to the latest National Long-Term Development Plan (2005–2025) (Law No. 17/2007). Overall, national development plans do not seem to consider climate change and development linkages in any depth, nor do they discuss or assess in any detail the potential trade-offs between development plans and climate change objectives. However, 2019 saw an extensive policy development effort to integrate mitigation into development plans. The Ministry of National Development Planning published the Low Carbon Development Initiative (LCDI) assessment, which foresees such integration into the 2020–2024 five-year development plan. One of the seven main development agendas of the document relates to the environment, and features both disaster and climate change resilience and low carbon development as main goals together with environmental quality (Bappenas 2019; PerPres 18 2020).

Before the 2020–2024 development plan, the only main policy document that focused on the integration of adaptation into development was the 2012 Strategy for Mainstreaming Adaptation into National Development Planning (Bappenas 2012). The strategy aimed at developing an integrated and cross-sectoral plan for mainstreaming adaptation in planning, implementation, and evaluation of development processes. It also served as a background study for the 2015–2019 Mid-Term Development Plan and was to be followed by several sectoral strategies on mainstreaming adaptation. The 2010 Road Map

Strategy of the Agricultural Sector Addressing Climate Change and the Indonesia Climate Change Sectoral Roadmap Synthesis Report aimed to provide a starting point for more extensive integration. Through these policy processes, the Ministry of National Development Planning has established itself as the lead agency in external climate policy integration efforts in Indonesia. Slow progress in the past has accelerated in recent years. Yet, how effectively climate change objectives will be integrated in the implementation of the 2020–2024 five-year development plan remains to be seen.

The political economy of climate change policy integration

Achieving effective climate policy integration is not just a matter of careful policy formulation and planning aimed at facilitating both vertical and horizontal linkages. The political-economic processes within the land use sector are key determinants of whether and how integration occurs.

They include: i) the interests and associated incentives facing different policy actors and how they exercise power within the land use sector; ii) the role of formal and informal institutions that shape social, political, and economic interaction; iii) and the impacts of values and ideas, such as political ideology and culture on political behaviour of these policy actors (Hall 1997). Recent approaches to the political economy of climate change have called for more attention to the role of the complex nature of climate change as an environmental and societal problem, and how interests and institutions and resulting power relations mediate competing claims to resources within multi-level climate change policy processes. In turn, ideas – in our case the social construction of climate change itself and the very understanding of climate change policies and responses – in turn shape ideology, incentives, and power relations (Tanner and Allouche 2011). The outcomes of these processes vary from the reproduction of existing institutions to institutional transformation. Below, we discuss how, in Indonesia, the political economy conditions have impacted the integration of climate change objectives in land use policy and practice.

The politics of internal climate policy integration

Interests: The scramble for mitigation control

Mitigation and adaptation have different characteristics, including their distinct nature as goods. Mitigation is a typical global public good, while adaptation resembles a more private good (Hasson, Löfgren & Visser 2010; Nordhaus 1994). At the international level, mitigation and adaptation compete for resources. Developed countries have a stronger interest in funding mitigation compared to adaptation action in developing countries, because mitigation benefits are shared across all countries, while adaptation benefits are experienced mainly locally (Bustamante et al. 2014; Ciplet, Roberts &

Khan 2013). The incentives for the Global North to focus international support on mitigation as opposed to adaptation are even stronger if developed countries can offset their own emissions in return for such funding (Kaul, Grunberg & Stern 1999). It is no secret that the push for land-based mitigation measures to be undertaken in the Global South has been supported by the hope of countries from the Global North to use these measures to be able to offset part of their own emissions (Dooley & Gupta 2017).

In 2010, Indonesia's major climate change mitigation funder, Norway, and the Indonesian President found a common interest in setting up autonomous agencies to lead climate change mitigation policy (the National Council on Climate Change and the REDD+ Agency) instead of relying on the state bureaucracy. Norway was wary to relinquish control on mitigation decisions to the Indonesian bureaucracy and was concerned of special interests putting pressure on ministries in order to hamper progress on climate change mitigation. At the same time, the use of semi-independent agencies allowed the President to retain substantial control and oversight to ensure compliance by ministries. However, this approach led to increased resentment on the part of major ministries who felt sidelined, in particular the Ministry of Forestry and the Ministry of Environment (Di Gregorio et al. 2015b). The balance of power shifted dramatically in 2014 with Widodo's election, as his interests did not include climate change objectives (Alisjahban and Busch 2017). In the government restructuring that followed, the newly merged Ministry of Environment and Forestry gained control over the climate mitigation agenda. The loss of presidential oversight on climate change weakened vertical climate policy integration as climate change policy progress increasingly moved under the control of the bureaucracy (Di Gregorio et al. 2015b).

Institutions: Hampering and facilitating internal climate policy integration

The focus of global financial institutions on supporting mitigation as opposed to adaptation actions in the land use sector is reflected nationally, as the Ministry of Environment and Forestry has an almost exclusive focus on mitigation, without explicitly recognizing or seeking internal climate policy coherence with adaptation policies. Within the ministry, mitigation and adaptation activities remain largely disconnected and independently pursued by different teams (Di Gregorio et al. 2019; Di Gregorio et al. 2015b). The work on forestry remains largely shaped by the earlier institutional structure developed under the former Ministry of Forestry, which was an exclusively sectoral line ministry working primarily through vertical integration efforts, with limited expertise in coordinating horizontal integration, whether internally across departments or externally across sectors.

Instead, the main efforts to facilitate internal climate policy integration in land use are undertaken by the leading climate adaptation actor, the Ministry of National Development Planning. A coordinating ministry, it has substantial expertise and is institutionally structured to bring together other line ministries and facilitate policy integration. As the responsible agency for the coordination of the spatial planning development process, it has a prominent role in coordinating the integration of climate mitigation policy processes across sectors and across governance levels. In particular, it has become a major supporter of internal climate policy integration. The Ministry's climate change policies discuss synergies between mitigation and adaptation in some detail. They refer to REDD+ activities, which are traditionally understood as climate change mitigation measures, as main climate change adaptation measures. These include forest management, forest conservation and rehabilitation, and the control of forest and land fires aimed at maintaining forest ecosystems, avoiding ecosystem and biodiversity loss, and ensuring the sustainability of water supply in land use (Di Gregorio et al. 2017). However, effective climate change integration will not be achieved in land use without active engagement of the main mitigation actor, the Ministry of Environment and Forestry.

Ideas: Mitigation and adaptation as separate discourses

Another consideration that makes it difficult to effectively integrate mitigation and adaptation policies relates to the differences in ideas – in our case, the very understanding of climate change – across levels of governance. At the level of global climate change institutions, mitigation and adaptation are largely pursued separately (Locatelli et al. 2016). In the land use sector, climate mitigation is usually associated with tropical forests, while adaptation needs in forests are not extensively acknowledged. In part this is due to the enduring wilderness myth around tropical forests, which wrongly envisions them as devoid of or of little use to people (Oates 1999; Clement et al. 2015; Denevan 1992). Conversely, adaptation is associated with agriculture, largely because of its link to food security. Mitigation in agriculture is less discussed, in part because of the trade-offs with food security (IPCC 2019). This separation between mitigation and adaptation in land use discourses at the global level is mirrored also at the national level (Somorin et al. 2012).

In Indonesia, national level policy processes have been driven by two separate discourse coalitions. The dominant coalition focuses on climate mitigation, while a less powerful coalition focuses on adaptation; and while there is some cross-coalition interaction, they operate largely independently. At the local level, however, climate change discourses are more holistic, closely integrated, and consider mitigation demands, adaptation needs, and economic development objectives together. Here, mitigation is largely seen as a co-benefit from sustainable development and climate change adaptation policies. Yet, because national level actors control funding and dominate policy decisions, local concerns about climate adaptation and integration with local development aims remain largely neglected (Di Gregorio et al. 2019). The Ministry of National Development Planning, which leads the integration between mitigation and adaptation, takes an explicitly ecosystem-based adaptation approach that pursues synergies and facilitates policy integration of adaptation with mitigation efforts (Pramova et al. 2012). But such approaches are not widespread among other key policy actors.

Thus, imbalances in interests in mitigation and adaptation of dominant actors, institutional path dependencies established by global and national organizational structures, and the discursive separation of mitigation and adaptation hampered internal climate policy integration in Indonesia. Next, we analyse how the political economy of the land use sector has impacted external climate policy integration.

The politics of external climate policy integration

Interests: Drivers of deforestation and agrifood systems

The politics of external climate policy integration in land use relates primarily to mainstreaming either of the two climate change responses of mitigation or adaptation into the forest and agriculture sectors. In Indonesia, this politics is closely linked to the political economy of the drivers of deforestation and of agrifood systems. GHG emissions from land use are primarily due to emissions from the conversion of natural forests into agriculture. Between 2000 and 2010, Indonesia lost over 14.5 million hectares of forests. Just under half of the loss occurred in industrial concessions, respectively in pulpwood (12.8 percent), logging (12.5 percent), and oil palm (11 percent), which led to estimated emissions of 11–19 million tons of CO_2. More than a quarter of pulpwood plantations and one-fifth of pulpwood concessions are located on peatlands, and around half of these are deep peat, which puts them at high risk of long-term CO_2 emissions and forest fires (Abood et al. 2015). Half of the land planted with oil palm is held by large-scale plantation operations, controlled by Indonesian and foreign (in particular Malaysian) capital owners. These powerful industrial interests, linked to forest exploitation and conversion of forest into large-scale agriculture, pose major challenges to climate mitigation policy action. Their structural power explains slow progress and political inaction in addressing the drivers of deforestation head on (Di Gregorio et al. 2012).

Evidence of such structural power includes the weakening of the 2011 Moratorium on the Clearing of Primary Forests, through lobbying by the timber, pulp, and paper and oil palm industries to exclude existing concessions and secondary forests. This leaves as many as 90,000 km^2 of logged and relatively threatened forest unprotected, with 46.7 million hectares of degraded forest that can still be designated for new concessions (Afiff 2016; Sloan, Edwards & Laurance 2012; Murdiyarso et al. 2011). Local simulations indicate that the implementation of the moratorium is unlikely to be effective at substantially reducing carbon emissions, because companies convert secondary forests instead of primary forests (Suwarno et al. 2018). Conveniently for businesses, millions of hectares of new concessions were granted on 31st December 2010, the day before the moratorium was first planned to come into force (Edwards, Koh & Laurance

2012). Similarly, while the moratorium on oil palm concessions was under discussion, the Ministry of Environment and Forestry released several tracts of state forest lands, classified as conversion forest, to oil palm companies, with one program allocating close to 13 million hectares for conversion into agriculture (Mongabay 2018; Purnomo et al. 2020). In 2017, the Ministry of Environment and Forestry passed new regulations stating that companies with timber plantation concessions holding land affected by the moratorium could apply for land swaps (PermenLHK 17 2017; PermenLHK 40 2017). While this might reduce some emissions from peatlands, land swaps open up further secondary forests to exploitation, which means that part of the emissions will only be displaced. They also have equity implications as they are likely to overlap with existing local level land use permits (Jong 2018). The end result is the limitation of the ability of the moratoria to substantially contribute to reducing carbon emissions from forests.

Furthermore, oil palm plantation companies have long appointed advisors and board members who are retired high-level civil servants, politicians, and military personnel, at both national and local level, to build patronage networks, which allow them to both resist and disregard enforcement of environmental policies (Varkkey 2013). Indonesia's latest development plan includes further expansion for palm oil for both food and biofuel production. The target for biofuel expansion alone is an additional 4 million hectares between 2016–2025, while long-term plans for oil palm development reach 20 or 30 million hectares (Brockhaus et al. 2012; Li 2018; Cramb & McCarthy 2016). The Indonesian government has long internalized the interests of the timber industry and of oil palm agribusinesses in their development plans. Such structural power is exercised even without lobbying on their part and has major influence on both policy design and implementation.

Institutions: Breakdowns and lock-ins

Historically, interlocking mechanisms that have accelerated forest exploitation have also been linked to collusion between the forest bureaucracy and the private sector (Barr 1998). Rent seizing, or the practice of state actors gaining the right to allocate rent, and rent-seeking behaviour, where private actors pursue economic benefits through the political system, led to the institutional breakdown of sustainable forest management during the Suharto era (Ross 2001). Rent seizing opportunities are facilitated by the structure of property rights, the historical control of the Ministry of Forestry over state forest lands, its control over the allocation of concessions for forest exploitation, and the resulting insecurity of land tenure of small-scale farmers. An oligarchy that emerged in the 1980s as timber barons with logging and plywood export interests maintained strong military ties (Barr 1998) later gave way to industrial forest plantations for large-scale pulp production (Afrizal & Anderson 2010) as demand from China and Japan drove illegal logging (Gellert 2005; Brockhaus et al. 2012). Not even the IMF's efforts to liberalize the forest sector after the 1997 financial crises would dismantle this system of forest exploitation (Barr and Setiono 2001; Hapsari 2011). In the 1990s, clear-cutting of forest areas for oil palm development increased dramatically (Casson 2000) leading to over 14 million hectares of oil palm by 2017, and making Indonesia the largest exporter of palm oil (Varkkey, Tyson & Choiruzzad 2018; Jelsma et al. 2017). In the early 2000s, decentralization, which was envisioned as a way to empower localities, also decentralized rent-seeking activities as oil palm development came increasingly under the control of local governments. Oil palm transformations of forest landscapes are much more extensive than the impact of logging, because they are "deep, massive and permanent" (Li 2018, p.330). Critics suggest that these institutional lock-ins are geared toward disenfranchising local people to the detriment of their local livelihoods and restricting their ability to stage forms of resistance to the benefits of investors and of the state (McCarthy and Robinson 2016; Li 2018; Peluso 1992).

Ideas: Win-win rhetoric and development discourse

The win-win ecological modernization rhetoric, which envisions economic growth alongside sustainability is used by a variety of actors in the climate change domain as a positive and unproblematic portrayal of climate action, particularly among state actors (Di Gregorio et al. 2015a). The Indonesian government discourse around climate mitigation showed this dual approach. Under the Yudhoyono Presidency, the government portrayed Indonesia as a regional climate leader with high ambitions for emission reductions (Aspinall, Mietzner & Tomsa 2015). Under the Widodo Presidency, discourse about climate change has continued to embrace notions of 'green growth' and 'low emission development'. Yet, on closer inspection, this rhetoric is accompanied by contradictory policies and reveals an attitude of "having your cake and eating it too" (Cronin et al. 2016). Such an approach does not openly recognize the existence of trade-offs between climate change responses and land use practices and plans (IPCC 2019), which is crucial in pursuing external climate policy integration (Di Gregorio et al. 2017). At the same time, despite the rhetoric, actors seem to believe that growth and emission reductions might not always go hand in hand. For example, during the 2020 NDC enhanced process debates, government officials stated that Indonesia will not compromise on economic growth ambitions for the sake of climate action (Jong 2020). Within this frame, older rhetoric of industry and planners presenting expansion of plantations as a unique opportunity to bring rural people into modernity and transform under-utilized resources into high productivity agriculture continue to resonate and remain influential (Li 2018). Policy actors who underline the importance of climate justice in mainstreaming mitigation remain a minority voice in Indonesia and yield less influence in climate policy (Di Gregorio et al. 2017). Under these circumstances, drawing attention to the need to reduce vulnerability of local populations to climate impacts and to strengthen local adaptive capacity remains a challenge. Yet, for effective climate policy integration in the land use sector, Indonesia needs an integrated approach that mainstreams climate change responses into both forestry and agriculture. This will require the integration and pursuit of both REDD+ mitigation and climate-smart agriculture (Carter et al. 2018).

Conclusion

In Indonesia, the politics behind the processes of climate policy integration in the land use sector are influenced by three interacting processes: First, the international focus and funding availability from development aid for mitigation, as opposed to adaptation; second, the domestic political economy of the drivers of greenhouse gas emissions from the land use sector; and third, the different priorities and understandings of climate change and development linkages across policy actors and levels of governance. Indonesia has taken substantial steps to formulate climate change mitigation policies to tackle emissions from deforestation. The use of moratoria on new forest concessions, peatlands, and new palm oil concessions are examples of vertical climate policy integration processes in which the Office of the President took the lead to mandate sectoral policy changes in forestry and agriculture. But overall, the land use mitigation agenda is firmly in the hands of the Ministry of Environment and Forestry, which controls most of the forest estate as well as local forestry services. At the same time, business interests behind the drivers of deforestation have been able to use their structural power to limit the impacts of these policies on their balance sheets, in practice reducing potential mitigation gains. With regards to land use and climate change adaptation, the lack of substantive policy initiatives is evident. It reveals major political inaction linked to limited international funding, lack of scientific knowledge, and lack of capacity to address local adaptation needs. This imbalance between climate mitigation and adaptation policies and actions in the land use sector limits the ability to achieve effective climate policy integration. Trade-offs between the two climate change responses remain largely ignored, while potential synergies are not fully exploited.

A variety of policy actors need to take action to enhance progress on the integration and mainstreaming of climate change in Indonesia's land use sector. First, multilateral and bilateral funding needs to provide more resources for adaptation, so that it is possible to rebalance the attention and efforts across mitigation and adaptation responses. Second, the Indonesian government should use the funding for climate adaptation to first expand knowledge about adaptation. This requires investments in local vulnerability assessments in rural communities and collaborative research between national and local universities on locally appropriate adaptation options. Third, local governments need to be endowed with resources for capacity building around adaptation to be able to enhance expertise on climate change within local government agencies. Resources should be used to effectively mainstream climate change adaptation into disaster risk reduction. This will first require training, capacity building, and enhanced expertise within local disaster management agencies. Only then can plans to build local adaptive capacity be put in place. Land use planning processes are among the most important venues to pursue the integration of mitigation, adaptation, and development objectives. These should explicitly investigate potential trade-offs between the three objectives, and policy formulation should explore ways to achieve synergies and reduce trade-offs wherever possible. Inevitably, policy decisions, especially where trade-offs cannot be avoided, will have distributional consequences. The best way to address these is to invest resources in facilitating inclusive decision-making processes with representation of different interests from the land use sector. Given the unbalanced playing field within the land use sector, mechanisms need to be put in place to effectively represent weaker interests, such as small-holders or marginalized groups. Finally, Indonesia's governance system needs to be able to envision how to break with past institutional and development trajectories favoring special interests in the land use sector and develop a new long-term vision that empowers local people to access and use land resources in climate-resilient ways.

References

Abood, S. A., Lee, J. S. H., Burivalova, Z., Garcia-Ulloa, J. & Koh, L. P. 2015, 'Relative Contributions of the Logging, Fiber, Oil Palm, and Mining Industries to Forest Loss in Indonesia', *Conservation Letters*, vol. 8, pp. 58–67.

Afiff, S. A. 2016, 'REDD, Land Management and the Politics of Forest and Land Tenure Reform with Special Reference to the Case of Central Kalimantan province', in J.F. McCarthy & K. Robinson (eds), *Land and Development in Indonesia: Searching for the People's Sovereignty*, ISEAS, Singapore.

Afrizal & Anderson, P. 2010, 'Industrial Plantations and Community Rights: Conflict and Solutions', in J.F. McCarthy & K. Robinson (eds), *Land and Development in Indonesia: Searching for the People's Sovereignty*, ISEAS, Singapore.

Alisjahban, A. S. & Busch, J. M. 2017, 'Forestry, Forest Fires, and Climate Change in Indonesia', *Bulletin of Indonesian Economic Studies*, vol. 53, pp. 111–136.

Aspinall, E., Mietzner, M. & Tomsa, D. 2015, *The Yudhoyono Presidency: Indonesia's Decade of Stability and Stagnation*, ISEAS, Singapore.

Bappenas (Ministry of National Development Planning) 2010, *Regulation of the President of the Republic of Indonesia Number 5 of 2010 Regarding the National Medium-term Development Plan (RPJMN) 2010–2014: Book 1, National Priorities*, Ministry of National Development Planning, Jakarta, Indonesia.

Bappenas (Ministry of National Development Planning) 2011, *Guideline for Implementing Greenhouse Gas Emission Reduction Action Plan*, Bappenas, Jakarta, Indonesia.

Bappenas (Ministry of National Development Planning) 2012, *The Strategy for Mainstreaming Adaptation in National Development Planning: Integration Framework*, Bappenas, Jakarta, Indonesia.

Bappenas (Menteri Negara Perencanaan Pembangunan Nasional) 2013, *Rencana Aksi Nasional Adaptasi Perubahan Iklim*, Menteri Negara Perencanaan Pembangunan Nasional, Jakarta, Indonesia.

Bappenas (Ministry of National Development Planning) 2014, *National Action Plan for Climate Change Adaptation (RAN-API)*, Bappenas, Jakarta, Indonesia.

Bappenas (Menteri Negara Perencanaan Pembangunan Nasional) 2019, *Pembangunan Rendah Karbon: Pergeseran Paradigma Menuju Ekonomi Hijau di Indonesia*, Menteri Negara Perencanaan Pembangunan Nasional, Jakarta, Indonesia.

Barr, C. M. 1998, 'Bob Hasan, the Rise of Apkindo, and the Shifting Dynamics of Control in Indonesia's Timber Sector. *Indonesia*, vol. 65, pp. 1–36.

Barr, C. M. & Setiono, B. 2001, 'Corporate Debt and Moral Hazard in Indonesia's Forestry Sector Industries, in C. M. Barr (ed.), *Banking on Sustainability: Structural Adjustment and Forestry Reform in Post-Suharto Indonesia*, CIFOR, Bogor, Indonesia.

Boer, R. & Perdinan 2008, '*Adaptation to Climate Variability and Climate Change: Its Socio-Economic Aspects*', EEPSEA Conference on Climate Change: Impacts, Adaptation, and Policy in South East Asia with a Focus on Economics, Socio-Economics and Institutional Aspects, Economic and Environmental Program for Southeast Asia,Bali, Indonesia.

Boer, R. & Subbiah, A. R. 2005, in V. K. Boken, A. P. Cracknell & R. L. Heathcote (eds.), *Monitoring and Predicting Agricultural Drought: A Global Study*, Oxford University Press, Oxford.

Brockhaus, M., Obidzinski, K., Dermawan, A., Laumonier, Y. &Luttrell, C. 2012, 'An Overview of Forest and Land Allocation Policies in Indonesia: Is the Current Framework Sufficient to Meet the Needs of REDD+?', *Forest Policy and Economics*, vol. 18, pp. 30–37.

Bustamante, M., Robledo-Abad, C., Harper, R., Mbow, C., Ravindranat, N. H., Sperling, F., Haberl, H., Pinto, A. S. & Smith, P. 2014, 'Co-benefits, Trade-Offs, Barriers and Policies for Greenhouse Gas Mitigation in the Agriculture, Forestry and Other Land Use (AFOLU) Sector, *Global Change Biology*, vol. 20, pp. 3270–3290.

Carter, S., Arts, B., Giller, K. E., Golcher, C. S., Kok, K., de Koning, J., van Noordwijk, M., Reidsma, P., Rufino, M. C., Salvini, G., Verchot, L., Wollenberg, E. & Herold, M. 2018, 'Climate-Smart Land Use Requires Local Solutions, Transdisciplinary Research, Policy Coherence and Transparency', *Carbon Management*, vol. 9, pp. 291–301.

Casson, A. 2000, *The Hesitant Boom: Indonesia's Oil Palm Sub-sector in an Era of Economic Crisis and Political Change*, Occasional Paper No. 29, CIFOR, Bogor, Indonesia.

Ciplet, D., Roberts, J. T. & Khan, M. 2013, 'The Politics of International Climate Adaptation Funding: Justice and Divisions in the Greenhouse', *Global Environmental Politics*, vol. 13, pp. 49–68.

Clement, C. R., Denevan, W. M., Heckenberger, M. J., Junqueira, A. B., Neves, E. G., Teixeira, W. G. & Woods, W. I. 2015, 'The Domestication of Amazonia Before European Conquest', *Proceedings of the Royal Society B: Biological Sciences*, vol. 282, no. 20150813.

Cramb, R. A. & McCarthy, J. F. 2016, *The Oil Palm Complex: Smallholders, Agribusiness and the State in Indonesia and Malaysia*, NUS Press, Singapore.

Cronin, T., Santoso, L., Di Gregorio, M., Brockhaus, M., Mardiah, S. & Muharrom, E. 2016, 'Moving Consensus and Managing Expectations: Media and REDD+ in Indonesia', *Climatic Change*, vol. 137, pp. 57–70.

Denevan, W. M. 1992, 'The Pristine Myth: The Landscape of the Americas in 1492', *Annals of the Association of American Geographers*, vol. 82, pp. 369–385.

Di Gregorio, M., Brockhaus, M., Cronin, T. & Muharrom, E. 2012, 'Politics and Power in National REDD+ Policy Processes, in A. Angelsen, M. Brockhaus, W. D. Sunderlin & L. V. Verchot (eds), *Analysing REDD+: Challenges and Choices,* Center for International Forestry Research, Bogor, Indonesia, pp. 69–90.

Di Gregorio, M., Brockhaus, M., Cronin, T., Muharrom, E., Mardiah, S. & Santoso, L. 2015a, 'Deadlock or Transformational Change? Exploring Public Discourse on REDD+ Across Seven Countries', *Global Environmental Politics*, vol. 15, pp. 63–84.

Di Gregorio, M., Nurrochmat, D. R., Fatorelli, L., Pramova, E., Locatelli, B., Brockhaus, M.& Sari, I. M. 2015b, *Integrating Mitigation and Adaptation in Climate and Land Use Policies in Indonesia*, CCCEP Working Paper and CIFOR Working Paper, University of Leeds and Center for International Policy Research, Leeds, UK and Bogor, Indonesia.

Di Gregorio, M., Nurrochmat, D. R., Paavola, J., Sari, I. M., Fatorelli, L., Pramova, E., Locatelli, B., Brockhaus, M. & Kusumadewi, S. D. 2017, 'Climate Policy Integration in the Land Use Sector: Mitigation, Adaptation and Sustainable Development Linkages' *Environmental Science & Policy*, 67, 35–43.

Di Gregorio, M., Fatorelli, L., Paavola, J., Locatelli, B., Pramova, E., Nurrochmat, D. R., May, P. H., Brockhaus, M., Sari, I. M. & Kusumadewi, S. D. 2019, 'Multi-level Governance and Power in Climate Change Policy Networks', *Global Environmental Change*, vol. 54, pp. 64–77.

Dooley, K. & Gupta, A. 2017, 'Governing by Expertise: The Contested Politics of (Accounting for) Land-based Mitigation in a New Climate Agreement', *International Environmental Agreements: Politics, Law and Economics*, vol. 17, pp. 483–500.

Edwards, D. P., Koh, L. P. & Laurance, W. F. 2012, 'Indonesia's REDD+ Pact: Saving Imperilled Forests or Business as Usual?' *Biological Conservation*, vol.151, pp. 41–44.

Edwards, G. A. S. 2019, 'Coal and Climate Change', *WIREs Climate Change*, vol. 10, e607.

Fünfgeld, A. 2019, 'Just Energy? Structures of Energy (In)Justice and the Indonesian Coal Sector' in T. E. Jafry, K. E. Helwig & M. E. Mikulewicz (eds), *Routledge Handbook of Climate Justice*, Routledge, London.

Gaveau, D., Locatelli, B., Salim, M., Husnayaen, H., Manurung, T., Descals, A., Angelsen, A., Meijaard, E. & Sheil, D. 2021, 'Slowing Deforestation in Indonesia Follows Declining Oil Palm Expansion and Lower Oil Prices', *Nature Climate Change*, in review.

Gellert, P. K. 2005, 'Oligarchy in the Timber Markets of Indonesia: From Apkindo to IBRA to the Future of the Forests' in B. P. Resosudarmo (ed.), *The Politics and Economics of Indonesia's Natural Resources*, ISEAS, Singapore.

Global Forest Watch 2020, 'Forest Monitoring Designed for Action', viewed 7 August 2020 at, <https://www.globalforestwatch.org >.

Hall, P. A. 1997, 'The Role of Interests, Institutions, and Ideas in the Comparative Political Economy of the Industrialized Nations', *Comparative Politics: Rationality, Culture, and Structure*, pp. 174–207.

Hapsari, M. 2011, 'The Political Economy of Forest Governance in Post-Suharto Indonesia', in H. Kimura, Suharko, A. B.Javier & A. Tangsupvattana (eds.), *Limits of Good Governance in Developing Countries*, Gadjah Mada University Press, Yogyakarta, Indonesia.

Hasson, R., Löfgren, Å. & Visser, M. 2010, 'Climate Change in a Public Goods Game: Investment Decision in Mitigation versus Adaptation', *Ecological Economics*, vol. 70, pp. 331–338.

InPres6 (President of the Republic of Indonesia) 2017, *Presidential Instruction No. 6 Year 2017 on Forest Moratorium*, Republic of Indonesia, Jakarta, Indonesia.

InPres8 (President of the Republic of Indonesia) 2018, *Instruction of the President of the Republic of Indonesia 8/2018 on the Delay and Evaluation of Licensing of Oil Palm Plantations and the Increase in Productivity of Oil Palm Plantations*, Republic of Indonesia, Jakarta, Indonesia.

IPCC 2019, *Climate Change and Land: An IPCC Special Report on Climate Change, Desertification, Land Degradation, Sustainable Land Management, Food Security, and Greenhouse Gas Fluxes in Terrestrial Ecosystems* [P.R. Shukla, J. Skea, E. Calvo Buendia, V. Masson-Delmotte, H.-O. Pörtner, D. C. Roberts, P. Zhai, R. Slade, S. Connors, R. van Diemen, M. Ferrat, E. Haughey, S. Luz, S. Neogi, M. Pathak, J. Petzold, J. Portugal Pereira, P. Vyas, E. Huntley, K. Kissick, M. Belkacemi, J. Malley, (eds.)], In Press.

Jelsma, I., Schoneveld, G. C., Zoomers, A. & van Westen, A. C. M. 2017, 'Unpacking Indonesia's Independent Oil Palm Smallholders: An Actor-Disaggregated Approach to Identifying Environmental and Social Performance Challenges', *Land Use Policy*, vol. 69, pp. 281–297.

Jong, H. N. 2018, 'Indonesia's Land Swap Program Puts Communities, Companies in a Bind' [Online], *Mongabay*, 27 August 2018, viewed 3 August 2020 at, <https://news.mongabay.com/2018/08/indonesias-land-swap-program-puts-communities-companies-in-a-bind/>.

Jong, H. N. 2020, 'Indonesia Won't "Sacrifice Economy" for More Ambitious Emissions Cuts'[Online], *Mongabay*, 14 April 2020, viewed 10 August 2020 at, <https://news.mongabay.com/2020/04/indonesia-emissions-reduction-climate-carbon-economy-growth/>.

Jong, H. N. 2021, 'Deforestation in Indonesia Hits Record Low, but Experts Fear a Rebound' [Online], *Mongabay*, 9 March 2021, viewed 9 March 2021 at, <https://news.mongabay.com/2021/03/2021-deforestation-in-indonesia-hits-record-low-but-experts-fear-a-rebound>.

Kaul, I., Grunberg, I. & Stern, M. A. 1999, *Global Public Goods: International Cooperation in the 21st Century*, Oxford University Press and UNDP, Oxford and New York.

Kep.38/M.PPN/HK/03/2012, Menteri Negara Perencanaan Pembangunan Nasional 2012, *Keputusan Menteri Negara Perencanaan Pembangunan Nasional, Kepala Badan Perencanaan Pembangunan Nasional Nomor Kep.38/M.PPN.HK/03/2012 Tentang Pembentukan Tim Koordinasi Penangannan Perubahan Iklim*, Menteri Negara Perencanaan Pembangunan Nasional, Jakarta, Indonesia.

KepPres 10/2011, President of the Republic of Indonesia 2011, *Presidential Instruction No. 10/2011 on the Postponement of Issuance of New Forest Licences and Improving Governance of Primary Natural Forest and Peatland*, Republic of Indonesia, Jakarta, Indonesia.

KepPres 25/2011, President of the Republic of Indonesia 2011, *Presidential Regulation of the Republic of Indonesia No 25 Year 2011 Concerning the Task Force for preparing the establishment of the REDD+ Agency*, Republic of Indonesia, Jakarta, Indonesia.

KepPres 61/2011, President of the Republic of Indonesia 2011, *Presidential Regulation of the Republic of Indonesia No 61 Year 2011 on The National Action Plan for Greenhouse Gas Emissions Reduction*, Republic of Indonesia, Jakarta, Indonesia.

Klein, R. J. T., Schipper, E. L. F. & Dessai, S. 2005, 'Integrating Mitigation and Adaptation into Climate and Development Policy: Three Research Questions,' *Environmental Science & Policy*, vol. 8, pp. 579–588.

Kok, M. T. J. & de Coninck, H. C. 2007, 'Widening the Scope of Policies to Address Climate Change: Directions for Mainstreaming', *Environmental Science & Policy*, vol. 10, pp. 587–599.

Koplitz, S. N., Mickley, L. J., Marlier, M. E., Buonocore, J. J., Kim, P. S., Liu, T. J., Sulprizio, M. P., DeFries, R. S., Jacob, D. J., Schwartz, J., Pongsiri, M. & Myers, S. S. 2016, 'Public Health Impacts of the Severe Haze in Equatorial Asia in September-October 2015: Demonstration of a New Framework for Informing Fire Management Strategies to Reduce Downwind Smoke Exposure', *Environmental Research Letters*, vol. 11.

Li, T. M. 2018, 'After the Land Grab: Infrastructural Violence and the "Mafia System" in Indonesia's Oil Palm Plantation Zones', *Geoforum*, vol. 96, pp. 328–337.

Locatelli, B., Fedele, G., Fayolle, V. & Baglee, A. 2016, 'Synergies Between Adaptation and Mitigation in Climate Change Finance', *International Journal of Climate Change Strategies and Management*, vol. 8, pp. 112–128.

Locatelli, B., Pavageau, C., Pramova, E. & Di Gregorio, M. 2015, 'Integrating Climate Change Mitigation and Adaptation in Agriculture and Forestry: Opportunities and Trade-Offs', *WIREs Climate Change*, vol. 6, pp. 585–598.

Marlier, M. E., DeFries, R. S., Kim, P. S., Gaveau, D. L. A., Koplitz, S. N., Mickley, L. J., Margono, B. A. & Myers, S. S. 2015a, 'Regional Air Quality Impacts of Future Fire Emissions in Sumatra and Kalimantan', *Environmental Research Letters*, vol. 10, 054010.

Marlier, M. E., DeFries, R. S., Kim, P. S., Koplitz, S. N., Jacob, D. J., Mickley, L. J. & Myers, S. S. 2015b, 'Fire Emissions and Regional Air Quality Impacts from Fires in Oil Palm, Timber, and Logging Concessions in Indonesia', *Environmental Research Letters*, vol. 10, 085005.

McCarthy, J. F. & Robinson, K. 2016, *Land and Development in Indonesia: Searching for the People's Sovereignty*, ISEAS, Singapore.

Mongabay 2018, 'Indonesian President Signs 3-year Freeze on New Oil Palm Licenses', [Online], 20 September 2018, viewed 10 August 2019 at, <https://news.mongabay.com/2018/09/indonesian-president-signs-3-year-freeze-on-new-oil-palm-licenses/>.

Murdiyarso, D., Dewi, S., Lawrence, D. & Seymour, F. 2011, *Indonesia's Forest Moratorium: A Stepping Stone to Better Forest Governance?* Working Paper 76, CIFOR, Bogor, Indonesia.

Nordhaus, W. D. 1994, *Managing the Global Commons: The Economics of Climate Change*, MIT, Cambridge, Mass. and London.

Nunan, F. (ed.) 2017, *Making Climate Compatible Development Happen*, Routledge, London.

Oates, J. F. 1999, *Myth and Reality in the Rain Forest: How Conservation Strategies are Failing in West Africa*, University of California Press, Berkeley.

Peluso, N. L. 1992, *Rich Forests, Poor People: Resource Control and Resistance in Java*, University of California Press, Berkeley.

PerMenLHK 17 2017, *Ministerial Regulation No 17/2017 On Changes of the Regulation of the Ministry of the Environment and Forestry No. P.12/Menlhk-II/2015 About Industrial Forest Plantation Development*, Ministry of the Environment and Forestry, Jakarta, Indonesia.

PerMenLHK 40 2017, *Ministerial Regulation No 40/2017 on Government Assistance for Industrial Forest Plantation Businesses for Peat Ecosystem Protection and Management*, Ministry of the Environment and Forestry, Jakarta, Indonesia.

PerPres 1, (President Republik of Indonesia) 2016, *Presidential Degree 1/2016 on the Peatland Restoration Agency*, Republic of Indonesia, Jakarta, Indonesia.

PerPres 16, (President Republik of Indonesia) 2015, *Presidential Decree 16/2015 on the Ministry of Environment and Forestry*, Republic of Indonesia, Jakarta, Indonesia.

PerPres 18 (President Republik of Indonesia) 2020, *National Medium-Term Development Plan (RPJMN) 2020–2024*, Republic of Indonesia, Jakarta, Indonesia.

PerPres 62 (President Republik of Indonesia) 2013, *Regulation of the President of the Republic of Indonesia Number 62 of 2013 About the Management of the Reduction of Greenhouse Gas Emissions from Deforestation, Forest Degradation and Peatland*, Republic of Indonesia, Jakarta, Indonesia.

PerPres 71 (President Republik of Indonesia) 2011, *Regulation of the President of the Republic of Indonesia Number 71/2011 Concerning the Provision of Inventory of National Greenhouse Gases*, Republic of Indonesia, Jakarta, Indonesia.

Pramova, E., Locatelli, B., Djoudi, H. & Somorin, O. A. 2012, 'Forests and Trees for Social Adaptation to Climate Variability and Change', *Wiley Interdisciplinary Reviews: Climate Change*, vol. 3, pp. 581–596.

Purnomo, A., Katili-Niode, A., Melisa, E., Helmy, F., Sukadri, D. & Sitorus, S. 2013, *Evolution of Indonesia's Climate Change Policy: From Bali to Durban*, National Council on Climate Change, Jakarta, Indonesia.

Purnomo, H., Okarda, B., Dermawan, A., Ilham, Q. P., Pacheco, P., Nurfatriani, F. & Suhendang, E. 2020, 'Reconciling Oil Palm Economic Development and Environmental Conservation in Indonesia: A Value Chain Dynamic Approach', *Forest Policy and Economics*, vol. 111, 102089.

Rizaldi, B. & Suharnoto, Y. 2012, *Climate Change and Its Impact on Indonesia's Food Crop Sector*, Sixth Executive Forum on Natural Resource Management: Water & Food in a Changing Environment on 11–13 April 2012 at SEARCA headquarters, Los Baños, Philippines.

RoI (Republic of Indonesia) 2007, *National Action Plan Addressing Climate Change*, State Ministry of Environment, Jakarta, Indonesia.

RoI (Republic of Indonesia) 2016, *First Nationally Determined Contribution*, Republic of Indonesia, Jakarta, Indonesia.

Ross, M. L. 2001, *Timber Booms and Institutional Breakdown in Southeast Asia*, Cambridge University Press, Cambridge, UK.

Sari, A. P., Dohong, A. & Wardhana, B. 2021, 'Innovative Financing for Peatland Restoration in Indonesia', in R. Djalante, J. Jupesta & E. Aldrian (eds.), *Climate Change Research, Policy and Actions in Indonesia: Science, Adaptation and Mitigation*, Springer International Publishing, Cham, Switzerland.

Sloan, S., Edwards, D. P. & Laurance, W. F. 2012, 'Does Indonesia's REDD+ Moratorium on New Concessions Spare Imminently Threatened Forests?' *Conservation Letters*, vol. 5, pp. 222–231.

Somorin, O. A., Brown, H. C. P., Visseren-Hamakers, I. J., Sonwa, D. J., Arts, B. & Nkem, J. 2012, 'The Congo Basin Forests in a Changing Climate: Policy Discourses on Adaptation and Mitigation (REDD+)', *Global Environmental Change*, vol. 22, pp. 288–298.

Suwarno, A., van Noordwijk, M., Weikard, H.-P. & Suyamto, D. 2018, 'Indonesia's Forest Conversion Moratorium Assessed with an Agent-based Model of Land-Use

Change and Ecosystem Services (LUCES)', *Mitigation and Adaptation Strategies for Global Change*, vol. 23, pp. 211–229.

Tanner, T. & Allouche, J. 2011, 'Towards a New Political Economy of Climate Change and Development', *IDS Bulletin-Institute of Development Studies*, vol. 42, pp. 1–14.

Todung, M. L. 2020, 'A Decade of Partnership Between Indonesia, Norway to Combat Climate Change', *Jakarta Post*.

van Noordwijk, M., Farida, Saipothong, P., Agus, F., Hairiah, K. & Suprayogo, D. 2006, 'Wateshed Functions in Productive Agricultural Landscapes with Trees', in D. P. Garrity (ed.), *World Agroforestry into the Future*, World Agroforestry Centre, Nairobi.

Varkkey, H. 2013, 'Patronage Politics, Plantation Fires and Transboundary Haze', *Environmental Hazards*, vol. 12, pp. 200–217.

Varkkey, H., Tyson, A. & Choiruzzad, S. A. 2018, 'Palm Oil Intensification and Expansion in Indonesia and Malaysia: Environmental and Socio-Political Factors Influencing Policy', *Forest Policy and Economics*, vol. 92, pp. 148–159.

Widiaryanto, P. 2015, 'Merging REDD+ into Ministry Should Be More Effective', *Jakarta Post*, 31 January 2015.

Wijaya, A., Chrysolite, H., Ge, M., Wibowo, C. K., Pradana, A., Utami, A. F. & Austin, K. 2017, *How Can Indonesia Achieve Its Climate Change Mitigation Goal? An Analysis of Potential Emissions Reductions from Energy and Land-Use Policies*, Working Paper, World Resources Institute, Washington D.C.

Zhida, J. 2012, 'Indonesia's Confidence Diplomacy Under the Yudhoyono Government', *China International Studies*, vol. 37, pp. 126–141.

5 Everyday climate politics in Laos

Miles Kenney-Lazar

Introduction

The Mekong River saw dramatic changes in 2019. During what is normally the wet season, the river swells with the monsoon rains. There is sufficient rain to supply water to millions of paddy fields across Laos, Myanmar, Thailand, Cambodia, and Vietnam. In July 2019, however, there was an extreme drought that brought the Mekong River to its lowest level in 100 years, delaying farmers from planting their rice crop (Lovgren 2019). By September, the situation had reversed as the banks of the Mekong and its tributaries were overflowing, leading to flooding in Thailand, Laos, and Cambodia (MRC 2019). The city of Pakse, the capital of Champassak Province, was inundated, and across the rest of southern Laos at least 14 people died while 102,000 were displaced (RFA 2019a). Yet, by late October, drought had returned, with a recorded river height of only 1.5 meters and riverbed dunes emerging near Nakhon Phanom in northeastern Thailand (Sripiachai 2019).

International media and organizations have questioned the impact that extensive development of hydropower dams on the mainstream Mekong and its tributaries has had upon flooding and drought in the region (Lovgren 2019; Eyler & Weatherby 2020; Osborne 2020). There are now 13 active dams along the mainstream Mekong, two of which are in Laos and largely supply energy to Thailand. With more planned, their cumulative impact upon water flow is increasingly feared, because they can exacerbate the ebbs and flows of the water table (Kijewski 2019). During droughts, hydropower operators have an economic incentive to retain water in their reservoirs to preserve their capacity to produce electricity, thus exacerbating the effects of drought. During periods of heavy rainfall, they are more likely to release extra water to prevent the reservoir from overfilling and compromising the structural integrity of the dam. This was especially relevant in 2019, considering the collapse of the Xe Pian Xe Namnoy Saddle Dam in southern Laos in 2018 (Baird 2020). As weather patterns become increasingly extreme and unpredictable due to climate change, the role of hydropower in exacerbating patterns of drought and flooding becomes ever paramount.

DOI: 10.4324/9780429324680-7

However, the government of Laos has pinned the blame for the recent drought exclusively on lighter rainfall due to climate change with no mention of the potential role of hydropower, even as an exacerbating factor rather than a direct cause (Xinhua 2019). Precise climate events, such as a delay in the start of the rainy season at the time of the first drought in July or two tropical storms at the time of the flooding in September, cannot be directly linked to climate change. While there have yet to be any scientific studies conducted demonstrating that the 2019 floods in southern Laos were caused by climate change, they do match a pattern of variable and extreme weather that has been linked to a changing climate (Lefroy, Collet & Grovermann 2010; GIZ & MoNRE 2014; USAID 2014; Lao PDR 2015). Particularly in southern Laos, the rainy season has been delayed and there has been a higher likelihood of extreme dry and wet periods, droughts, and floods. Thus, climate change is an easy scapegoat for the rainy season's havoc upon the Mekong region in 2019. Although it is caused by human activity, it is presented as an almost natural process. It is seemingly outside the control of humans living and working in the Mekong region. As it is not easy to separate the impacts of climate change from hydropower, the science of flooding and drought in the region has become highly political.

Yet, this is a type of technological and scientific politics that is far removed from the lives of those who are dealing with the everyday impacts of climate change. While the politics of climate among states and international organizations is important, this chapter focuses on the politics of those who might not be aware of such debates but are intensely conscious of how weather patterns that they have come to expect over generations have been changing and impacting their lives. These are the average citizens whose lives are affected by unpredictable changes in river water levels or the impacts of changing weather patterns on crop cycles. While their engagement with climate change may seem peripheral, they are the ones who will be most impacted by a changing climate, and thus their interaction with these changes is as important as those of scientists, activists, and policymakers. Furthermore, the average citizen is more directly subjected to a different register of domestic politics in Laos, which is dominated by the government and the Marxist-Leninist Lao People's Revolutionary Party (LPRP) than international actors.

The domestic, everyday politics of climate change are best illustrated by the story of Ms. Houayheuang Xayabouly, nicknamed Mouay. On 12 September 2019, Mouay was arrested for 'defaming' the Lao People's Democratic Republic (Lao PDR) (RFA 2019b). A Lao woman residing in Phonthong District, Champassak Province, Mouay had posted a Facebook Live video the previous week criticizing the government's efforts to provide relief to victims of the flooding that devastated the six southernmost provinces of Laos. Impassioned by the impacts of the flooding on her family and friends, she asked why the government had done so little. She recognized the government's limited budget, but she questioned why that was the case considering the inflated personal wealth of individual officials, implying that they had enriched themselves at the expense of public resources that could have been used to save lives and support livelihoods. In late November, Mouay was sentenced to five years in prison, despite confessing in an attempt to secure a lighter sentence (RFA 2019c).

Mouay's expression of frustration via social media and her subsequent arrest were not explicitly concerned with climate change, but they spoke volumes about how climate politics operate in Laos. Mouay directly commented on the government's response to a climate-related disaster, yet climate change was not mentioned, and she may not have considered the connection. This is exemplary of how Lao people are engaging in climate politics: rather than taking on the topic directly, they are interacting with the impacts of climate change and mitigation and adaptation programs on their everyday lives, responding to threats to their material interests and concerns, and reacting in ways that are shaped by living and operating in Laos's restrictive political system. Furthermore, it demonstrates how the politics of nature-society-state relations in Laos, such as the government's disaster response, are not focused on climate change, per se. In contrast, references to it are used to blunt criticism of human transformations of the environment, as exemplified in the government's use of climate change to deflect criticism of hydropower development.

This chapter is a commentary on the intersections of various forms of politics with climate change in Laos, as a biophysical material transformation, a discourse, and an object to be worked on through policies, plans, and programs. Most of the scholarship on climate change in Laos has been in the form of scientific studies focusing on its impacts (Hett et al. 2012; Huan et al. 2013; Ingxay, Yokoyama & Hirota 2015) or analyses of the policies and politics of climate change mitigation and adaptation programs developed by international donors in cooperation with the Lao government (Lange & Jensen 2013). Studies focused on the multi-scalar politics of climate change in Laos have tended to look more at forestry programs, but with an institutional and techno-political orientation (Ingalls and Dwyer 2016; Cole et al. 2017; Ingalls et al. 2018; Ramcilovik-Suominen 2019). While these aspects of climate politics are important and feature heavily in this chapter, they miss out on a crucially important element of the politics that pervade climate change across multiple sectors – how Lao people are engaging with it in their everyday lives and how they are expressing concerns about these issues. Such politics operate in ways similar to the politics of land dispossession and resource extraction (McAllister 2015; Baird 2017; Kenney-Lazar, Suhardiman & Dwyer 2018). They do so by engaging in the politics of the specific changes that they see in their surrounding environments and livelihoods. Furthermore, they seek to navigate the boundaries between acceptable and censored political expression, where their concerns can be heard without putting them in jail, like Mouay.

Climate change: An emerging political object

In many ways, Laos is an odd choice as a case study for investigating climate change politics. Its carbon dioxide (CO_2) emissions from the burning of fossil fuels are negligible in global comparison – 2.8 megatons in 2017, which is 0.01% of global emissions (Muntean et al. 2018). As a landlocked country, Laos is not vulnerable to the impacts of sea-level rise that low-lying coastal countries face. Furthermore, no significant political movements related to climate change have emerged in the country, likely due to the limited freedoms of outward, contentious political expression, especially organized protests or demonstrations. The wave of climate strikes and rallies held around the world at the end of September 2019, which saw over 6 million people in the streets (Taylor et al. 2019), did not include any events in Laos.[1]

Yet, climate change is affecting Laos, just as it has touched every other part of the globe in one way or another, and people are responding politically, in ways that are unexpected but relevant to global climate politics. High rates of deforestation and forest degradation have meant that Laos's effect upon global emissions and capacity to sequester carbon is higher than might have been estimated (Lestrelin et al. 2013). Climate change is altering weather patterns, delaying the wet season by almost a month, and leading to more wet and dry extremes, creating droughts and floods, and posing threats to the livelihoods of Lao farmers (USAID 2014). Increases in rainfall and higher temperatures are also increasing the risk of the spread of diseases and pests, which threaten livestock, fisheries, and human health (ibid.). Laos has been selected as one of the key countries for the implementation of programs that seek to mitigate emissions and increase sequestration through a reduction of deforestation and forest degradation, and a significant amount of international aid has been secured for these goals (Lestrelin et al. 2013; Cole et al. 2017; Ramcilovik-Suominen 2019).

The limits of climate action and politics in Laos reflect how it has not historically emerged as a domestic object of political action, contention, or movement, whether for the average citizen or the Lao government. Laos's political landscape is still very much an outcome of over two decades of nationalist and socialist struggle for liberation and the construction of a prosperous, equal, and just society. This included struggles against colonial France during the First Indochina War (1946–1954) and against the imperial United States during the Second Indochina War (1955–1975). The victory of the Pathet Lao, with the support of the Vietnamese communists, and the establishment of the Lao PDR in 1975 has been framed by the government as a people's victory as it was a struggle fought by workers and peasants (Evans 1998). Additionally, the Lao People's Revolutionary Party (LPRP) has been constructed as a party that represents the people's interests in the model of Leninist democratic centralism, whereby inputs and even votes are allowed to filter from the bottom up, but decisions subsequently made are binding for all members. After the LPRP came to power in 1975, its vision, which is the driving force of political ideas in the country, has focused on industrial modernization to generate prosperity that can be shared broadly across the population (Stuart-Fox 1997).

Within this context, the concerns of the international development community have been peripheral to those of the Lao government, both in terms of political ideology and practice. Yet, openings were created for the interests of the international community, eventually including climate change, as the country opened its doors to international aid as part of a broader program of economic reform that began in the mid-1980s. In 1986, the government established the New Economic Mechanism (NEM), permitting and integrating market forces into the economy to address the economic failures of its centrally planned state-run economy. Such changes were also initiated in response to decreasing levels of aid from the Soviet Union and the rest of the Eastern Bloc as their economies faltered toward the end of the Cold War. Thus, the government was open to new sources of aid, such as from multilateral institutions, like the United Nations, World Bank, and Asian Development Bank, as well as Western countries.

With increasing aid from Western-oriented international development institutions, new discourses of development began to circulate throughout Laos, particularly in the national capital, Vientiane. In the 1990s, Laos increasingly became a premier destination for development donors and projects working on environmental issues because it was perceived to have one of the most pristine environments in the Mekong region (Hett et al. 2012). Furthermore, as one of the region's poorest countries, and having just opened its economy to market forces, it was eligible for significant amounts of development support. Such aid, which focused on sustainable development, has had significant impacts upon Laos's development scene, leading to the reshuffling of institutional arrangements to strengthen environmental departments and facilitate the creation of new environmental protection laws (Goldman 2001; Singh 2014). Meanwhile, Lao government officials, consultants, and NGO staff have quickly learned the new jargon of sustainable development, including terms such as 'biodiversity conservation' and 'environmental impact assessments'.

Yet, it is important to not assume that legal, institutional, and cultural adaptations of sustainable development have transformed Laos's development strategy into one that is environmentally friendly. Goldman (2001) has written about how the pairing of environment-related donor funds and state development objectives has transformed Laos into an 'environmental state', that seeks to link resource extraction, agro-industrial expansion, and hydropower development with environmental conservation initiatives. Conservation could be used to legitimize resource industries that have significant environmental impacts and risks. Laos's engagement with climate change is part of this same strategy of legitimizing its pre-existing development strategy while using climate change rhetoric to seek donor funds for projects it already sought to implement (Lange and Jensen 2013). Laos has readily ratified international conventions on climate change, such as the United Nations Framework Convention on Climate Change in 1995 and the Kyoto Protocol in 2003 and was reportedly the first ASEAN country to ratify the Paris Agreement.[2] It has also formulated a National Adaptation Programme of Action to Climate Change (Lao PDR 2009) and a National Strategy on Climate Change (Lao

PDR 2015), which were drafted with support from international donors. Furthermore, it has fully supported the United Nations program for Reducing Emissions from Deforestation and Forest Degradation (REDD+). Yet, as Lange and Jensen (2013) have argued, climate change action in Laos is still largely a donor-driven affair and the governmental institutions that have been set up to work on climate change within the Ministry of Natural Resources and Environment have relatively little power in comparison with those facilitating the extraction, processing, and commodification of the country's resources, such as the Ministry of Energy and Mines.

The politics of adaptation: Mediated impacts

Despite being a landlocked and mountainous country, the material impacts of a changing climate for rural people are significant, as the 2019 droughts and floods described above so clearly illustrate. Thus, there has been an expanding program of work in Laos funded by international donors on adaptation to the impacts of climate change. The Lao government has largely focused its climate change strategy on adaptation. It sees itself as a victim of climate change, considering that the country is a net sink of carbon due to its extensive forest cover and low degree of industrialization and consumption (Lange and Jensen 2013). As a result, climate change is often imagined to be an external force bearing down upon the country, mainly caused by wealthier, more industrialized countries who should be funding adaptation to it. In this sense, the government's position is aligned with many other countries in the Global South.

The Lao government is right to assert that the impacts of climate change are not driven by activities originating from within the country. Thus, it is afflicted by climate injustice. Nonetheless, such a position ignores how climate change impacts are not direct, but instead are mediated through political-economic and socio-ecological relations, as political ecologists have argued (Tschakert 2012; Taylor 2014; Eriksen et al. 2015). Although adaptation is often framed in technical terms, it is a highly political practice because of the variable ways in which it could occur, affect people's livelihoods, and relate to government development plans and projects. This is evident in the example of hydropower described above – the effects of climate change upon flooding and drought are mediated by hydropower's alteration of water flows in the Mekong and its tributaries. Yet even such a perspective tends to focus on the techno-politics of adaptation, which include the politics of scientific analyses of impacts and the technological solutions employed in response. However, there is another type of politics at play: the everyday politics of engaging with a transforming environment and the types of changes that farmers seek to secure their livelihoods in general, including against adverse environmental impacts and the socio-economic costs of adaptation programs.

The Lao government is pursuing a range of different strategies to adapt to current and perceived future climate change impacts. Adaptation is a key component of the government's National Strategy on Climate Change (NSCC), specifically set out in the 2008 National Adaptation Programme of Action (Lao PDR 2009). The main aim of the government is to "increase resilience of key economic sectors and natural resources to climate change and its impacts" (Lao PDR 2015, p.5). These key economic sectors are set out by the government as agriculture, forestry and land-use change, water resources, transport and urban development, and public health. The remainder of this section focuses on agriculture and related natural resources, as changes to these due to flooding, drought, and increased temperature will affect the majority of the country's largely rural population.

Donor-driven and government reports that recommend making agriculture more resilient to climate change focus on technical-managerial approaches to farming systems. For example, USAID (2014) suggests adopting improved rice varieties that are more resistant to flooding and extreme heat or shifting cropping calendars to avoid harvests during periods of high rainfall. They further recommend improving soil erosion control techniques and practising intercropping and other forms of agricultural diversification to reduce reliance on monoculture. For livestock, diets should be improved to minimize the potential for malnourishment, vaccines are needed to prevent the contraction of diseases that will spread more quickly, and access to markets need to be improved to reduce input costs and increase prices received. For aquaculture, pond aeration should be used to mitigate impacts of higher temperatures, on-site water storage is needed to reduce risks of reduced water availability in the dry season, and embankments and diversionary canals should be built to deal with flooding.

It is remarkable that all of these recommendations are purely technical and do not deal with the political-economic issues that impact impoverished rural Lao people and make them particularly vulnerable to climate change in the first place. These recommendations could potentially be useful and are not necessarily problematic, per se. However, they avoid the real issues that constrain farmers. For example, it has been documented that farmers most often identify a lack of agricultural land as the most significant problem for livelihood development (Chamberlain 2007; Arnst 2010). Since the National Assembly opened a telephone hotline over a decade ago, one of the few direct ways in which citizens could petition their concerns directly to the central government without getting into trouble, conflicts over land have consistently been one of the most prominent topics of concern (*Vientiane Times* 2018). Additionally, there are increasing numbers of farmers finding ways to creatively resist the acquisition of their lands for development projects like large-scale plantations (McAllister 2015; Baird 2017; Kenney-Lazar, Suhardiman & Dwyer 2018). Securely accessing enough land for all households is a critically important feature of resilient rural livelihoods which can help farmers weather the impacts of climate change, including allowing them to pursue many of the technical solutions that the government and USAID propose.

Yet, access to land has been decreasing in many areas across Laos. Land and forest zoning and allocation programs have increased forest land areas at the expense of swidden agriculture, reducing the area that can be used for shifting cultivation (Vandergeest 2003; Ducourtieux, Laffort & Sacklokham 2005; Fujita and Phanvilay 2008). As part of government programs to move upland people into lowland areas, or due to hydropower and other types of development projects, resettlement has forced people into areas where they have less agricultural land than where they previously lived (Baird and Shoemaker 2007; Delang and Toro 2011). Large-scale land concessions for farming and tree plantations or infrastructural development, special economic zones, and mineral extraction have decreased land access without replacement (Baird 2011; Kenney-Lazar 2012).

Access to land is a perfect example of a critical development issue in Laos that needs significant attention in and of itself but also has important links to climate change. However, because land access is not directly connected to the impacts of climate change, it is often not considered as seriously as impacts of increased temperatures, flooding, and droughts upon agricultural production. Yet, having sufficient land access is as important, if not more, than the technical programs intended to allow agriculture to continue. Furthermore, access to land is an issue that Lao farmers deeply care about and it is one that they have a political perspective on.

The politics of mitigation: Material interests

Lao people are directly feeling the impacts of climate change and there is a politics to how they react to those direct changes. Furthermore, adaptation projects are being pursued with implications for their livelihoods and they are responding politically to these as well. However, one of the most prominent ways in which climate change has been acted upon in Laos is through projects that aim to mitigate carbon emissions, either by reducing forest loss or by pursuing ostensibly renewable and low-carbon forms of energy production, such as hydropower (*Vientiane Times* 2013). Here, again, there is a politics of reaction to the impacts of mitigation programs. However, the ways in which such mitigation programs are run have roots in the Lao government's political goals and projects that were established long before climate change had become a political object in Laos. Thus, the politics of mitigation efforts often have to do with these prior projects and goals and their immediate impacts upon rural, environmental-based livelihoods than climate change as an object, per se. Nonetheless, they are all tied into climate change mitigation strategies and discourses and thus can be considered a form of climate politics.

Hydropower is one of the government's strategies for pursuing mitigation – the bulk of the Clean Development Mechanism (CDM) projects in Laos are for hydropower dams.[3] However, constructing hydropower as a tool of climate change mitigation is a secondary justification for the pursuit of dam development that is primarily driven by political-economic interests of profit, modernization, and economic development (Williams 2020). Thus, this section instead focuses on projects that are set up explicitly for climate mitigation, principally the United Nations' REDD+ program. Laos contributes to global climate change more due to emissions from deforestation and forest degradation than the combustion of fossil fuels from industry and consumption. Thus, REDD+ seeks to reduce deforestation in the country. However, as agreed upon at the 2005 UNFCCC meeting, reducing such emissions could hamper the financial benefits to developing countries that have not contributed much to global emissions. Therefore, they should be compensated for such activities by wealthier nations.

In Laos, preparations for REDD+ began in 2008, while REDD+ projects have been in operation since 2009 (Dwyer & Ingalls 2015). REDD+ was initially envisioned as a relatively straightforward mechanisms for paying developing countries to not engage in activities that would lead to carbon emissions from deforestation or forest degradation. However, with REDD+'s progression, it has run into a range of difficulties associated with emissions from forests and the realities of development. These include illegal logging, insecure and unclear land tenure arrangements in forested areas, and large-scale development plans with significant impacts on forest areas that would be expensive to pay governments not to pursue. As Dwyer and Ingalls (2015) point out, this is not necessarily a bad thing as it means that REDD+ can be used as a tool for dealing with and addressing these complex issues. However, it has also meant that in Laos, as elsewhere, REDD+ has aimed for easy targets, which tend to be smallholder farmers, especially swidden cultivators in the Lao context (Ramcilovik-Suominen 2019). This is partly due to the lower cost of providing incentives for them not to cultivate lands compared to larger resource extraction and infrastructure projects. However, it also has to do with how REDD+ maps onto pre-existing national development strategies and plans.

Thus, REDD+ has a politics of techniques, methods, and assessments for sequestering carbon. These include questions concerning the most efficient, effective way to sequester carbon and how to implement REDD+ in a way that will also be good for smallholders' livelihoods and their land tenure security. Because of how REDD+ combines science, technology, expertise, and politics, Dwyer and Ingalls (2015, p.1) write that "REDD was born techno-political … and has remained so ever since." Yet, these techno-politics occur in close circles of UN meetings, in-country development workshops, and ministry roundtables among UN bureaucrats, bilateral donor representatives, target country government officials, scientists, and development practitioners. In the Lao case, this means that, geographically, such politics largely remain within Vientiane, or other regional and global cities, held in five-star hotels, government ministry offices, or the offices of NGOs and bilateral and multilateral donors.

Yet, the politics of REDD+ are not constrained to such places – they do occur in rural areas, among upland, forest-using peasant communities, but are focused on their direct material interests, rather than the techno-politics of REDD+. There is a lack of scientific evidence showing that swidden cultivation is a significant cause of emissions from deforestation and forest degradation, especially compared to other forms of land use and land-use change (Ziegler et al. 2012). Nonetheless, it has been the main target of REDD+ programs, because of how REDD+ maps onto longstanding government programs, to stabilize and eventually eliminate swidden cultivation (Kenney-Lazar 2013). Thus, the government has continued to limit swidden cultivation in some areas and ban it in others.

Although farmers have not outwardly protested such policies, knowing the political risks of defying government objectives, they have engaged in a politics of maintaining access to swidden lands in other ways. First, they have engaged in various forms of everyday resistance or "weapons of the weak" (Scott 1985), agreeing to government policies and programs on the surface but defying them in practice. After being resettled from upland to lowland areas, villagers have returned to farm their uplands using swidden techniques (Évrard and Baird 2017). Villagers have ignored the classifications of land as forest, as part of the land and forest allocation programs, continuing their practices as they had before (Dwyer 2014). In other cases, when swidden village land was granted as concession land to plantation companies, they have expanded swidden into new areas, including old-growth forests (Baird and Fox 2015).

Second, Lao farmers have made the case that although they agree with government policies on swidden and forest preservation, they need access to land, and in the uplands, there are few other alternatives to swidden. In this way, farmers have engaged in a form of "resisting with the state" (McAllister 2015; Kenney-Lazar, Suhardiman & Dwyer 2018), mobilizing the contradictions inherent in government ideologies to their advantage, such as the socialist rhetoric of allocating "land to the tiller" (Vandergeest 2003). Without making an argument about the merits of swidden as a sustainable system of production, or one that can be carbon-neutral, as is done in technocratic circles of REDD+ politics, upland villagers have simply made the case that they do not have enough land for agriculture (Chamberlain 2007; Arnst 2010). In their case, this means that they cannot rotate their fields and leave them fallow long enough to be fertile and to limit the growth of weeds and the spread of pests.

Conclusions: Centring everyday climate politics

Assessing the politics of climate change in Laos is not an easy endeavor. There are no climate protests, movements, or demonstrations. With only one political party legally allowed in the country, electoral politics are non-existent, especially when it comes to addressing climate change. The lack of contentious politics is not only due to the limiting political system but also because the current and future impacts of climate change in Laos, while significant and worrying, are hardly catastrophic, such as in other places like small island states or impoverished countries with extensive, low-lying coastlines.

Thus, unlike those countries' governments, the Lao government is not particularly vocal about climate change and environmental injustice at the international level.

Yet, there is a politics of climate change in Laos. At one level, there are techno-politics concerning climate change impacts, adaptation strategies, and mitigation approaches. These are the politics of whether hydropower is a carbon-neutral form of development and to what degree it is affecting flooding and drought in the region; or how rainfall patterns will be altered and which farmers they will affect, as well as what types of agricultural techniques should be adopted to best deal with erratic rainfall; or what types of human activities are causing deforestation and forest degradation, and which of these should be curtailed. However, the techno-politics of climate change are far removed from the lives of everyday Lao citizens, especially peasant farmers. Yet, these are the same people who are most affected by climate change and by programs to adapt to or mitigate climate change. Thus, it is essential to center the politics of how farmers engage with climate change, even if the connection is indirect. Farmers are primarily focused on their immediate material interests, particularly when such interests are threatened or impeded. Centring farmers' material interests is essential for working towards a climate politics that is socially just for the most marginalized groups.

This chapter has also shown that a direct focus on climate change often has the effect of evading or delegitimizing the political-economic, social, or human processes that are at play in shaping people's relationship with the environment. Focusing exclusively on climate change as the political object of concern leads to a concern with only an external threat, rather than the internal social and political relations that mediate climate change or that make people more vulnerable to it and other forms of impoverishment. In terms of impacts, climate change is a convenient external cause of environmental change that can be used to ignore the impacts of human-driven activities in Laos, such as hydropower dams. Similarly, adaptation approaches are almost exclusively focused on technical solutions that respond directly to climate change threats, rather than an interest in working with farmers to negotiate the socio-political relations that are impoverishing them, such as a lack of access to land. Additionally, mitigation strategies prioritize low-hanging fruits – making changes to behaviors that are cheapest and will generate the greatest carbon gains, thus creating a disproportionate focus on swidden cultivators. In contrast, thinking more broadly about the diverse forms of climate politics that the rural poor engage in is instructive for evading the regressive politics that center climate change as global and apolitical. Instead, it focuses on the social natures that matter to rural people.

Notes

1 The School Strike for Climate included events in many countries, but not Laos, as can be seen in their map of events: https://www.fridaysforfuture.org/events/map. Last accessed 21 April 2021.

2 See http://www.mofa.gov.la/index.php/activities/state-leaders/1406-laos-becomes-the-first-country-from-asean-to-ratify-paris-climate-agreement. Last accessed 5 November 2020.
3 See www.cdmpipeline.org/. Last accessed 12 November 2020.

References

Arnst, R. 2010, *Farmers' Voices*, LEAP, Vientiane, Laos.

Baird, I. G. 2011, 'Turning Land into Capital, Turning People into Labour: Primitive Accumulation and the Arrival of Large-Scale Economic Land Concessions in the Lao People's Democratic Republic', *New Proposals: Journal of Marxism and Interdisciplinary Inquiry*, vol. 5 no. 1, pp. 10–26.

Baird, I. G. 2017, 'Resistance and Contingent Contestations to Large-Scale Land Concessions in Southern Laos and Northeastern Cambodia', *Land*, vol. 6, no. 1, pp. 1–19.

Baird, I. G. 2020, 'Catastrophic and Slow Violence: Thinking About the Impacts of the Xe Pian Xe Namnoy Dam in Southern Laos', *The Journal of Peasant Studies*, early online publication, doi:10.1080/03066150.2020.1824181.

Baird, I. G. & Fox, J. 2015, 'How Land Concessions Affect Places Elsewhere: Telecoupling, Political Ecology, and Large-Scale Plantations in Southern Laos and Northeastern Cambodia', *Land*, vol. 4, pp. 436–453.

Baird, I. G. & Shoemaker, B. 2007, 'Unsettling Experiences: Internal Resettlement and International Aid Agencies in Laos', *Development and Change*, vol. 38, no. 5, pp. 865–888.

Chamberlain, J. 2007, *Participatory Poverty Assessment II (2006): Lao People's Democratic Republic*, National Statistics Center and Asian Development Bank, Vientiane, Laos.

Cole, R., Wong, G., Brockhaus, M., Moeliono, M. & Kallio, M. 2017, 'Objectives, Ownership and Engagement in Lao PDR's REDD+ Policy Landscape', *Geoforum*, vol. 83, pp. 91–100.

Delang, C. & Toro, M. 2011, 'Hydropower-Induced Displacement and Resettlement in the Lao PDR', *Journal of South East Asia Research*, vol. 19, no. 3, pp. 567–594.

Ducourtieux, O., Laffort, J.-R. & Sacklokham, S. 2005, 'Land Policy and Farming Practices in Laos', *Development and Change*, vol. 36, no. 3, pp. 499–526.

Dwyer, M. B. 2014, 'Micro-Geopolitics: Capitalising Security in Laos's Golden Quadrangle', *Geopolitics*, vol. 19, pp. 377–405.

Dwyer, M. B. & Ingalls, M. 2015, *REDD+ at the Crossroads: Choices and Tradeoffs for 2015–2020 in Laos*, Working Paper 179, CIFOR, Bogor, Indonesia.

Eriksen, S. H., Nightingale, A. J. & Eakin, H. 2015, 'Reframing Adaptation: The Political Nature of Climate Change Adaptation', *Global Environmental Change*, vol. 35, pp. 523–533.

Evans, G. 1998, *The Politics of Ritual and Remembrance: Laos Since 1975*, University of Hawai'i Press, Honolulu.

Évrard, O. & Baird, I.G. 2017, 'The Political Ecology of Upland/Lowland Relationships in Laos Since 1975', in V. Bouté & V. Pholsena (eds.), *Changing Lives in Laos: Society, Politics, and Culture in a Post-Socialist State*, NUS Press, Singapore.

Eyler, B. & Weatherby, C. 2020, 'How China Turned Off the Tap on the Mekong River', *The Stimson Center*, 13 April 2020, viewed 12 November 2020 at, <https://www.stimson.org/2020/new-evidence-how-china-turned-off-the-mekong-tap/>.

Fujita, Y. & Phanvilay, K. 2008, 'Land and Forest Allocation in Lao People's Democratic Republic: Comparison of Case Studies from Community-Based Natural Resource Management Research', *Society & Natural Resources*, vol. 21, no. 2, pp. 120–133.

GIZ & MoNRE 2014, *Ten Facts on Climate Change in Lao PDR*, Lao PDR, Vientiane, Laos.

Goldman, M. 2001, *Imperial Nature: The World Bank and Struggles for Social Justice in the Age of Globalization*, Yale University Press, New Haven.

Hett, C., Heinimann, A., Epprecht, M., Messerli, P. & Hurni, K. 2012, 'Carbon Pools and Poverty Peaks in the Lao PDR', *Mountain Research and Development*, vol. 32, no. 4, pp. 390–399.

Huan, S., de Rouw, A., Bonté, P., Robain, H., Valentin, C., Lefèvre, I., Girardin, C., Le Troquer, Y., Podwojewski, P. & Sengtaheuanghoung, O. 2013, 'Long-term Soil Carbon Loss and Accumulation in a Catchment Following the Conversion of Forest to Arable Land in Northern Laos', *Agriculture, Ecosystems and Environment*, vol. 169, pp. 43–57.

Ingalls, M. and Dwyer, M. B. 2016, 'Missing the Forest for the Trees? Navigating the Trade-Offs Between Mitigation and Adaptation Under REDD, *Climatic Change*, vol. 136, no. 2, pp. 353–366.

Ingalls, M., Meyfroidt, P., To, P. X., Kenney-Lazar, M. & Epprecht, M. 2018, 'The Transboundary Displacement of Deforestation Under REDD+: Problematic Intersections Between the Trade of Forest-Risk Commodities and Land Grabbing in the Mekong Region', *Global Environmental Change*, vol. 50, pp. 255–267.

Ingxay, P., Yokoyama, S. & Hirota, I. 2015, 'Livelihood Factors and Household Strategies for an Unexpected Climate Event in Upland Northern Laos', *Journal of Mountain Science*, vol. 12, no. 2, pp. 483–500.

Kenney-Lazar, M. 2012, 'Plantation Rubber, Land Grabbing and Social-Property Transformation in Southern Laos', *Journal of Peasant Studies*, vol. 39, no. 3–4, pp. 1017–1037.

Kenney-Lazar, M. 2013, *Shifting Cultivation in Laos: Transitions in Policy and Perspective*, Ministry of Agriculture and Forestry, Vientiane, Laos.

Kenney-Lazar, M., Suhardiman, D. & Dwyer, M. 2018, 'State Spaces of Resistance: Industrial Tree Plantations and the Struggle for Land in Laos', *Antipode*, vol. 50, no. 5, pp. 1290–1310.

Kijewski, L. 2019, 'Officials to Meet on Mekong Crisis as Fishing Communities Suffer, *Al Jazeera*, 25 November 2019.

Lange, R. B. & Jensen, K. M. 2013, *Climate Politics in the Lower Mekong Basin: National Interests and Transboundary Cooperation on Climate Change*, Danish Institute for International Studies, Copenhagen.

Lao PDR 2009, *National Adaptation Programme of Action to Climate Change*, Lao PDR, Vientiane, Laos.

Lao PDR 2015, *National Strategy on Climate Change: Intended Nationally Determined Contribution*, Lao PDR, Vientiane, Laos.

Lefroy, R., Collet, L. & Grovermann, C. 2010, *Potential Impacts of Climate Change on Land Use in the Lao PDR*, International Center for Tropical Agriculture (CIAT), Palmira, Colombia.

Lestrelin, M., Trockenbrodt, M., Phanvilay, K., Thongmanivong, S., Vongvisouk, T., Thuy, P.T., Castella, J.-C. 2013, *The Context of REDD+ in the Lao People's Democratic Republic: Drivers, Agents and Institutions*, Occasional Paper 92, CIFOR, Bogor, Indonesia.

Lovgren, S. 2019, 'Mekong River at Its Lowest in 100 years, Threatening Food Supply', *National Geographic*, 31 July 2019.

McAllister, K. 2015, 'Rubber, Rights, and Resistance: The Evolution of Local Struggles Against a Chinese Rubber Concession in Northern Laos', *Journal of Peasant Studies*, vol. 42, no. 3/4, pp. 817–837.

MRC (Mekong River Commission) 2019, 'First Mekong Flood Reaches Thailand and Lao PDR Before Landing on Cambodia Over the Next Few Days', viewed at, <http://www.mrcmekong.org/news-and-events/news/new-event-page-3/>.

Muntean, M., Guizzardi, D., Schaaf, E., Crippa, M., Solazzo, E., Olivier, J. G. J. & Vignati, E. 2018, *Fossil CO2 Emissions of All World Countries – 2018 Report*, Publications Office of the European Union, Luxembourg.

Osborne, M. 2020, 'Chinese Dams and the Mekong Drought', *The Interpreter*, 11 August 2020.

Ramcilovik-Suominen, S. 2019, 'REDD+ as a Tool for State Territorialization: Managing Forests and People in Laos', *Journal of Political Ecology*, vol. 26, no. 1, pp. 263–281.

RFA 2019a, 'Floods Kill At Least 14 in Southern Laos, Hundreds of Thousands Displaced', 11 September 2019.

RFA 2019b, 'Lao Authorities Arrest Woman for Criticizing Flood Relief Efforts on Facebook', 16 September 2019.

RFA 2019c, 'Lao Woman Gets Five Years for Criticizing Government on Facebook', 25 November 2019.

Scott, J. C. 1985, *Weapons of the Weak: Everyday Forms of Peasant Resistance*, Yale University Press, New Haven.

Singh, S. 2014, 'Developing Bureaucracies for Environmental Governance: State Authority and World Bank Conditionality in Laos', *Journal of Contemporary Asia*, vol. 44, no. 2, pp. 322–341.

Sripiachai, P. 2019, 'Mekong River Falls to Critical Level, Sand Dunes Emerge', *Bangkok Post*, 29 October 2019.

Stuart-Fox, M. 1997, *A History of Laos*, Cambridge University Press, Cambridge, UK.

Taylor, M. (ed.) 2014, *The Political Ecology of Climate Change Adaptation: Livelihoods, Agrarian Change and the Conflicts of Development*, Routledge, London.

Taylor, M., Watts, J. & Bartlett, J. 2019, 'Climate Crisis: 6 Million People Join Latest Wave Of Global Protests', *The Guardian*, 27 September 2019.

Tschakert, P. 2012, 'From Impacts to Embodied Experiences: Tracing Political Ecology in Climate Change Research', *Geografisk Tidsskrift-Danish Journal of Geography*, vol. 112, no. 2, pp. 144–158.

USAID 2014, *Lao PDR Climate Change Vulnerability Profile*, USAID, Bangkok, Thailand.

Vandergeest, P. 2003, 'Land to Some Tillers: Development-Induced Displacement in Laos', *International Social Science Journal*, vol. 55, no. 175, pp. 47–56.

Vientiane Times 2013, 'Laos Explains its Hydropower Policy', *Vientiane Times*, 6 June 2013.

Vientiane Times 2018, 'Bad Roads, Land Disputes, Drug Trade Top NA Hotline Calls', *Vientiane Times*, 26 December 2018.

Williams, J. M. 2020, 'The Hydropower Myth', *Environmental Science and Pollution Research*, vol. 27, pp. 12882–12888.

Xinhua 2019, 'Climate Change, Global Warming Cause Drought in Laos: Experts', 16 December 2019.

Ziegler, A. D., Phelps, J., Yuen, J. Q., Webb, E. L., Lawrence, D., Fox, J. M., Bruun, T. B., Leisz, S. J., Ryan, C. M., Dressler, W., Mertz, O., Pascual, U., Padoch, C. & Koh, L.P. 2012, 'Carbon Outcomes of Major Land-Cover Transitions in SE Asia: Great Uncertainties and REDD+ Policy Implications', *Global Change Biology*, vol. 18, no. 10, pp. 3087–3099.

6 Malaysia's complex multi-level climate governance between institutionalization and non-state actor interventions

Irina Safitri Zen and Zeeda Fatimah Mohamad

Introduction

Similar to other emerging economies, Malaysia is confronted by the challenging task of climate change mitigation and adaptation[1] measures with the goal of rapid industrialisation and catch-up development (Rasiah et al. 2017). Since its independence in 1963, Malaysia's economic growth has been accompanied by increasing energy demand, which is heavily dependent on fossil fuels – resulting in significant greenhouse gas (GHG) emissions. In 2017, Malaysia emitted almost 250 million tons of CO_2 equivalents, which accounts for ca. 0.7 percent of global CO2 emissions. Major GHG emissions in Malaysia come from the energy sector (including transport), forestry, and waste (MESTECC 2018).

At the same time, Malaysia is highly affected by the impact of climate change (Rahman 2009; Palermo & Hernandez 2020; Al-Amin & Filho 2011), most notably in the fields of water and coastal resources, food security and agriculture, forestry and biodiversity, infrastructure, energy, and public health (MESTECC 2018). Malaysia's average temperatures and sea-levels are also rising year after year, with an increasing variability in rainfall patterns and a growing frequency of extreme weather events. Future projections up to 2100 indicate that these trends will continue (Tang 2019), raising concerns about higher climate change-related risks in the coming years.

Following cumulative scientific evidence compiled in the National Communication report, the economic impact of climate change has caused Malaysia to realize the stark reality that climate change – if left unchecked – will pose significant ecological, social, and economic risks. Since then, Malaysian governments have advocated for climate change measures at the international, national, and local levels, and there have been efforts to harmonize climate change concerns with those of national interests and development priorities (Al-Amin and Filho 2011). In response to this challenge, the country has made a clear commitment since 2009 to address climate change at the institutional level by introducing a dedicated climate change policy and multi-level governance strategy with a voluntary reduction

DOI: 10.4324/9780429324680-8

target. Despite the promises of its climate change strategies, Malaysia's low carbon transition over the years has been stymied by various limiting factors.

The complexity of the implementation of climate governance continues at the sub-national government. The state of Melaka, for instance, records active involvement in the regional economic development through foreign organizations and in the Global Platform for Sustainable Cities facilitated by the federal agencies for the vertical axis mode of climate governance. This mode of Melaka climate governance is absorbed by the state structure up to the local councils. The horizontal axis of Melaka climate governance has been dominated by the Melaka green technology corporation who works closely with various government agencies, local stakeholder and local authorities. The involvement of the local university is considered as an independent organization who distributes knowledge and expertise as part of the research grant secured from the Ministry of Higher Education, Malaysia. In contrast, the low carbon society of Iskandar Malaysia in Johor state was initiated by the local university who secured the international funds from Japan. It helps the federal agencies, Iskandar Regional Development Authority, IRDA, and the local authority in developing the Blueprint implementation of a low carbon society for Iskandar Malaysia.

How Malaysia's climate change policies and multi-level governance evolved

Historically, Malaysia has had a concrete environmental policy framework in place since the early 1970s. Hence, any analysis of its current strategy on environmental governance, including climate change, needs to reflect the evolution of environmental policymaking within the socio-economic context of the country.

During colonial rule and for the first two decades after independence, Malaysia already had laws to control pollution, but these were not specifically aimed at associated environmental problems. Instead, they were intended to promote 'wise use' and 'sustainable yield' of resources. They were single-issue regulations, fragmented and inconsistent, and generally ineffective in environmental management (Hezri and Nordin Hasan 2006). However, a more integrated policy commitment for environmental protection was made due to the federal government's recognition at the time that an industrializing Malaysia needed to develop "appropriate policies and programs so that accelerated industrial development goes hand in hand with sound management of the environment" during the Mid-Term Review of the Second Malaysia Plan in 1973 (Government of Malaysia 1973). From then onwards, environmental protection has been included in Malaysia's subsequent development plans. On this basis, the Environmental Quality Act (EQA) was passed in 1974 which, until today, is the key environmental legislation in the country. The elaboration of legislations and related programs outlined in the Act demonstrated the federal government's commitment to deal with environmental issues since the 1970s. Since then, a host of environmental-related policies have been introduced in the country (Ambali 2011).

Yet, Malaysia's environmental agenda is also very much influenced by the international policy landscape. There have been four main waves of Malaysia's environmental policy agenda in line with global environmental agendas (see Table 6.1). The first wave occurred in the 1970s to 1980s, especially after the United Nations Conference on the Human Environment in 1972. This watershed event resulted in the worldwide call for more effective environmental protection in both the Global North and the Global South. Malaysia responded quite swiftly, as mentioned earlier, by raising the issue of environmental sustainability in its development plan in 1973, the subsequent introduction of concrete statutory provisions under the EQA 1974, and the setting up of a dedicated implementing agency: the Department of Environment. At that time, Malaysia was considered a frontrunner among developing countries. Yet, the institutional set-up was inadequately resourced, working at the margins of public policy because of limited resources and influence in government (Ambali 2011; Hezri, 2011).

The second wave occurred in the 1990s where global environmentalism shifted from crisis to reform agenda for sustainable development. An important policy turn was provided by the Brundtland Report (WCED 1987) suggesting that economic growth can continue with a reduced impact on the environment. Malaysia's response to this second wave has been described as patchy (Hezri & Nordin Hasan 2006) – where neither prominent structures and processes, nor statutory review were introduced to equip the policy system in addressing the challenges of environmental sustainability. Given that Malaysia was one of the pioneers in establishing a framework for environmental governance in the 1970s, its response to the post-1992 sustainable development agenda was considered lacking in comparison (Hezri 2011).

As the world entered the millennium, the third wave of environmental institutionalization was increasingly aimed at changing patterns of production – an industrial revolution where 'greening' was the focus. In terms of development strategy, there was a new enthusiasm that developing countries like Malaysia would continue developing via the ecological modernization trail, following Europe and North East Asian countries, or develop its own green growth model (Mol, Sonnenfeld & Spaargaren 2020). Economy and

Table 6.1 Four waves of environmental institutionalization

First wave	Second wave	Third wave	Fourth wave
1970s–present	1990s–present	2006–present	Since 2015
Environmental protection	*Sustainable development*	*Green investment*	*Sustainable development goals (Agenda 2030)*

Source: based on Hezri 2011

nature could be favourably combined through the greening of economic growth and industrial development. One of the key drivers behind this green growth agenda is the development and adoption of 'green technology choices' that meet local ecological and human development needs (Jacobs 2013).

In Malaysia, a basic institutional architecture responding to this green growth agenda was swiftly undertaken through the incorporation of a green technology portfolio within the Ministry of Energy, Green Technology, and Water (MEGTW) in 2009. In addition, this was soon after equipped with the launch of the National Green Technology Policy (NGTP) and the Malaysian Green Technology Corporation (formerly National Energy Centre) in the same year. Green technology was defined in the NGTP as "the development and application of products, equipment and systems used to conserve the natural environment and resources, which minimizes and reduces the negative impact of human activities" (MEGTW 2009). The speed of institutional set-up in this third wave was a stark contrast to the relatively slower pace during the second wave (Chua and Oh 2011; Hezri 2012; Kasayanond, Umam & Jermsittiparsert 2019).

Interestingly, enthusiasm on climate change mitigation and low-carbon technology was also heightened within this period of national green growth agenda. The National Climate Change Policy (NCCP) was also introduced in the same year (2009) as the NGTP. The formulation of the National Policy on Climate Change (NPCC) was an important milestone for Malaysia in its efforts at addressing the challenges of climate change. The policy was an outcome of a study entitled 'Policy Study on Climate Change' under the Ninth Malaysia Plan (2006–2010). The main purpose of the study was to develop a national policy and strategies to address the issues of climate change at the national and international level (Pin, Pereira & Aziz 2013). The NPCC was approved by the Malaysian Cabinet in 2009 and the document was published the following year. It delineates five principles to strengthen its ten strategic thrusts that direct national responses on climate change in all sectors (MNRE 2010). A total of 43 key actions have been developed for its implementation. The NPCC also specified key sectors for mitigation and adaptation based on the main sectors that are impacted by climate change in the country. Hence, NPCC is not a stand-alone policy, but must be viewed as a cross-cutting policy framework that needs to be coordinated with a complex set of policy domains and implementing agencies that directly or indirectly support its implementation.

During this period, interest in mainstreaming environmental protection was heightened through the explicit inclusion of 'sustainability' in the government of Malaysia's 2010 New Economic Model (NEM). The National Green Technology and Climate Change Council was also established in September 2009 to formulate policies and identify strategic issues to implement both the NGTP and the NCCP. It was chaired by the Prime Minister and consisted of a number of key Cabinet Ministers as a high-level decision-making platform on climate change (MESTECC 2018).

Following this rather comprehensive institutional set-up, the Malaysian Prime Minister at the time, Dato' Sri Najib Razak, made a commitment in December 2009 during the UNFCCC Climate Change Summit in Copenhagen (CoP 15) that Malaysia would be adopting a clear climate change mitigation target: Malaysia proposed a voluntary reduction of up to 40 percent in terms of emissions intensity of GDP by 2020 compared to 2005 levels, contingent on adequate technology transfer and financing from the developed world. By the end of 2014 and 2015, the government officially announced that Malaysia had achieved reducing the GHG emissions intensity of GDP by 33 percent and 35 percent, respectively – well on its way towards the 40 percent by the year 2020 as targeted (Gee 2015; Razak 2016). In addition, the government often emphasized that as Malaysia strived to become a high-income economy, it was critical for the country to adapt towards an economic strategy that struck the right balance between green and pro-business approaches (STAR 2015). Again, the connection between climate change mitigation and industry-driven green growth was emphasized.

Albeit the ambitious sounding target, it is important to emphasize that Malaysia's 40 percent carbon intensity reduction target is per unit of GDP, and has received some criticism (Susskind et al. 2020). This means that as Malaysia's GDP increases, the country's total GHG emissions can increase along with it. Therefore, Malaysia could reach its intensity target while total GHG emissions may increase (Zhu et al. 2014). Measuring emissions per unit of GDP reflects the challenge that many emerging economies face when doing their fair share to reduce global GHG emissions while also encouraging rapid industrialization and economic growth (Rasiah et al. 2017). This explains Malaysia's leaning to 'green growth' as its main decarbonization strategy as it allows efforts on environmental protection and climate mitigation without curbing economic growth.

Finally, the fourth wave is the period in which current implementation of climate change policymaking and governance are being implemented, but within the historical context of the previous three waves. The fourth wave occurred since the Sustainable Development Goals (SDGs) were introduced in 2015. Following this global commitment, Malaysia has explicitly aligned the country's Eleventh Malaysia Plan (2016–2020) alongside the SDGs (EPU 2017), with the green growth agenda continuing to be a main driver across different SDGs (EPU 2017). 2015 was also the year in which the Paris Agreement – the latest national treaty to cap global temperature rise to 1.5 Celsius over the next 100 years – was ratified. The agreement upholds the Common but Differentiated Responsibilities (CBDR) principle of the Kyoto Protocol, which acknowledges that all states have a shared obligation to address environmental destruction but denies equal responsibility of all states with regards to environmental protection, including climate change. Hence, although the Paris Agreement has increased the obligation for developing countries, including Malaysia, in terms of carbon reductions, many of its provisions also continue to safeguard the development rights of developing countries. This includes maintaining the continued commitment of developed countries to provide financial, technological, and capacity building assistance to developing countries for climate mitigation and adaptation (Fei 2018).

Following this, Malaysia's latest commitment, reflected in its Nationally Determined Contribution (NDC) to the UNFCCC, is to reduce its emissions intensity by 45 percent by 2030 (relative to its emissions intensity per unit of GDP in 2005), which was changed to 2050 following the Marrakech Action Proclamation in 2016 (Rasiah et al. 2017). More targeted plans to concretize policy actions have also been introduced in this current wave which include: Green Technology Master Plan (GTPM) (2016–2030), Malaysia's Roadmap Towards Zero Single-Use Plastics (2018–2030), Clean Air Action Plan (2010), National Haze Action Plan (2019), and National Open Burning Action Plan (2019), among others.

Despite a promising institutional framework and governance strategy by the state actors, Malaysia's low-carbon transition has been stymied by various socio-economic and socio-political factors. In a recent study, Susskind et al. (2020) outlined six 'categories' of constraints: federal-state friction; limited government capacity and willingness to regulate; absence of a dedicated decarbonization agency; lack of international funding and support; nascent public concern about environmental issues (like climate change); and barriers to renewable energy adoption. However, that statement has not happened at the sub-national government, i.e., the state of Melaka and Iskandar Malaysia, and the state of Johor who are able to tap foreign funding by the help of non-state actors (NSAs). Further, non-partisan efforts and cooperation by NSAs such as the private sector, academia, and non-governmental organizations have played an important bridging and supporting role in sustaining the process before and during this period of political uncertainty. In the next section, the role of non-state actors in supporting the role of state actors in climate change governance and their contribution in filling the void or addressing weaknesses left by state actors in certain aspects of the governance will be explored – with empirical illustration from two case studies in the implementation of low-carbon cities in the state of Johor and Melaka, Malaysia.

State and sub-national climate action

Climate change efforts at the sub-national level in Melaka and Iskandar Malaysia, Johor respectively, capture the proactive action by the sub-national government and higher education institutes (HEI). The formation of a network of city governments with many international organizations, as well as the involvement of local stakeholder and HEIs are examples of NSA practices. For example, their active involvement in the Indonesia-Malaysia-Thailand Growth Triangle (IMT-GT) regional development network was facilitated by the Asian Development Bank (ADB). Later on, the federal government helped the state to achieve its aims through the involvement of the Malaysian Industry-Government Group for High Technology (MIGHT), an independent non-profit technology think tank under the Prime Minister's office. Their participation helped to accelerate the green technology initiative and served as a counterbalance to the Green State claimed by the Penang state government in 2012. Melaka Solar Valley, smart meters for buildings and housing, and the Melaka Sustainable City Program under the Global Platform for Sustainable Cities (GPSC) were among the important strategic projects. Later, through the formulation of the Melaka State Climate Action Plan, the HEI's engagement aided in the translation of ICLEI's greenhouse gas (GHG) inventory programme. ICLEI, a global network for local government for sustainability, facilitates GHG inventory for Melaka state. GHG inventories are essential in developing effective low-carbon strategies, tracking GHG reductions, responding to regulations and local GHG program requirements, and securing climate finance. Some cities also believe that tracking emissions can eventually conserve financial and other resources (Fong, Sotos & Kotorac 2013).

The proactive action of the local HEI, Universiti Teknologi Malaysia (UTM), in the early establishment of Iskandar Malaysia's low-carbon society agenda opened up a flexible mode of collaboration between the Iskandar Malaysia Regional Authority (IRDA), the local council, and stakeholders under the international collaborative project of Iskandar Malaysia's low-carbon society with several HEI in Japan. Iskandar Malaysia, formerly known as Iskandar Development Region, and South Johor Economic Region, is the key southern development corridor in Johor, Malaysia (SJER). As a government agency, IRDA is responsible for overseeing and regulating regional economic development in Iskandar Malaysia, as well as accelerating it. Behind the establishment of Iskandar Malaysia is Khazanah Nasional Berhad – the government of Malaysia's strategic investment fund. Khazanah plays a critical role in encouraging economic growth and making strategic investments on behalf of the government. These two states exemplify how sub-national actors are working to promote low-carbon cities (Ministry of Natural Resources and Environment 2015).

Melaka Green Technology State

The Melaka state aims to become the country's first green technology city or 'Green' State by 2020. In 2010, the state adopted the Urban Environmental Accords (UEA) indicators as a basis for the Melaka Green Technology State Blueprint (2011–2020). This was followed by the development of the Melaka Green City Action Plan (GCAP) in 2014 through facilitation from the ADB. The Melaka GCAP (IMT-GT 2014) consists of 35 thematic actions in the areas of water management, energy efficiency and renewable energy, green transportation, zero waste, cultural heritage and tourism, and urban forestry and agriculture which serve as a basis for the green initiative. It is part of the network of regional development of the IMT-GT through the ADB's 3E strategies (Environment, Economic Competitiveness, and Equity). The initiative was part of the ADB Urban Operational Plan (ADB 2011, 2012) and the Urban Management Partnerships (UMPs) for city-to-city peer learning and knowledge sharing. Melaka's long-term commitment to low-carbon growth was recognized as one of the first initiatives taken by sub-national actors in Malaysia to support the NDC's goal to reduce GHG emissions (Krishnan et al. 2014). Later on, the regional collaboration was documented in IMT-GT Blueprint 2017-2021 (CIMT 2017).

Melaka's green technology city state was supported by institutional arrangements to facilitate implementation of the GCAP. Established in 2010, the Melaka Green Technology Council is chaired by the Chief Minister of Melaka and comprises of State Executive Committee members, local councillors, and related government agencies. The council's objective is to oversee the planning and monitoring of green technology developments. In 2013, the Melaka Green Technology Corporation was established to execute the green development agenda and also function as the secretariat for the Melaka Green Technology Council. The corporation aims to lead green development initiatives, implement green policies, plan and monitor green technology development, and enhance business and investment activities for green technologies. The structure extends to the local councils by establishing Green Technology Action Committees, which require the four councils to report to the Melaka Green Technology Council. The four local councils are: Historical Malacca City Council, Hang Tuah Jaya Municipal Council, Alor Gajah Municipal Council, and Jasin Municipal Council.

Despite ADB and IMT-GT participation through public-private partnerships (PPPs) Melaka maintains its structure by establishing the Melaka Green Technology Corporation as the government's arm for green technology-related initiatives. The corporation is structurally under the Melaka state EPU and indicates a state-centric mode to control the green initiative in this state, which shows a shift from 'government' to 'governance' in the Melaka journey towards becoming a green technology state. This is an indication of a new approach of global governance in anticipating climate change challenges which has been captured in several studies (Andonova & Levy 2003; Bäckstrand 2006), as well as a shift from hierarchical setting to networked governance, which consists of state and non-state actors (Keck & Sikkink 1998). This new mode of governance is characterized by decentralized, voluntary, market-oriented interaction between public and private actors (Zen, Hashim & Sinniah 2019).

As the first demonstration site of GCAP, Melaka set an example for replication in another two cities within the sub-region (IMT-GT 2014). It aimed to spur economic development in participating provinces and states in the three countries involved in the IMT-GT; the cities of Melaka (in Malaysia), Songkhla (in Thailand), and Medan (in Indonesia). These Green Cities

Initiative flagship programs demonstrate a new form of learning through the networking city (Osborne, Kearns & Yang 2013). Moreover, the coalitions cross national borders of the IMT-GT, demonstrating the worth of the international cooperative initiative (Widerberg & Pattberg 2015), where non-state or sub-national actors from at least two different countries share the same public goals across borders (Andonova, Hale & Roger 2017) and engage in a form of "transnational climate governance" (Andonova, Betsill & Bulkeley 2009). This initiative captures the trend of learning cities for climate action as practised by several Nordic countries (Osborne, Kearns & Yang 2013).

Moreover, Melaka state received facilitation from the Malaysian Industry-Government Group for High Technology (MIGHT) to accelerate several key projects under the Malaysia Sustainable Cities program. The project is part of the GPSC network to support Melaka's ambitious plan to become the first state in Malaysia to adopt green technology and become a green 'city-state' by 2020. As a co-founding project between the Global Environment Facility's (GEF) Sustainable Cities Integrated Approach Pilot (SC-IAP) and the United Nations Industrial Development Organization (UNIDO) in collaboration with the Ministry of Urban Well-being, Housing, and Local Government, as well as the MIGHT, the project demonstrates a PPP mechanism for global governance captured in several studies (Andonova & Levy 2003; Bäckstrand 2006). GPSC comprises 28 cities across 11 countries and provides a network of practitioners and leaders worldwide developing solutions for sustainable urban growth which allows peer-to-peer sharing and emphasis on capacity building.

Since the mechanism is not yet established to measure the achievement of Melaka as a green technology state through the implementation of green initiative, the Global Protocol for Community-Scale Greenhouse Gas Emission Inventories (GPC) framework was adopted for that purpose. The GPC framework allows the community-scale GHG inventory deployed for city-to-city comparison (GHG Protocol 2014). The Center for IMT–GT Subregional Cooperation and ICLEI facilitate the development of two GHG inventory reports (ICLEI 2016; ICLEI 2017). Melaka Green Technology Corporation plays a central role in data gathering for the state government agencies and its four urban local authorities. Since 2018, the three years of collaboration by UTM,the local university facilitated co-development of the Melaka State Climate Action Plan 2020–2030 translates the GHG inventory report into an action plan for long-term monitoring purposes (Zen, Hashim & Sinniah 2019). The document has become the main reference for the state to execute the climate change initiative and has been used as a basis to monitor green initiatives at the state level. Key initiatives in Melaka include energy efficiency projects and energy audits for nine government buildings, installing street lighting using light emitting diode (LED) lights, the GHG inventory with ICLEI, green city benchmarking and baseline indexing, and the International Green Training Center (IGTC).

We have so far described the continuing effort of Melaka in the regional and global climate governance, which functions as learning network platforms and provides a platform for the involvement of the private sector and HEI as NSAs. However, there has been no formal announcement by the state on the achievement of Melaka as a green technology state.

Low Carbon Society of Iskandar Malaysia

The Iskandar Malaysia low carbon society initiative was influenced by the involvement of HEI in an international collaborative research project, 'Development of Low Carbon Society Scenarios for Asian Regions'. Led by the local university, Universiti Teknologi Malaysia (UTM), this initiative involves Kyoto University, Okayama University, Japan's National Institute for Environmental Studies (NIES), and also the Iskandar Malaysia Development Authority (IRDA). A five-year project had secured the grant from the Japan Science and Technology Agency and Japan International Cooperation Agency, under the Science and Technology Research Partnership for Sustainable Development, or SATREPS (Low Carbon Asia Research Center 2015). This modus operandi demonstrate the flexible platform of the international cooperative initiative through non-state or sub-national actors to collaborate across borders towards one goal: the 'Low Carbon Society of Iskandar Malaysia'. As a result, a type of 'transnational climate governance' is indicated.

Iskandar Malaysia is a regional economic conurbation in the southern part of Peninsular Malaysia, which consists of the five districts; Johor Bahru District, Kulai District, Pontian District, Kota Tinggi District, and Kluang District. Overseen by the IRDA, the urban region development of Iskandar Malaysia has two aims: first, to develop a prosperous, resilient, robust, and globally competitive economy (the 'strong' dimension) and second, to nurture a healthy, knowledgeable and globally competitive society that subscribes to low carbon living (the 'sustainability' dimension). Simultaneously, it aims to develop a total urban-regional environment that enables rapid economic growth but reduces growth's energy demand (Iskandar Malaysia 2021).

The integration of two competing goals – strong and sustainable – in a single development vision poses a great challenge to Iskandar Malaysia's growth policies and development planning. As a multi-disciplinary project to developed Iskandar's future LCS scenarios, propose actions, quantify the GHG emission reduction potential of the proposed LCS actions, and continuously engage local stakeholders in a series of focus group discussions (Ho et al. 2016). During the development of LCS's future scenario, several engagements to develop a Blueprint action plan reflected the climate change governance initiative combining networks of multi-universities – local and Japanese – as the NSAs.

A Low Carbon Society aims to accelerate the common carbon development initiative and produce the Low Carbon Society Blueprint of Iskandar Malaysia 2025 (LCSBP-IM2025). The LCSBP-IM 2025 presents comprehensive climate change mitigation (carbon emission reduction) policies (LCS actions and sub-actions) and detailed strategies (measures and programs) to guide the development of Iskandar Malaysia. The four local councils adopted the Blueprint which provides guidance for the implementation of low carbon initiatives with the help of the LCS team. This is to show the monitoring control mechanism for a low carbon initiative towards achieving its vision of "a strong, sustainable metropolis of international standing" by 2025 (Ho et al. 2013).

The participatory approach of a series of workshops and focus group discussions (FGD) for Blueprint development conducted by the LCS team functions as an educational approach and serves as training for government officers at the sub-national and local council level on how to successfully take action on climate change. This process exhibits the inclusive approach in educating a society facilitated by NSAs where a network of multi-university collaboration functions as a team of experts. Even though the Blueprint for each local council has been adopted, the challenge is sustaining the LCS initiative financially in the long term. The local government, not the federal agency, has the legal authority to execute the LCS (i.e., IRDA).

Non-state climate action

Private institution and companies

In 2013, the Malaysian Ministry of Natural Resources and Environment, NRE, and the United Nations Development Program, UNDP Malaysia, initiated a two-year program for a National Corporate Greenhouse Gas Reporting Programme for Malaysia, which is known as *MyCarbon*. The program calls for a national disclosure with the objectives to set up a globally recognized, standard corporate GHG accounting and reporting program, provide a data source for analysis and development of local emissions factors, and encourage corporate level carbon accounting and emissions reductions, as well as providing standards, guidance, and support measures. This reporting helps the ministry to measure the private institution involvement to achieve the national voluntary carbon emissions reduction target. Hence, the establishment of GHG reporting standards or carbon disclosure supports low carbon pathways in supporting the National Policy on Climate Change and the National Green Technology Policy. This initiative aligns with the Tenth Malaysia Plan (RMK 10), emphasizing mitigation strategies to reduce the emission of GHGs. This initiative stimulates more substantial incentives for the investment of renewable energy, energy efficiency, improved waste management, conserving forest, and reduced emissions to improve air quality.

As of 2014, 23 private enterprises had pledged to voluntarily disclose their GHG emissions in order to reduce their emissions. The effort provided a platform for knowledge sharing and targeted 50 companies to join the program by the end of 2015. Several local companies making GHG reporting voluntarily as part of their corporate sustainability reporting or social responsibility reporting are: local telecommunication network, such as DIGI telecommunication companies like Maybank, UEM Environ, Puncak Niaga, etc. This also follows the global standard of the GHG Protocol as a corporate accounting and reporting standards by World Resources Institute (WRI) and the World Business Council for Sustainable Development.

This initiative provides vital networking from private companies and creates its own pathway for private companies, helping in strengthening the decarbonization pathway for Malaysia. By linking 'orchestrator' (i.e., international organizations and governments) and 'intermediaries' (i.e., private companies), networks could harness additional contributions by building catalytic linkages that create an enabling and conducive environment for a growing number of climate actions. The two terms, 'orchestrator' and 'intermediaries', are referred to in Hsu et al. (2018). The corporate level emission data from different sectors helps to strengthen the identification of local emission factors in Malaysia. *MyCarbon* data helps to synergize data collection and exchange for GHG inventory, which helps in strengthening national communication to UNFCCC and ensures higher tier national reporting is more accurate.

Recently, the establishment of a climate governance initiative, or CGI, provides a platform to assist the business sector on the longer-term risks of climate change (www.cgmalaysia.com/). It is a platform to inform, engage, and encourage non-executive directors (NEDs) of listed company boards, as well as provide oversight and inform board conversations to assist businesses in mitigating climate change. CGI, a project hosted by the World Economic Forum (WEF) helps in strengthening the capacity of NEDs in developing climate change risk management strategies. The CGI Principles, which were established by the WEF, are consistent with the aims of the Task Force on Climate-Related Financial Disclosures (TCFD) led by Michael Bloomberg. The task force sets out a comprehensive and ambitious standard for directors to address what the 1,000 top global experts from business, academia, and civil society have now rated as the top risk-facing society for 2019. The CGI Malaysian Chapter will be the first country chapter in the Asian region.

Higher education institutions

HEIs are involved in a number of major efforts in Malaysia, including as demonstration sites for energy efficiency programs and net zero building programs, the low carbon cities program, as well as low carbon initiatives and assessments. Compared to a network of NSAs, climate actions are captured in several actor groups such as cities, states, and regions, companies and investors, and banks but there is less involvement of HEIs (Hsu et al. 2018). However, an increasing trend of HEIs' involvement in climate actions exists in the United States – since 2015, there have been 700 schools and universities involved in climate actions (Hsu et al. 2018). Further, The emergence of climate action initiatives by HEIs as part of the NSAs in Malaysia records several interesting patterns of involvements.

At the national level, the creation of a Low Carbon Cities Framework (LCCF) by Malaysia Green Technology Corporation and Climate Change Center (MGTC) provides guidance and assesses the low carbon development of cities. Malaysia Green Technology Corporation and Climate Change Center, MGTC or *GreenTech Malaysia*, as the government agency under the purview of the Ministry of Environment, has a mandate to lead green growth, climate change mitigation, and climate resilience and adaptation. The program focuses on five elements – energy, water, waste, mobility, and greenery – which is used for carbon sequestration track and target for carbon reduction initiatives (KeTTHA 2011). Involvement of HEIs in the early stage of LCCF development helps to foster the online carbon calculator and provides capacity building for admin staff and academia, which aligns with the spirit of a low carbon campus or campus sustainability initiative (Zen et al. 2019). This effort represents a transnational society moving towards a common goal (Andonova 2005).

Albeit, HEIs' involvement as NSAs in several energy efficiency programs at the regional level, and the development of homegrown assessment tools for energy efficiency, indicates a good local response for long-term use. At an early stage, Universiti Teknologi Malaysia (UTM) participated in the ASEAN Energy Management Scheme (AEMAS) in 2014 and was awarded the Energy Management Gold Standard, 3rd Gold Star rating for in-house energy efficiency management, which included the training for a certified energy manager (UTM Newshub 2014). Later on, the Voluntary Sustainable Energy Low Carbon Building Assessment, or *GreenPASS*, was developed by the Sustainable Energy Development Authority of Malaysia (SEDA). It is an effort encourage energy efficiency programs for all building types in Malaysia. The *GreenPASS* program attracted HEIs, which have large-scale building complexes with a high consumption of energy. Universiti Teknologi Melaka (UTeM) was the first university to receive the category of Zero Energy Building (ZEB) from SEDA in 2018, followed by the Management and Science University (MSU) achieving a Diamond-2 rating for *GreenPASS* with 24.21 percent less carbon emissions against the baseline measurements in 2019. This works out at an annual saving of 5,646,661 kWh or 3,918.78 tonnes less CO_2 in 2020 (Management and Science University 2020). Later, *GreenPASS* developed self-assessment through an online system – Building Energy Data Online System (BEDOS) – to encourage self-monitoring of their performance. Recently, under the PPP financing mechanism, the International Islamic University Malaysia (IIUM) received a National Energy Award 2020 (IIUM News 2021) which fosters results from the energy efficiency program for the climate mitigation action.

Another rather compelling example is demonstrated in a large-scale solar panel initiative by Universiti Teknologi MARA (UiTM) with innovative financing mechanisms. UiTM as the public chain of local universities, which has 34 satellite-campuses scattered around the country, set about the installation of 1.17 sq. m of solar power plants, which cost RM278 million and was targeted to generate over RM650 million revenue for the next 21 years (Shahar 2018). A PPP between UiTM, under the indirect subsidiary of UiTM Solar Power Pte. Ltd., and a private bank, Affin Hwang Investment Bank Berhad, was constructed to create the Green Sustainable and Responsible Investment: SRI Sukuk. The large scale of solar power plants was set up to finance the RM240 million development and operation of the 50MW utility solar power plant in Gambang, Pahang (Wahab & Naim 2019). Hence, it has become the first HEI in the world to issue a Green SRI Sukuk (Chonghui 2018).

The implementation of Green SRI Sukuk was supported by the release of the Sustainable and Responsible Investment (SRI) Sukuk Framework by the Securities Commission Malaysia in 2014. This movement develop connection between financial sector and HEIs which showed the readiness of the financial sector to support climate action and create a conducive financial ecosystem for the country. Not only that,the solar power plant functions as a house for solar research centers: the Centre of Excellence for Innovation, Development, and Commercialization of Renewable Energy. The partnerships among NSAs and HEIs with private companies and government agencies demonstrates networked governance consisting of state and non-state actors (Keck & Sikkink 1998). It is characterized by voluntary, market-oriented interaction between public and private actors (Bäckstrand 2008). This form of PPP demonstrates a new global governance on climate action. Nevertheless, UiTM added its original function as an education institution where the initiative includes educational training and research related to renewable energy.

An earlier theoretical definition of non-state actors describe them as having a "state-centric approach … there to structurally enhance the economic performance with less efficacy" (Halliday 2001). NSAs evolve into four different forms of reconceptualization of the 'non-state'. First, an earlier definition of NSAs evolved from only covering NGOs in the narrow, post-1960 definition to encompass all that is 'non-state', i.e., business and banks, religious movements, social movements, and criminal organizations. Second, the NGOs in the contemporary world, where the state is being eroded, rethink how the state should be involved. Third, the influence of NGOs that promote a good effort would affect the state form. Fourth, NSAs that promote the normative cajole the state (Archibugi, Held & Kohler 1998), as part of the global civil society movement, hence the climate action initiative. In this context, NSAs' diverse involvement in climate change pledges various mechanisms for climate action initiatives. NSAs involve the city, state, and regional governments as well as companies, investors, HEIs, and civil society organizations.

Conclusion

Melaka Green Technology State showcases a network of regional actors of climate governance. It involves the regional economic development of the

IMT-GT and the GPSC's worldwide network by providing a chance for a peer-to-peer city learning platform through the networking city which demonstrates the international cooperative initiative. However, state-centric control for this green initiative is captured with multi-sectoral collaboration and a shift from 'government' to 'governance' with active involvement of NSAs. It is characterized by decentralized, voluntary, market-oriented interaction between public and private actors. The Iskandar Malaysia low carbon society initiative remains the main agenda of the regional development authority, IRDA, especially for the foreign investment to spur up the Southern regional economic development. However, the implementation of the LCS Blueprint initiative remains at the local authority level under the purview of the state of Johor. Besides, the carbon disclosure mechanism and reporting adopted the Japanese standard, such as Comprehensive Assessment System for Built Environment Efficiency (CASBEE), rather than the worldwide GPC framework.

The involvement of private institution illustrates proactive action by the federal agency in developing the *MyCarbon* program, backed by the United Nations. The program adopts the disclosure and standard reporting of GHG Protocol. On the other hand, NSAs' involvement with the HEIs shows a wide range of involvement. It includes planning for climate action initiatives, finding the possible funding, and planning for the implementation through engagement of local councils or stakeholders. The interesting initiative captured in the energy efficiency program demonstrates a shift from building internal capacity of certified energy managers such as in UTM, creating a PPP mechanism to foster the results, such as in UTeM and IIUM, and an innovative financial mechanism through the Green Sukuk SRI for a large-scale solar power plant in UiTM. This leads to better energy conservation and lower electricity bills, which helps to reduce the University's massive operational and maintenance expenditures. Under the freedom of partnership and the loose definition function of HEIs, which cover research, education training, and campus operation, it seems like an attractive package of climate governance being offered under NSAs' terms. Hence, HEI demonstrate an attractive institutional innovation package in enhancing climate change governance and demonstrating a better position for decarbonization growth pathways in this country under a control ecosystem of higher learning institutions.

Note

1 Mitigation involves options and strategies for reducing GHG emissions and increasing GHG uptakes by the Earth's system – which involve actions based on direct reduction of anthropogenic GHG emissions, or the enhancement of carbon sinks to limit long-term climate damage. Adaptation refers to adapting actions in a changing climate – which involves adjusting to expected or actual future climate to agriculture. Appropriate climate change policy needs an efficient mix of mitigation and adaptation options/actions that limits the overall impacts of climate change (IPCC 2014).

References

Al-Amin, A. Q. & Filho, W. L. 2011, 'An Overview of Prospects and Challenges in the Field of Climate Change In Malaysia', *International Journal of Global Warming*, vol. 3, no. 4, pp. 390–402.

Ambali, A. R. 2011, 'Policy of Sustainable Environment: Malaysian Experience', *European Journal of Scientific Research*, vol. 48, no. 3, pp. 466–492.

Andonova, L. B. 2005, '*International Institutions, Inc: The Rise of Public-Private Partnerships in Global Governance*', Conference on the Human Dimensions of Global Environmental Change, Berlin, pp. 2–3.

Andonova, L. B., Betsill, M. M. & Bulkeley, H. 2009, 'Transnational Climate Governance', *Global Environmental Politics*, vol. 9, no. 2, pp. 52–73.

Andonova, L. B., Hale, T. N. & Roger, C. B. 2017, 'National Policy and Transnational Governance of Climate Change: Substitutes or Complements', *International Studies Quarterly*, vol. 61, no. 2, pp. 253–268.

Andonova, L.B. & Levy, M.A. 2003, 'Franchising Global Governance: Making Sense of the Johannesburg Type II partnerships', in O. S. Stokke & O. B. Thommessen (eds), *Yearbook of International Co-operation on Environment and Development, 2004*, pp. 19–31.

Archibugi, D., Held, D. & Kohler, M. 1998, *Reimagining Political Community*, Stanford University Press, Stanford, California.

Asian Development Bank (ADB) 2011, *Technical Assistance for the Master Plan on ASEAN Connectivity Implementation*, ADB, Manila.

Asian Development Bank (ADB) 2012, *Regional Technical Assistance Green Cities: A Sustainable Urban Future in Southeast Asia*, ADB, Manila.

Asian Development Bank (ADB) 2014, 'IMT-GT 2014, A Framework for GrEEEn Actions: Green City Action Plan for Melaka', viewed at <https://www.adb.org/sites/default/files/related/41571/imt-gt-green-city-action-plan-melaka-april-2014.pdf/>.

Bäckstrand, K. 2006, 'Multi-Stakeholder Partnerships for Sustainable Development: Rethinking Legitimacy, Accountability and Effectiveness', *European Environment*, vol. 16, no. 5, pp. 290–306.

Bäckstrand, K., 2008, 'Accountability of Networked Climate Governance: The Rise of Transnational Climate Partnerships', *Global Environmental Politics*, vol. 8, no. 3, pp. 74–102.

Centre for IMT-GT Subregional Cooperation (CIMT) 2017, 'IMT-GT Implementation Blueprint 2017-2021', viewed at, <https://www.adb.org/sites/default/files/related/41543/imt-gt-implementation-blueprint-2017-2021.pdf/>.

Chonghui, L. 2018, 'UiTM's Solar Power Plant Being Built', *The Star Newspaper*, 29 April 2018.

Chua, S. C. & Oh, T. H. 2011, 'Green Progress and Prospect in Malaysia', *Renewable and Sustainable Energy Reviews*, vol. 15, no. 6, pp. 2850–2861.

Climate Governance Malaysia 2020, 'Climate Governance Malaysia', viewed at, <www.cgmalaysia.com/>.

Economic Planning Unit (EPU) 2017, 'Malaysia: Sustainable Development Goals Voluntary National Review 2017: High-level Political Forum', *EPU Publication*, viewed at, <https://sustainabledevelopment.un.org/content/documents/15881Malaysia.pdf>.

Economic Planning Unit 2015, *Eleventh Malaysia Plan, 2016–2020*, Percetakan Nasional.

Facer, K. & Buchczyk, M. 2019, 'Understanding Learning Cities as Discursive, Material and Affective Infrastructures', *Oxford Review of Education*, vol. 5, no. 2, pp. 168–187.

Fei, L. L. 2018, 'The Implications of the Paris Climate Agreement for Malaysia', *International Journal of Science Arts and Commerce* vol. 3, no. 2, pp. 27–39.

Fong, W. K., Sotos, M. E. & Kotorac, S. 2013, 'For the First Time, a Common Framework for Cities' Greenhouse Gas Inventories', World Research Institute, viewed at, <https://www.wri.org/insights/first-time-common-framework-cities-greenhouse-gas-inventories/>.

Fulton, L., Mejia, A., Arioli, M., Dematera, K. & Lah, O. 2017, '*Climate Change Mitigation Pathways for Southeast Asia: CO2 Emissions Reduction Policies for the Energy and Transport Sectors*', *Sustainability*, vol. 9, no. 7, p.1160.

Gee, L. T. 2015, 'Implementation of Green Technology Policy in Malaysia', Ministry of Energy, Green Technology & Water Malaysia, viewed at, <https://www-iam.nies.go.jp/aim/event_meeting/2015_cop21_japan2/file/03_malaysia.pdf>.

Government of Malaysia 1973, *The Midterm Review of Second Malaysia Plan, 1971–1975*, Government Printer, Kuala Lumpur.

GHG Protocol 2014, 'Global Protocol for Community-Scale Greenhouse Gas Emission Inventories', viewed at <http://www.ghgprotocol.org/city-accounting/>.

Halliday, F. 2001, 'The Romance of Non-state Actors', in D. Josselin & W. Wallace, *Non-state Actors in World Politics*, Palgrave Macmillan, London, pp. 21–37.

Hezri, A. A. 2011, 'Sustainable Shift: Institutional Challenges for the Environment in Malaysia', *Akademika*, vol. 81, no. 2.

Hezri, A. A. & Nordin Hasan, M. 2006, 'Towards Sustainable Development? The Evolution of Environmental Policy in Malaysia', *Natural Resources Forum*, vol. 30, no. 1, pp. 37–50, Oxford, UK: Blackwell Publishing.

Ho, C. S., Matsuoka, Y., Simson, J. & Gomi, K. 2013, 'Low Carbon Urban Development Strategy in Malaysia–The Case of Iskandar Malaysia Development Corridor', *Habitat International*, vol. 37, pp. 43–51.

Hsu, A., Widerberg, O., Weinfurter, A., Chan, S., Roelfsema, M., Lütkehermöller, K. & Bakhtiari, F. 2018, 'Bridging the Emissions Gap: The Role of Non-state and Subnational Actors', Pre-release version of a chapter of the forthcoming UN Environment Emissions Gap Report 2018, in *The Emissions Gap Report 2018. A UN Environment Synthesis Report*.

ICLEI 2016, 'Melaka State: Greenhouse Gas Emission Inventory Report 2013', Melaka Green Technology Corporation, ICLEI – Local Governments for Sustainability, South Asia.

ICLEI 2017, 'Melaka State: Greenhouse Gas Emission Inventory Report 2015', Melaka Green Technology Corporation, ICLEI – Local Governments for Sustainability, South Asia.

ICLEI, Local Government for Sustainability 2021, 'Melaka GHG Emissions Inventory Report', released at CoP 22, ICLEI South Asia.

IEA 2013, 'World Energy Outlook 2013', viewed at: <http://www.worldenergyoutlook.org/publications/weo-2013/>.

IIUM News 2021, 'Congratulations! IIUM Won National Energy Award (NEA) 2020', viewed at, <www.iium.edu.my/v2/congratulations-iium-won-national-energy-award-nea-2020/>.

Intergovernmental Panel on Climate Change 2018, 'Global Warming of 1.5° C: An IPCC Special Report on the Impacts of Global Warming of 1.5° C Above Pre-

108 Zen and Mohamad

Industrial Levels and Related Global Greenhouse Gas Emission Pathways, in The Context of Strengthening the Global Response to the Threat of Climate Change, Sustainable Development, and Efforts to Eradicate Poverty'.

International Energy Agency & Birol, F. 2013, *World Energy Outlook 2013*, International Energy Agency, Paris, viewed at, <http://www.worldenergyoutlook.org/publications/weo-2013/>.

IPCC 2014, *Climate Change 2014: Synthesis Report*, Contribution of Working Groups I, II and III to the Fifth Assessment Report of the Intergovernmental Panel on Climate Change [Core Writing Team, R.K. Pachauri & L.A. Meyer (eds.)], IPCC, Geneva, Switzerland, p.151.

Iskandar Malaysia 2021, 'Home', viewed at, <https://iskandarmalaysia.com.my>.

Jacobs, M. 2013, 'Green Growth', In Falkner, R. (Ed.). (2016). *The handbook of global climate and environment policy.* John Wiley & Sons.

Jalil, A. 2020, 'Govt Plans to Assess the Need for Climate Change Legislation', *The Malaysian Reserve.*

Johnson, O. W. 2020, 'Learning from Nordic Cities on Climate Action', *One Earth*, vol. 2, no. 2, pp. 128–131.

Kasayanond, A., Umam, R. & Jermsittiparsert, K., 2019, 'Environmental Sustainability and Its Growth in Malaysia by Elaborating the Green Economy and Environmental Efficiency', *International Journal of Energy Economics and Policy*, vol. 9, no. 5, p.465.

Keck, M.E. & Sikkink, K. 1998, *Activists Beyond Borders: Advocacy Networks in International Politics*, Cornell University Press, Ithaca, New York.

Kementerian Tenaga, Teknologi Hijau dan Air (KeTTHA) 2011, 'Low Carbon Cities Framework and Assessment System', viewed at < http://lccftrack.greentownship.my/files/LCCF-Book.pdf/>.

Kementerian Tenaga, Teknologi Hijau dan Air (KeTTHA) 2015, 'Implementation of Green Technology Policy in Malaysia' [Powerpoint presentation], viewed at, <https://www-iam.nies.go.jp/aim/event_meeting/2015_cop21_japan2/file/03_malaysia.pdf>.

Kementerian Tenaga, Teknologi Hijau dan Air (KeTTHA) 2017, 'Green Technology Master Plan Malaysia 2017–2030', viewed at, <https://www.pmo.gov.my/wp-content/uploads/2019/07/Green-Technology-Master-Plan-Malaysia-2017-2030.pdf>.

Kingsbury, B. & Casini, L. 2009, 'Global Administrative Law Dimensions of International Organizations Law', *International Organizations Law Review*, vol. 6, no. 2, pp. 319–358.

Kok Seng, Y. 2016, 'Malaysia's Response to Climate Change', 'The 25th Asia-Pacific Seminar on Law', *International Organizations Law Review*, vol. 6, no. 2, pp. 319–358.

Krishnan, G., Sandhu, S. C., Prothi, A. R. Singru, R. & van Dijk, N. 2014, 'Green City Action Plan: A Framework for GrEEEn Actions Melaka, Malaysia', viewed at, < https://www.adb.org/sites/default/files/related/41571/imt-gt-green-city-action-plan-melaka-april-2014.pdf/>.

Low Carbon Asia Research Center 2015, 'SATREPS-Low Carbon Society', viewed at, <http://www.utm.my/partners/satreps-lcs/about/>.

Malaysia, UTM. 2013, *Low Carbon Society Blueprint for Iskandar Malaysia 2025*, UTM-Low Carbon Asia Research Center, Johor Bahru, Malaysia.

Management and Science University 2018, 'A Green Milestone for MSU', viewed at, <https://www.msu.edu.my/MSUnews/Apr2021-sustainability-SEDA-GreenPass/>.

Ministry of Energy, Green Technology and Water (MEGTW) 2009, 'National Green Technology Policy', viewed at <https://www.kasa.gov.my/resources/alam-sekitar/national-green-technology-policy-2009.pdf>

Ministry of Energy, Science, Technology, Environment & Climate Change (MESTECC), 2018, 'Malaysia's Third National Communication and Second Biennial Update Report to the UNFCCC', *MESTECC Publication*, viewed at, <https://unfccc.int/sites/default/files/resource/Malaysia%20NC3%20BUR2_final%20high%20res.pdf>.

Ministry of Energy, Green Technology and Water 2017, 'Green Technology Masterplan Malaysia 2017–2030', viewed at, <Green-Technology-Master-Plan-Malaysia-2017-2030.pdf>

Ministry of Natural Resources and Environment 2009, 'National Policy on Climate Change', viewed at, <http://www.nre.gov.my/English/Environment/Pages/environment.aspx>.

Ministry of Natural Resources and Environment 2014, 'A Roadmap of Emissions Intensity Reduction in Malaysia', viewed at, <https://enviro2.doe.gov.my/ekmc/wp-content/uploads/2020/04/Final_A-Roadmap-of-GHG-Intensity-Reduction-Malaysia.pdf>.

Ministry of Natural Resources and Environment 2015, 'Malaysia Biennial Update Report to the United Nations', Framework Convention on Climate Change, UNFCCC, viewed at, <https://unfccc.int/sites/default/files/resource/MALBUR1.pdf>.

Ministry of Natural Resources and Environment 2020, 'Third Malaysia Biennial Update Report to the United Nations', Framework Convention on Climate Change, UNFCCC, viewed at, <https://unfccc.int/sites/default/files/resource/MALAYSIA_BUR3-UNFCCC_Submission.pdf>.

MNRE, 2010, '*National Policy on Climate Change*', Workshop on Climate Change & Biodiversity: Mobilizing the Research Agenda, 13–14 December 2010, UKM, Bangi.

Mol, A.P., Sonnenfeld, D.A. & Spaargaren, G. (eds.) 2020, *The Ecological Modernisation Reader: Environmental Reform in Theory and Practice*, Routledge, Oxford.

MOSTI 2021, 'About Us', viewed at, <https://www.mosti.gov.my/web/en/corporate-profile/about-us/>.

Osborne, M., Kearns, P. & Yang, J. 2013, 'Learning Cities: Developing Inclusive, Prosperous and Sustainable Urban Communities', *International Review of Education*, vol. 59, no. 4, pp. 409–423.

Palermo, V. & Hernandez, Y. 2020, 'Group Discussions on How to Implement a Participatory Process in Climate Adaptation Planning: A Case Study in Malaysia', *Ecological Economics*, vol. 177, p.106791.

Phang, F. A., Wong, W. Y., Ho, C. S., Musa, A. N., Fujino, J. & Suda, M. 2016, 'Iskandar Malaysia Planned Behaviour of Malaysian Citizens', *Journal of Cleaner Production*, vol. 235, pp. 125–1264.

Phang, F. A., Wong, W. Y., Ho, C. S. & Musa, A. N. 2017, 'Achieving Low Carbon Society Through Primary School Ecolife Challenge in Iskandar Malaysia', *Chemical Engineering Transactions*, 56, 415-420.

Pin, K. F., Pereira, J. J. & Aziz, S. 2013, 'Platforms of Climate Change: An Evolutionary Perspective and Lessons for Malaysia', *Sains Malaysiana*, vol. 42, no. 8, 1027–1040.

Rahman, H.A. 2009, 'Global Climate Change and Its Effects on Human Habitat and Environment in Malaysia', *Malaysian Journal of Environmental Management*, vol. 10, no. 2, pp. 17–32.

Rahman, S. 2020, 'Malaysia and the Pursuit of Sustainability', *Southeast Asian Affairs*, pp. 209–232.

Rasiah, R., Al-Amin, A. Q., Habib, N. M., Chowdhury, A. H., Ramu, S. C., Ahmed, F. & Leal Filho, W. 2017, 'Assessing Climate Change Mitigation Proposals for Malaysia: Implications for Emissions and Abatement Costs', *Journal of Cleaner Production*, vol. 167, pp. 163–173.

Razak, N., 2016. 'Speech at Opening of Paloh Hinai Green Technology Park', Paloh Hinai.

Sandhu, S. C. & Singru, R. N. 2014, 'Enabling Green Cities: An Operational Framework for Integrated Urban Development in Southeast Asia', *Scientific Research*, vol. 48, no. 3, pp. 466–492.

Shahar, F. M. 2018, 'UITM's Solar Photovoltaic Plant in Gambang Will Generate Over RM650m Revenue Over 21 Years', *New Strait Times*.

Susskind, L., Chun, J., Goldberg, S., Gordon, J. A., Smith, G. & Zaerpoor, Y. 2020, 'Breaking Out of Sustainable Urban Communities', *International Review of Education*, vol. 59, no. 4, pp. 409–423.

Tang, K. H. D. 2019, 'Climate Change in Malaysia: Trends, Contributions, Impacts, Mitigation and Adaptations', *Science of the Total Environment*, vol. 650, pp. 1858–1871.

The Star 2015, 'PM: Malaysia to Achieve 35% Cut in GDP's Greenhouse Gas Intensity by Year End', *The Star*, 11 September 2013, viewed at, <https://www.thestar.com.my/news/nation/2015/09/11/pm-35pc-cut-in-gdp-greenhouse-gas-intensity-2015/>.

The World Bank Data n.d., 'CO2 Emissions (Metric Tons per Capita)', viewed at, <https://data.worldbank.org/indicator/EN.ATM.CO2E.PC>.

United Nations (UN) n.d., 'Framework Convention on Climate Change', UNFCCC, MESTECC Publication.

United Nations (UN) 2015, 'Paris Agreement', United Nations Framework Convention on Climate Change, viewed at, <https://unfccc.int/sites/default/files/english_paris_agreement.pdf>.

United Nations (UN) 2018, 'Paris Agreement', United Nations Framework Convention on Climate Change, viewed at, <https://unfccc.int/sites/default/files/english_paris_agreement.pdf>.

UTM Newshub 2014, 'UTM Awarded the Energy Management Gold Standard (EMGS) 3rd Gold Star', viewed at, <https://news.utm.my/2014/12/utm-awarded-the-energy-management-gold-standard-emgs-3rd-gold-star/>.

Varkkey, H. 2019, 'Winds of Change in Malaysia: The Government and the Climate', Heinrich Böll version of a chapter of the forthcoming UN Environment Emissions Gap Report 2018. The Emissions Gap Report 2018. A UN Environment Synthesis Report.

Wahab, M. Z. & Naim, A. M. 2019, 'Malaysian Initiatives to Support Sustainable and Responsible Investment (SRI) Especially Through Sukuk Approach', *Journal of Emerging Economies & Islamic Research*, vol. 7, no. 3, pp. 1–11.

Widerberg, O. & Pattberg, P. 2015, 'International Cooperative Initiatives in Global Climate Governance: Raising the Ambition Level or Delegitimizing the UNFCCC?', *Global Policy*, vol. 6, no. 1, pp. 45–56.

World Commission on Environment and Development (WCED) 1987, *Our Common Future*, Oxford University Press, New York.

Zaid, S. M., Myeda, N. E., Mahyuddin, N. & Sulaiman, R. 2015, 'Malaysia's Rising GHG Emissions and Carbon "Lock-In" Risk: A Review of Malaysian Building

Sector Legislation and Policy', *Journal of Surveying, Construction and Property*, vol. 6, no. 1, pp. 1–13.

Zen, I. S., Al-Amin, A. Q. & Doberstein, B. 2019, 'Mainstreaming Climate Adaptation and Mitigation Policy: Towards Multi-Level Climate Governance in Melaka, Malaysia', *Urban Climate*, vol. 30, p.100501.

Zen, I.S., Hashim, H. & Sinniah, G. K. 2019, *Melaka State Climate Action Plan 2020–2030*, Melaka Green Technology Corporation Pub.

Zhu, Z.-S., Liao, H., Cao, H.-S., Wang, L., Wei, Y.-M. & Yan, J. 2014, 'The Differences of Carbon Intensity Reduction Rate Across 89 Countries In Recent Three Decades', *Applied Energy*, vol. 113, pp. 808–815, doi:10.1016/j.apenergy.2013.07.062.

7 Evolving climate change governance in Myanmar

Limitations and opportunities in a political crisis

Adam Simpson and Ashley South

Introduction

As one of the countries most affected by climate change over the last two decades (Eckstein et al. 2019; Simpson 2018a), Myanmar has a strong interest in investing in both climate mitigation and adaptation. Climate change results in delayed and shorter, but often more intense, monsoons, which causes more severe flooding and drought, but poor coordination and capacity across all levels and areas of government have resulted in a fragmented and inadequate response (see also South & Demartini 2020).

This incoherence has been exacerbated by an ongoing political crisis following a military coup in February 2021 (Simpson & Farrelly 2021d), which has derailed even the country's meagre attempts to deal with climate change. At the time of writing, eight months after the coup, Myanmar was still in the grip of a violent crackdown by the military against unarmed protesters as well as a widespread civil disobedience campaign which effectively froze government activities. This traumatic experience was exacerbated by a third wave Delta outbreak of COVID-19 in mid-2021, which caused untold suffering and likely tens of thousands of deaths (Simpson & Farrelly 2021f). The final outcome of the conflict between the military and opposition forces is yet to be determined but it is clear that, despite the urgency, climate change action will be a second-order priority for Myanmar for the foreseeable future due to the political crisis engulfing all aspects of the country (Simpson & Farrelly 2021a).

Until the coup, the Myanmar Climate Change Alliance (MCCA), supported by the United Nations, had been the key coordinating body in the country since 2013 and had assisted in the development of an emerging Myanmar Climate Change Strategy and Master Plan. In addition, various civil society organizations, centered in Yangon or ethnic nationality-populated areas and focused on renewable energy transitions and climate resilience, made significant contributions to 'activist environmental governance' (Simpson 2014).

Political developments in Myanmar had been largely overshadowed after 2017 by the Rohingya crisis with engagement by the international community

DOI: 10.4324/9780429324680-9

decreasing and domestic antipathy towards some international actors increasing in some areas due to a perceived bias in favor of the Rohingya, complicating the coordination of climate responses (Simpson & Farrelly 2021e). The military coup resulted in a further reduction of international engagement, with various countries applying targeted sanctions or redirecting their aid budgets to non-state actors (Payne 2021), reflecting a return to the conditions prior to the political liberalization of 2011. In the meantime, illiberal countries such as Russia and Myanmar's influential neighbor, China, were engaging the new junta, with Thailand and ASEAN playing difficult and ambiguous roles.

The coup also altered domestic perspectives of international actors such as the UN, with most civilians wanting more international engagement in Myanmar. However, for most actors this meant punishing and isolating the military junta, and empowering the increasingly progressive National Unity Government (NUG), the parallel government established in hiding and in exile (Simpson 2021c; 2021d; Simpson & Farrelly 2021g), and the opposition, Ethnic Armed Organizations (EAO). This shift has the potential to improve domestic and international climate coordination in the future but is highly dependent on the political trajectory of the country.

Overall, state-led climate change governance in Myanmar has historically been afflicted by the same political constraints that characterize all areas of policymaking: the cultural and institutional legacies of half a century of authoritarian governance and maldevelopment, which was reinforced by the military's ongoing and constitutionally enshrined dominance within key areas of the country's governance structures during 2011–2021 under the 2008 institution; and an aloof National League of Democracy (NLD) government until the coup that was unwilling to consult with civil society adequately and incapable of accepting criticism (Farrelly, Holliday & Simpson 2018; Fink & Simpson 2018; Simpson & Farrelly 2021c; Thant Myint-U 2020).

Most government and non-government activities on climate change were brought to a standstill by the military takeover in February 2021, so this chapter focuses on the period prior to the coup. Nevertheless, key local actors (including EAOs) remain committed and willing to engage on climate change mitigation and adaptation.

While there was historically positive support for international climate change conventions at the government level, there was an absence of strong governmental action and policy implementation on climate change even prior to the coup. Our analysis adopts a critical approach to climate governance and focuses on the activities of non-state actors, such as non-governmental organizations (NGOs) and civil society organizations (CSOs), which have attempted to provide the leadership and activism often lacking in government. This analysis demonstrates the ability of non-state actors to undertake climate change 'activist environmental governance' (Simpson 2014) in the absence of government leadership, although results can be mixed (Simpson & Smits 2018). It also parallels a form of 'hybrid governance' (South 2018) that already diffuses power and authority within the country. Historically, civil society actors have been in the forefront of progressive environmental policy change in Myanmar and the broader region (Simpson 2014), including in relation to climate change issues. Some of these non-state CSO actors are linked to EAOs that have for decades been struggling for self-determination vis-à-vis a centralized state (still) dominated by the military (South 2010). Sometimes they are the governance/administration or services delivery wings of EAOs, sometimes operating with international donor funding.

This chapter interrogates the politics, policies, and risks associated with climate change in Myanmar and then analyses two case studies of non-state and sub-national climate governance. First, we consider the Renewable Energy Association of Myanmar (REAM), which has been at the forefront of the small, but growing, civil society movement supporting the shift towards renewable and climate-friendly energy technologies, particularly in Myanmar's mountainous ethnic minority regions. Second, we consider the local responses and adaptations of villages in southern Karen (Kayin) State. We analyse the drivers and impediments in national and sub-national climate change governance in Myanmar, particularly in relation to civil society, and examine the prospects for progress in this area in the uncertain but tragic context of the military coup.

Politics and governance in Myanmar

For most of the half-century to 2011, Myanmar was governed by an authoritarian military regime, which severely constrained most civil and political rights, including the formation of NGOs (Doyle & Simpson 2006). Following elections in 2010, the military-backed USDP government and former-general President Thein Sein took office and, between 2011 and 2016, initiated a surprising political and economic liberalization process throughout the country. After the first genuinely competitive elections in 2015, a more democratic yet still militarily circumscribed government led by Aung San Suu Kyi and the NLD – who had boycotted the 2010 elections – entered government in April 2016. There was a largely liberalizing process in governance over this time. However, illiberal tendencies remained within the NLD government and the military retained overall control over security issues and constitutional change through a military-authored constitution (Simpson & Farrelly 2021b; Simpson, Farrelly & Holliday 2018). Even this heavily circumscribed civilian government was too much for the military to suffer and it was brought down by the illegal military coup of February 2021 which reversed all of the liberalizing and democratic reforms of the previous decade (Simpson 2021a).

The history of authoritarian rule and persistent civil wars between the Burman (Bamar) ethnic majority and various ethnic minority groups in the country's mountainous periphery has, in some areas, resulted in a form of 'hybrid governance' (South 2018), which has evolved to include a range of sub-national actors. These actors include NGOs and ethnic minority CSOs that sometimes work in partnership with EAOs – some of which still control extensive territory in relatively remote border regions and demonstrate extensive para-state governance systems. During military rule there was little opportunity for local dissent and domestic environmental activists, particularly those based in ethnic minority areas, who questioned the necessity or rationale for energy or development projects were harassed by the military and its intelligence service, including arrests and torture. As a result of this repression, and particularly a national crackdown in 1988, many activists removed themselves from Myanmar to the 'liberated' border regions controlled or influenced by EAOs, or to neighboring countries such as Thailand, where millions of Myanmar citizens live as economic migrants, in addition to refugees in camps along the border.

In the absence of effective state-led environmental governance, this 'activist diaspora' (Simpson 2013b), which included numerous ethnically based environmental NGOs, provided crucial 'activist environmental governance' of Myanmar's environment and natural resources during this period (Simpson 2014). These activists undertook dangerous covert research in Myanmar proper and the liberated areas to produce environmental reports and assessments. A strong justice focus on security and human rights was then used to pressure corporations and Western governments to divest from these resource projects. Historically they did not lobby Myanmar's military government itself, since it was antagonistic to such groups, but from 2011 to 2021 the Thein Sein and Aung San Suu Kyi governments offered new opportunities for civil society engagement, together with a more effective state-led governance regime (Simpson 2017; Vijge & Simpson 2020). Nevertheless, the scale of the risks associated with climate change and the limitations regarding state attention and capabilities have ensured that sub-national and non-state actors continue to play a crucial role in climate mitigation and adaptation (Simpson & Smits 2018).

Climate risk and policies

The risks from climate change in Myanmar are many and varied (see South & Demartini 2020). Although historically, Myanmar has made a very small contribution to climate change in terms of GHG emissions, its reliance on the monsoon and agriculture for its economy and society highlight the risks of inaction in relation to both mitigation and adaptation. Concerning climate mitigation, there is a strong connection between climate and energy policies. Myanmar has historically relied primarily on renewable energy for its electricity in the form of large dams, but these were often built in ethnic minority areas against the wishes of the local communities. Furthermore, the lack of reliable and affordable electricity beyond key urban centers has meant that many communities have little choice but to rely on biomass (firewood) for cooking and other needs.

During the decade of reform (2011-21), as a result of civil conflict and community opposition, the governments' extensive dam building program was held in abeyance – although several projects remained on the drawing boards. Due to increased demand for electricity the importation of liquified natural gas (LNG) was increasing at a time when the government needed to make important decisions about the country's future energy development. This was occurring just as more climate-friendly technologies, such as solar and wind, were maturing globally. In this section we analyse the risks and policy responses – in both climate and energy – to global and local climate change.

Climate risks

The Sixth Report of the Intergovernmental Panel on Climate Change (IPCC) argues that "it is unequivocal that human influence has warmed the atmosphere, ocean and land" and that the rate of change is unprecedented (IPCC 2021a). In relation to Southeast Asia, it notes that climate change has resulted in more extreme tropical cyclones across the region (IPCC 2021b). Myanmar has been deeply affected by these events. In the Global Climate Risk Index 2020, Myanmar was considered the second most affected country by climate change between 1999 and 2018, primarily due to Cyclone Nargis in 2008 and recurrent flooding and drought (Eckstein et al. 2019). Cyclone Nargis killed more than 140,000 people, destroyed 800,000 houses, and left millions of Irrawaddy (Ayeyarwady) Delta residents, mostly ethnic Karen, homeless and facing disease and malnutrition. The cyclone path along the Delta meant that it caused maximum destruction, but it was clear that the widespread degradation of forests and mangrove ecosystems exacerbated the impacts of the cyclone. Mangroves provide a natural barrier against storm surges. The growth of military-run shrimp and fish farms along the coast had weakened these natural barriers, while deforestation had intensified river flooding.

Data from the Department of Meteorology and Hydrology, analysed by the World Wildlife Fund (Horton et al. 2017), show that, between 1981–2010, daily temperatures increased by about 0.25°C per decade, which is higher than the global average rate of increase of 0.18°C per decade (Lindsey & Dahlman 2021). Temperatures are likely to increase by mid-century, by between 1.3°C to 2.7°C above historical levels. Severe changes in rainfall patterns are also expected, and sea levels will likely rise by 20–41 cm by 2050. Already, the monsoon duration over the last 50 years shows a significant reduction, from 140–150 days in the mid-1950s to less than 120 days in 2008 (South & Demartini 2020).

Climate change hazards disproportionately affect the poorest and most vulnerable, such as conflict-affected people in Karen State who are increasingly exposed to floods and landslides, fire and droughts. South and Demartini (2020) noted that approximately 70 percent of Myanmar's workforce is engaged in agriculture, forestry, and fisheries – all of which are vulnerable to climate change. Agricultural losses in the Southeast Asia region could extend to 50 percent of present rice yields by the end of the century (Norwegian Institute of International Affairs 2017)

In addition to extreme events such as Cyclone Nargis, widespread flooding is occurring more regularly throughout the country, such as that which devastated large swathes of Myanmar in mid-2015 and affected 9 million people (Thomas 2016). As with most countries, climate change will be the most significant environmental issue for Myanmar's future because it is likely to exacerbate most existing environmental problems and also create new ones: storms and cyclones such as Nargis are likely to become more frequent and more intense, together with a more unpredictable monsoon, higher sea-levels, and more prolonged and severe droughts. Crucial strategies to deal with these global environmental changes include increasing the amount of forested areas in fragile watersheds and repairing mangrove ecosystems that provide natural buffer zones. The ADB lists climate change and pollution as two of the key risks facing Myanmar, highlighting the importance of these issues to Myanmar's long-term development (Asian Development Bank 2012a, p.32–33).

As South and Demartini (2020, p.18) note, existing

> practices and policies do not prepare or support smallholder farmers to face the challenges of climate change. With limited savings and often high debts, smallholder farmers cannot afford to maximize the utilization of land, causing exposure to vulnerable climatic and financial conditions. This means that a season of poor yield constitutes a significant financial and existential challenge.

Furthermore, marginalized ethnic communities, particularly the 1 million Rohingya crammed into the Kutupalong and Nayapara refugee camps perched precariously on the Bangladeshi border region's denuded hills, have already endured almost unimaginable hardship and face significant flood risks every monsoon season (Simpson & Farrelly 2020).

Climate policies

Despite the risks of potentially devastating climate change impacts, governments in Myanmar have traditionally demonstrated little commitment to mitigation or adaptation. During the pre-2011 years of military rule, the limited contact with outside organizations, the lack of official development assistance, and the absence of relations with global finance and multilateral development banks (Simpson 2013d; Simpson and Park 2013) meant that the regime was somewhat insulated from global governance developments associated with climate change. Nevertheless, the military government's attempts to normalize its international relations in the early 1990s led it to sign the UN Framework Convention on Climate Change (UNFCCC) in 1992. With UNFCCC funding through the UN Global Environment Facility (GEF), Win Myo Thu, an activist, and his NGO, EcoDev, began a project of national

communication while the country was still under military rule, including orga-
nizing the country's first national climate change conference (Simpson 2015,
pp.56–57).

Under the reforming regime of Thein Sein, the Myanmar government was
brought in from the cold and submitted its Nationally Determined Contribution (NDC) to the UNFCCC as part of the Paris Agreement in September 2015 (Ministry of Environmental Conservation and Forestry 2015).
The government committed to reduce the country's per capita emissions of 2
tonnes of carbon dioxide equivalent (tCO_2eq) in 2010 by 6 percent by 2030.
This emissions level ranks Myanmar around 46th lowest out of 198 countries
(Australian-German Climate and Energy College 2015). Given the existing
low level of per capita emissions – compared with, for example, Australia at
25.3 tCO_2eq – any commitment not to increase per capita emissions appears
quite significant. Nevertheless, although politically beneficial, the 6 percent
reduction is also relatively insignificant in terms of contributions to climate
change – on a per capita basis, it represents a 0.5 percent reduction in Australia's per capita emissions (Simpson 2021b).

The Myanmar Climate Change Master Plan, initiated under Thein Sein,
was released by the NLD government in 2019 and provides a cross-sectoral
framework for achieving Myanmar's NDC under the Paris Agreement (Government of Myanmar 2019a). The Myanmar Climate Change Alliance
(MCCA) supported the development of key policy instruments for addressing
climate change, notably the 2019 Myanmar Climate Change Strategy
(MCCS) and Master Plan (MCCSMP) and the Myanmar Climate Change
Policy (MCCP). Like most formal climate change initiatives in Myanmar,
however, the MCCA is dominated by the central government and major
international actors. Roles for communities and CSOs tend to be limited to
information sharing, and Myanmar EAOs were not included in the MCCA.

A revised National Environmental Policy, which was funded by the UNDP
and underwent several drafts and national consultation rounds, was finally
released in 2019 (Government of Myanmar 2019b). It provided 23 National
Environmental Policy Principles, including that "the rights of indigenous
people and ethnic nationalities to their lands, territories, resources and cultural heritage, and their roles in environmental conservation and natural
resources management, are recognized and protected" (7(a)(6)). If ever
adhered to, this principle provides a resource-focused basis for ethnic minority
autonomy and federalism; a key goal of many ethnic minorities and a way for
them to improve their own climate adaptation and resilience (BEWG 2017).

Overall, disaster risk reduction (DRR) and other policies in Myanmar
remain highly centralized and dependent on the decision-making of Union
government officials in Naypyidaw, who are largely unaware of the situation
on the ground. In order to be effective, these measures need to be decentralized, with more decision-making at the sub-national level. This should
include Myanmar's 14 state and regional governments, who will need to
coordinate closely with civil society groups and EAOs, in whose homelands

most remaining forests are located. Such an approach would be commensu-rate with ethnic stakeholders' demands for federalism as a structural solution to decades of armed state and society conflict in Myanmar (South & Demartini 2020).

Forestry policies

While improvements in governance occurred during the Thein Sein era, there continued to be significant unregulated natural resource exploitation, parti-cularly linked to military or crony-operated companies, which may also undermine the NDCs. It is difficult to reduce carbon emissions and maintain carbon sinks while clear-felling native forests. According to the Global Forest Resources Assessment 2015, Myanmar had the third largest area of annual deforestation – after Brazil and Indonesia – between 2010 and 2015, losing 546,000 hectares of forest per annum, which works out as 1.7 percent of the country (Food and Agriculture Organization 2015, p.15). By 2015, the fores-ted area had decreased to approximately 44.5 percent (World Bank 2016). Despite an export ban on timber from 1 April 2014, discrepancies in figures between the Myanmar government and its trading partners indicated that the corrupt and illegal export of logs continued (Simpson 2018b, 2018c). According to the government, after the ban came into effect, Myanmar received US$44 million from timber exports in the 2014–2015 fiscal year compared with $900m the year before. These figures were not borne out by figures from its six major trading partners, which indicate Myanmar's log exports were only slightly down to $1.3bn from $1.5bn the year before, with timber trade with China increasing from $622m to $677m in the same periods (NCRA 2015). This result was consistent with research demonstrating that military-connected companies in the border regions often receive agricultural concessions as a cover for logging operations that appear to be the primary objective (Woods 2015). These figures are particularly troubling since most remaining forests in the country are located in ethnic minority-populated areas and demands for local (decentralized) control over natural resources are one of the key aspirations of ethnic nationality leaders in Myanmar.

Despite five years of reforms by the Thein Sein government, the NLD inherited a long list of governance issues that had built up during decades of military mismanagement and neglect. One of the NLD government's first decisions was to halve the number of government ministries, with MOECAF rebranded as the Ministry of Natural Resources and Environmental Con-servation (MoNREC). The most significant change to the Ministry was the addition of mining to the forestry and environmental conservation portfolios. The appointed minister, U Ohn Win, had a strong background in forestry and was one of the few ministers appointed from outside the NLD. While the minister was conservation-oriented, the sprawling and largely unregulated mining sector represented an enormous challenge. The main obstacle facing the government was to find a balance between exploiting natural resources for sustainable and equitable development in one of Asia's poorest countries and repairing some of the past environmental damage caused by unfettered deforestation, mining, and previous destructive development projects. Halting deforestation, together with vigorous reforesting of the country would assist with this goal, but as always, there were political and administrative hurdles. Reorganization of Myanmar's state-owned enterprises, particularly in the natural resources sector, was a key goal of broader governance reforms (Heller & Delesgues 2016). However, as noted above, the issue was complicated by the fact that most remaining forests are located in ethnic areas, where EAOs often constitute de facto governing authorities.

One of the minister's first announcements, in June 2016, was that logging would be suspended nationwide by the end of that fiscal year in April 2017. Building on the export ban in April 2014, the policy helped stem the haemorrhaging of Myanmar's forest cover that occurred over the previous decades. The prospects for this policy's success were substantially greater under the new government than if such a policy had been proposed under the Thein Sein government. The NLD government had few of the links to the military and their associated businesses that have stymied conservation attempts in the past (Woods 2015). While deforestation continued in pockets due to illegal logging for agriculture or mining, the country-wide unregulated destruction of forests was reduced with some reforestation of recently forested areas and afforestation of long-denuded areas occurring. However, such policies are unlikely to succeed unless EAOs are engaged in the decision-making and implementation processes fully. Furthermore, many local communities in still-forested areas have strong claims to stewardship and local land tenure, based on their ancestors' well-documented and sustainable livelihoods practices, which contributed towards maintaining forest cover in more remote parts of the country. If discussed and implemented in partnership with key local stakeholders (EAOs and communities), this forestry policy could have numerous beneficial flow-on effects, including reduced erosion of arable land and soils and more amenable microclimates.

A range of government actors outside of MoNREC and the climate change portfolio, such as the Health Ministry, which has expertise in disaster response, could take on a greater role in climate change responses and DRR (Gilfillan 2019). Other government departments with responsibilities for DRR include the Fire Department, Department of Metrology and Hydrology, Disaster Management Department, and Social Welfare Department (South & Demartini 2020). One of the key areas that needs to be addressed in this equation is energy policy, which has had a significant impact on climate change globally and is at a critical juncture in Myanmar as it looks to expand and develop its economy over the next few decades.

Energy policies

Many of the issues related to mitigating climate change across the world are driven by energy policy. Unlike many developing countries, Myanmar does not rely primarily on coal for producing electricity since its own reserves are limited and of poor quality, while it has significant hydropower potential. Most electricity capacity in the country, historically around 75 percent, was derived from large-scale hydropower, most of which was constructed in ethnic-minority areas without community participation or consent during military rule (Simpson and Smits 2018). As with other forms of governance, energy policies during this period contributed little towards national development or sustainability. Electricity access and usage throughout the country was extremely low, with the electrification rate estimated at 26 percent at the end of military rule in 2011 (Asian Development Bank 2012b, p.23). As with many other developing countries, simple biomass technologies such as fuelwood, charcoal, agricultural residue, and animal waste have historically provided the dominant fuel source, supplying almost 70 percent of the country's primary energy (Asian Development Bank 2012b, p.3).

Throughout the 2000s, country-wide electricity shortages existed in the context of the development of the country's energy sector for exports in return for foreign exchange, initially to Thailand but also to China (Haacke 2010; Simpson 2007). Natural gas was the most successful of Myanmar's forays into this area, with several projects being instigated and developed during military rule, although some were completed during the Thein Sein government: the Yadana and Yetagun Natural Gas Pipelines exported gas from Myanmar to Thailand from the turn of the century; the Zawtika Gas Pipeline sent gas in the same direction from 2014; and the Shwe Gas Pipeline exported gas from Rakhine State to Yunnan Province in China from 2013. Each of these projects was associated with various human rights abuses, including the displacement of local communities (Simpson 2014).

Although several hydropower dams for domestic use were built during military rule, the program to build dams and export the energy was less successful: the China Power Investment Corporation and its partners started building the $3.6 billion Myitsone Dam on the Irrawaddy (Ayeyarwady) River in Kachin State, which was expected to provide up to 6,000MW of electricity, with 90 percent to be exported to China, while a cascade of dams was planned for the Salween (Thanlwin) River in Shan, Karenni (Kayah) and Karen (Kayin) States to export electricity to both Thailand and China (Simpson 2013a).

All these major development projects were undertaken in the absence of any significant or rigorous governance processes and without any meaningful local consultation or participation. Environmental Impact Assessments (EIAs) were either not undertaken at all or lacked any formal procedures for public or civil society input or response. Nevertheless, while the gas pipelines all reached completion, progress on the dams was often slow due to civil conflict and opposition from local communities, however limited the formal modes of participation. These projects tended to cause various social and environmental problems, mainly in the ethnic minority areas of Myanmar's mountainous border regions. Hydropower and natural gas are less harmful in relation to climate change and local pollution than coal, but large-scale hydropower projects have dire ramifications for fisheries, downstream water security, and displaced local communities, as well as the destruction of some of Myanmar's remaining forests. In 2018, the International Finance Corporation (IFC), in collaboration with Australian Aid, released its Strategic Environmental Assessment of the Myanmar Hydropower Sector, which provided a comprehensive overview of the government's difficulties in pursuing hydropower projects in Myanmar (International Finance Corporation 2018). The Myitsone Dam, for example, caused significant dislocation and resulted in a nationwide campaign against the project, culminating in its suspension by President Thein Sein in September 2011 (Kirchherr, Charles & Walton 2016).

The exploitation of natural gas rather than coal or oil had less to do with conscious government policy and more to do with the geological serendipity of plentiful reserves. For the many destitute and energy-poor communities of Myanmar, the exporting of most of these energy resources provided little hope for improved energy and environmental security (Simpson 2013c). Two of the ethnic minority areas that hosted the gas pipelines, Tanintharyi (Tenasserim) Region and Rakhine (Arakan) State, had the two lowest per capita levels of electricity usage in the country (Simpson 2014, p.84). Despite plentiful gas reserves, Myanmar has signed contracts to export the vast majority of that which is extracted. As a result, in recent years, it has been required to import LNG to satisfy its rapidly growing energy consumption. By 2018, around 56 percent of electricity production was from hydropower, while 37 percent was from natural gas, with about 6 percent from coal (International Energy Agency 2020).

As the shadow NUG turns to address a range of chronic and acute issues threatening Myanmar, there is an opportunity (a 'critical juncture') to review key decisions and issues around climate change in Myanmar. Despite the many other challenges, addressing climate change issues in an accountable and inclusive manner should be a priority for any credible alternative government. Energy policies that minimize carbon emissions while protecting local ecosystems and cultural practices will promote sustainable local practices while assisting global efforts to reduce climate change. They are also more likely to be equitable if involving local people in decision-making and planning (not just mitigation projects, after key decisions have already been made). These policies can be drawn from a critical approach to energy security, which focuses on the pursuit of social and environmental justice for individuals and marginalized communities rather than the state (Simpson 2014, pp.191–96), and which would prohibit the use of mega-dams across large free-flowing rivers such as the Irrawaddy and Salween.

Before the 2021 coup abruptly changed the situation, there were expectations about significant shifts in Myanmar, following regional and global trends, towards the decentralization of energy policy, which could simultaneously improve the development of distributed, decentralized, and low-carbon energy systems and technologies (du Pont 2019). This would also be in line with the move towards federalism, as envisaged in the peace process. The government's decision to raise the artificially low prices for national grid electricity in July 2019, effectively a reduction in the large subsidy provided, was also likely make solar and other climate-friendly and ecologically sensitive technologies more competitive with gas and large-scale hydropower. Although this policy by the government may well be helpful in reducing carbon emissions, it was driven largely by budgetary concerns. It has been primarily non-state actors that have been driving climate governance through a range of small-scale activities at the local level.

Non-state climate governance

How do key non-government stakeholders in Myanmar understand and relate to climate change issues, including in the policy domain? This issue is particularly relevant, with the collapse of the legitimate government in Myanmar following the military coup potentially opening space for other governance actors.

In this section, we adopt a critical approach to climate governance by focusing on two groups of non-state subnational actors. In the first part, we examine the NGO Renewable Energy Association of Myanmar (REAM). REAM does not openly protest for climate action, like some other campaign groups such as Climate Strike Myanmar (Dowling 2020), but is instead focused on service delivery of climate-friendly mini-grids. In the second part, we delve into community-based action to examine the responses and adaptations to climate risks and impacts of communities and other stakeholders, including EAOs, in southern Kayin (Karen) State villages.

NGO activism and effectiveness: the Renewable Energy Association of Myanmar (REAM)

The Renewable Energy Association of Myanmar (REAM), under the direction of its founder, U Aung Myint, has been at the forefront of the push towards more climate-friendly energy technologies in Myanmar. Given the energy cross-roads it finds itself in, Myanmar has an opportunity to improve its low levels of energy and climate security by adopting low-carbon renewable technologies employing battery storage and mini-grids. To achieve its aims, however, it requires the government to provide a transparent and amenable regulatory environment (Simpson & Smits 2018).

During the extreme illiberalism of earlier military rule (from the organization's establishment in 1993 until 2011), REAM's experience contrasted with many other environmental NGOs that existed at that time. Many NGOs (especially those based in the 'liberated areas', or other countries) focused on campaigning against large energy projects, which inevitably resulted in critiques of military rule. Many of these NGOs were ethnically based since most large projects were slated for the mountainous ethnic minority areas. The military often conflated this ethnic environmental activism with the ethnic insurgencies that raged around the country's periphery, with the NGOs therefore banned and attacked by the government, if they were paid any attention at all.

In contrast, REAM was based in Yangon and focused on developing small-scale renewable energy technologies, even if many of these existed in ethnic areas such as Shan State. Aung Myint had previously worked within government which provided him with a pre-existing network of government contacts. During the 2011–2016 period, under the more liberal USDP government of Thein Sein, REAM began to develop close relationships with other NGOs from around the region, notably the Bangkok-based Mekong Energy and Ecology Network (MEE Net). MEE Net, a small but well-established regional NGO, recognized REAM's expertise and placed an activist in the REAM office to develop links with local communities and promote climate-friendly energy technologies in Myanmar. Similarly, the small international NGO, Green Empowerment, and Dipti Vaghela of the international Hydro Empowerment Network began working closely with REAM on mini and micro-hydro projects, including regular visits to REAM in Yangon and field trips to ethnic regions.

Despite the gradual establishment of more civil and political freedoms over this period and a resultant recognition of the role NGOs could play in society, the government was still tied to the military with a policy focus on large dams and other large-scale energy projects that benefited military companies and associated cronies. Similarly, the Japan International Cooperation Agency (JICA), the ADB, and the World Bank were seen as fixated on large-scale projects rather than the small-scale sustainable energy technologies promoted by REAM.

Although there remained limitations on NGOs at this time, they also experienced unprecedented freedoms and access to government. Energy-focused workshops funded by international donors proliferated during this period, with REAM a pivotal contributor to many, such as a Mini-Hydropower Workshop in Naypyitaw in 2015 funded by the US State Department. Nevertheless, the government refused to support regulations that would allow mini-grids to feed into the national electricity network, which was considered an essential reform if climate-friendly mini-grids were to compete with the subsidized electricity from the national grid.

With the coming to power of what was expected to be a more liberal NLD government in April 2016, REAM saw opportunities for real change in energy and climate policy. Although Aung Myint contributed to the earlier government's INDC submission and COP21 planning in 2015, he argued that their direct links to the military and associated cronies prohibited them from adequately supporting the necessary technological, social, and economic shift required to achieve sustainable energy and climate security. Under the NLD government, which was considered likely to be more sympathetic to the plight of poor communities, it was hoped there would be significant opportunities for influencing policy to expand small-scale renewable energy technologies. As an example, the government's reduction in the subsidy for the national grid in July 2019 made renewable energy and mini-grids much more competitive.

With new international support, REAM had the opportunity to expand its staff significantly but purposefully stayed small to focus on networking with local communities. Under the previous government, foreign NGOs were required to register with the (newly renamed) Ministry of Natural Resources and Environmental Conservation. The World Wildlife Fund (WWF), a new international partner for REAM, received its permanent registration as an NGO after the NLD election landslide in late 2015. Its Energy Director, working at WWF since mid-2015, previously worked in the Ministry of Electric Power and argued that under the oppressive and conservative influence of the military, they had been taught to reject any renewables except large hydropower.

Once the NLD came to power, former colleagues in the ministries became far more open to discussing both their opinions and policy alternatives with NGOs. While the multilateral development banks and JICA have tended to offer more traditional solutions to Myanmar's energy insecurity, the technical and social innovations currently being developed across the world provide an opportunity that, with the right support from government, could drive the transformation of Myanmar towards a more energy-secure, climate-friendly future that also empowers local communities and CSOs. With the impacts of the military coup still uncertain, it is likely that REAM will continue its activities assisting local communities to build renewable micro-grids in the medium term but with a reduction in the international partnerships that were growing under the NLD. In the renewed reality of military rule, REAM's relatively uncontroversial and apolitical grassroots activities make it less of a target for repression than more campaign-oriented NGOs (Simpson & Smits 2018).

Climate change risk and resilience in Karen (Kayin) State

Climate change hazards are especially serious for vulnerable and marginalized groups, such as conflict-affected people in Karen (Kayin) State. Karen communities have experienced widespread and systematic human rights abuses in armed conflicts between the Union government and the Karen National Union (KNU) since independence in 1948. Although the Myanmar military and the KNU agreed on a ceasefire in 2012, in 2020 the peace process had not resolved underlying issues driving decades of armed conflict. Particularly in recent years, Karen State has suffered serious impacts of climate change, including widespread flooding (particularly in 2018), and rising temperatures resulting in declining agricultural yields.

In the immediate aftermath of a disaster, local self-help and coping mechanisms are the most important elements of response, with external actors usually only arriving on the scene sometime later. The quality of individual local leaders is an important factor, with Christian and Buddhist religious leaders often playing leading roles. Other important elements of local response include the availability of relief items and communities' access to networks of information and distribution. At the local level, relationships of solidarity, patronage, and protection with government and/or EAOs are significant. Strong community networks, based on ethno-linguistic and religious identities ('social capital'), have sustained and supported absorptive capacities and foster social protection, despite the increasing severity of hazards. Indigenous Buddhist networks provide psychological and spiritual care and material protection and support, such as monks providing shelter during disasters and distributing donations from the lay community.

Some Karen farmers have begun to experiment with climate change-adapted agriculture, including adopting new crops (green beans) in areas where climate change negatively impacts rice cultivation. Local transformative capacity is strengthened when women take more significant roles in DRR. Nevertheless, women and other marginalized groups remain especially vulnerable to climate change impacts and often suffer disproportionately from disasters (including in 2020 the coronavirus pandemic). Despite these positives, however, climate change-impacted communities often lack a voice, and are generally not consulted in the development of policies and strategies to address their plight. In many cases, their strongest advocates are local civil society actors, and EAOs.

The plight of Karen and other ethnic nationality villages has become particularly acute in the period after the research was undertaken. Since December 2020, more than 20,000 Karen civilians have been displaced, mostly in northern Karen State and eastern Bago Region (well to the north of the area discussed here). This has occurred in the context of Myanmar Army attacks on Karen areas, which have been repeated in aerial and other assaults on Kachin communities in the north (Free Burma Rangers 2021).

Although during the period of research Karen and other climate change-impacted villages were generally coping, fears existed that they may struggle in the future – particularly in the more disastrous climate change scenarios, which just a few years ago were at the extreme end of projections and now seem more likely. Loss and damage resulting from climate change occur not only because of limited capacities for absorption and adaption, because coping capacities are being exhausted, but also due to the increasingly severe and unpredictable nature of hazards. Some communities may reach a 'tipping point', beyond which local adaptation strategies no longer work.

In order to be effective, responses to climate change should be undertaken in partnership between government, relevant EAOs and CSOs, and communities. Since 2009, the KNU has officially banned large-scale commercial logging in its areas of authority or control, with exceptions for community use. However, this instruction is not always respected by local leaders, some of whom may have benefited personally from natural resource extraction. Within the KNU, responsibility for climate change issues rests with the Department for Agriculture and the KNU Forestry Department, and the KNU Environmental Protection Committee (which consists of these two departments, plus key Karen CSOs). Most of the KNU's seven districts have established at least one protected nature reserve.

The KNU's Land and Forest Policies promote federalism and self-determination in relation to natural resource governance. Between 2008–2017 the KNU established 147 new community forests. For the KNU and Karen CSOs, land rights are essential to sustainable climate change mitigation and responses. Senior KNU officials pointed to the relevance of the organization's Land Policy, which supports and promotes customary law and other traditional community practices in relation to land. Despite occasional informal discussions, government officials have not recognized the KNU's progressive land policies or endorsed its land registration activities.

Many of Myanmar's remaining forested areas of biodiversity are located in areas controlled by the KNU and other EAOs. Given the crucial role of such natural resources in mitigating climate change, and providing local resources for adaptation, the KNU should play a key role in climate change governance in Southeast Myanmar, as acknowledged in the Nationwide Ceasefire Agreement (Article 25), which the KNU signed in 2015. However, Myanmar's existing climate change responses and architecture tend to be top-down and technocratic, with only limited consultation of local stakeholders. This centralized and state-centric approach reflects Myanmar's authoritarian political cultures and the historical marginalization of ethnic nationality communities. It is important to support local agency and social capital as part of 'building back better' after disasters, including by supporting the capacities and networks described above. There is also a strong argument for decentralizing DRR activities within a federal constitutional framework as envisaged in the peace process. This should be combined with recognition of and support to both EAOs' governance/administration and services delivery, and impressive community resilience capacities demonstrated through a range of climate change adaptation and other locally based (and demand-driven) activities (South & Demartini 2020). However, following the coup it is even more uncertain as to whether this kind of decentralization and federalism will occur in the near future.

Conclusion

Effective state-led climate governance is desperately needed in Myanmar to address the historical environmental degradation that has occurred during the mismanagement of military rule and the environmental strains that were unleashed during the political and economic reform process. As Thant Myint-U (2020, p.258) noted just before the coup, 'Burma is running out of time [and the] country needs a radical agenda to fight inequality and prepare for the climate emergency to come'. While the policy architecture to improve outcomes was embryonic – and has since been severely disrupted by the coup – there were clear signs that a process of comprehensive and integrated climate governance was emerging over the decade to 2021. As with all legislation and policies in Myanmar, however, it is the implementation of the policy that provides the challenge, especially in the current deadly and dynamic context. As these case studies have demonstrated, civil society actors and EAOs will play crucial roles in the development and effectiveness of climate governance architecture, particularly since most forested and mountainous areas are in ethnic nationality-populated areas.

Even prior to the coup, it was clear that non-state actors were playing key roles in climate 'activist environmental governance' in Myanmar through building sustainability and resilience in communities via the installation of small-scale renewable energy mini-grids or encouraging experimentation with climate change-adapted agriculture and community-based forestry conservation. The coup has only exacerbated and highlighted the state's lack of capacity and inability to address climate change by itself. While genuine federalism and decentralization was anathema to previous Myanmar governments, the NUG has made it clear that if a new Myanmar emerges from the ashes under its leadership it will require support, assistance, and contributions from a far greater diversity of actors within Myanmar society (Simpson 2021c, 2021d), providing civil society organizations and EAOs with significantly more influence in future climate change governance.

Acknowledgements

Dr Adam Simpson would like to thank REAM, which hosted him for a workshop in Naypyitaw and fieldtrips to mini-grid sites in Shan State during the preparation of this chapter. Dr Ashley South would like to thank friends and colleagues in ActionAid Myanmar, including particularly Liliana Demartini and Marianna Cifuentes, and local AAM Fellows. Thanks also to civil society, government and KNU personnel who provided insights and interviews. This chapter is dedicated to all the Myanmar people fighting against military rule and for the protection of human rights and democratic values within a new federal and democratic Myanmar.

References

Asian Development Bank 2012a, 'Myanmar in Transition: Opportunities and Challenges', Asian Development Bank (ADB), Manila, viewed 1 July 2017 at, <https://www.adb.org/sites/default/files/publication/29942/myanmar-transition.pdf>.
Asian Development Bank 2012b, *Myanmar: Energy Sector Initial Assessment*, Asian Development Bank (ADB), Manila.
Australian-German Climate and Energy College 2015, 'INDC Factsheets', Australian-German Climate and Energy College, Melbourne, viewed 10 June 2016 at, <http://www.climate-energy-college.net/indc-factsheets>.
BEWG 2017, 'Resource Federalism: A Roadmap for Decentralised Governance of Burma's Natural Heritage', The Burma Environmental Working Group (BEWG), Chiang Mai, Thailand, viewed 19 April 2021 at, <https://www.burmalibrary.org/en/resource-federalism-a-roadmap-for-decentralised-governance-of-burmas-natural-heritage>.

Dowling, T. D. 2020, 'Extinction Rebellion and Myanmar (Part One)', *Tea Circle*, 1 January 2020, viewed 31 August 2020 at, <https://teacircleoxford.com/2020/01/01/extinction-rebellion-and-myanmar-part-1/>.

Doyle, T. & Simpson, A. 2006, 'Traversing More Than Speed Bumps: Green Politics Under Authoritarian Regimes in Burma and Iran', *Environmental Politics*, vol. 15, no. 5, pp. 750–767.

du Pont, P. 2019, 'Decentralizing Power: The Role of State and Region Governments in Myanmar's Energy Sector', Asia Foundation, Yangon, viewed 25 August 2019 at, <https://asiafoundation.org/wp-content/uploads/2019/04/Myanmar-Decentralizing-Power_report_11-April-2019.pdf>.

Eckstein, D., Künzel, V., Schäfer, L. & Winges, M. 2019, 'Global Climate Risk Index 2020', Germanwatch, Bonn, viewed 31 August 2020 at, <https://www.germanwatch.org/en/17307>.

Environmental Investigation Agency 2021, *Myanmar's Brutal Junta Makes Another Grab for Cash with Major Auctions of Illicit Timber*, 15 September, viewed 24 September 2021 at <https://eia-international.org/news/myanmars-brutal-junta-makes-another-grab-for-cash-with-major-auctions-of-illicit-timber/>.

Farrelly, N., Holliday, I. & Simpson, A. 2018, 'Explaining Myanmar in Flux and Transition', in A. Simpson, N. Farrelly & I. Holliday (eds), *Routledge Handbook of Contemporary Myanmar*, Routledge, London and New York, pp. 3–11.

Fink, C. & Simpson, A. 2018, 'Civil Society', in A. Simpson, N. Farrelly & I. Holliday (eds), *Routledge Handbook of Contemporary Myanmar*, Routledge, London and New York, pp. 257–267.

Food and Agriculture Organization 2015, 'Global Forest Resources Assessment 2015', Food and Agriculture Organization of the United Nations, Rome, viewed 15 September 2015 at, <http://www.fao.org/3/a-i4793e.pdf>.

Free Burma Rangers 2021, 'FBR Update: Burma Army Wounds 11-year-old Girl, Kills Man, Wounds Five Others in Continued Attacks, with 24,000 Now Displaced in Northern Karen State'. 14 April 2021.

Gilfillan, D. 2019, 'The Health Sector's Role in Governance of Climate Change Adaptation in Myanmar', *Climate and Development*, vol. 11, no. 7, pp. 574–584.

Government of Myanmar 2019a, *Myanmar Climate Change Master Plan (2018–30)*, Environmental Conservation Department, Ministry of Natural Resources and Environmental Conservation, Naypyitaw.

Government of Myanmar 2019b, *National Environmental Policy of Myanmar*, Government of Myanmar, Naypyitaw.

Haacke, J. 2010, 'China's Role in the Pursuit of Security by Myanmar's State Peace and Development Council: Boon and Bane?', *The Pacific Review*, 23 (1): 113–137.

Heller, P. R. P. & Delesgues, L. 2016, 'Gilded Gatekeepers: Myanmar's State-Owned Oil, Gas and Mining Enterprises', Natural Resource Governance Institute (NRGI), New York, viewed 18 February 2016 at, <http://www.resourcegovernance.org/sites/default/files/nrgi_Myanmar-State-Owned-Enterprises_Full-Report.pdf>.

Horton, R., Mel, M. D., Peters, D., Lesk, C., Bartlett, R., Helsingen, H., Bader, D., Capizzi, P., Martin, S. & Rosenzweig, C. 2017, 'Assessing Climate Risk in Myanmar: Technical Report', Center for Climate Systems Research at Columbia University, WWF-US and WWF-Myanmar, New York, viewed 14 September 2020 at, <https://unhabitat.org/assessing-climate-risk-in-myanmar-technical-report-2017/>.

International Finance Corporation 2018, *Strategic Environmental Assessment of the Myanmar Hydropower Sector*, International Finance Corporation, Washington D.C.

International Energy Agency 2020, 'Myanmar', International Energy Agency (IEA), Paris, viewed 1 September 2020 at, <https://www.iea.org/countries/myanmar>.

IPCC 2021a, 'Sixth Assessment Report (AR6) Climate Change 2021: The Physical Science Basis', The Intergovernmental Panel on Climate Change, Cambridge University Press, Cambridge, viewed 11 August 2021 at, <https://www.ipcc.ch/report/ar6/wg1/>.

IPCC 2021b, 'Regional Fact Sheet: Asia', in the Intergovernmental Panel on Climate Change (ed.) *Climate Change 2021: The Physical Science Basis. Contribution of Working Group I to the Sixth Assessment Report of the Intergovernmental Panel on Climate Change*, Cambridge University Press, Cambridge.

Kirchherr, J., Charles, K. J. & Walton, M. J. 2016, 'The Interplay of Activists and Dam Developers: The Case of Myanmar's Mega-Dams', *International Journal of Water Resources Development*, vol. 33, no. 1, pp. 111–131.

Lindsey, R. & Dahlman, L. 2021, 'Climate Change: Global Temperature', National Oceanic and Atmospheric Administration, US Department of Commerce, viewed 20 April 2021 at, <https://www.climate.gov/news-features/understanding-climate/climate-change-global-temperature>.

Ministry of Environmental Conservation and Forestry 2015, 'Myanmar's Intended Nationally Determined Contribution – INDC', Ministry of Environmental Conservation and Forestry (MOECAF), The Republic of the Union of Myanmar, Yangon, 25 August 2015.

Myint-U, T. 2020, *The Hidden History of Burma: Race, Capitalism, and the Crisis of Democracy in the 21st Century*. W W Norton & Company, New York.

NCRA 2015, 'Myanmar Business Update', New Crossroads Asia, Yangon, 15 September 2015, viewed 1 October 2015 at, <http://newcrossroadsasia.com/docs/Volume%2037.pdf>/.

Norwegian Institute of International Affairs 2017, 'Impact of Climate Change on ASEAN International Affairs: Risk and Opportunity Multiplier', Norwegian Institute of International Affairs (NUPI) and Myanmar Institute of International and Strategic Studies, Oslo, viewed 14 September 2020 at, <https://reliefweb.int/report/world/impact-climate-change-asean-international-affairs-risk-and-opportunity-multiplier>.

Payne, M. 2021, 'Statement on Myanmar', Australian Government, Minister for Foreign Affairs, Canberra, 7 March 2021, viewed 19 April 2021 at <https://www.foreignminister.gov.au/minister/marise-payne/media-release/statement-myanmar-0>.

Putnam, R., Leonardi, R. & Nanetti, R. 1993, *Making Democracy Work: Civic Traditions in Modern Italy*, Princeton University Press, Princeton, NJ.

Simpson, A. 2007, 'The Environment-Energy Security Nexus: Critical Analysis of an Energy 'Love Triangle' In Southeast Asia', *Third World Quarterly*, vol. 28, no. 3, pp. 539–554.

Simpson, A. 2013a, 'An "Activist Diaspora" as a Response to Authoritarianism in Myanmar: The Role of Transnational Activism in Promoting Political Reform', in F. Cavatorta (ed.), *Civil Society Activism under Authoritarian Rule: A Comparative Perspective*, Routledge/ECPR Studies in European Political Science, London and New York, pp. 181–218.

Simpson, A. 2013b, 'Challenging Hydropower Development in Myanmar (Burma): Cross-Border Activism Under a Regime in Transition', *The Pacific Review*, vol. 26, no, 2, 129–152.

Simpson, A. 2013c, 'Challenging Inequality and Injustice: A Critical Approach to Energy Security', in R. Floyd and R. Matthew (eds), *Environmental Security: Approaches and Issues*, Routledge, London and New York, pp. 248–263.

Simpson, A. 2013d, 'Market Building and Risk Under a Regime in Transition: The Asian Development Bank in Myanmar (Burma)', in T. Carroll & D. S. L. Jarvis (eds), *The Politics of Marketising Asia*, Palgrave Macmillan, Basingstoke and New York.

Simpson, A. 2014, *Energy, Governance and Security in Thailand and Myanmar (Burma): A Critical Approach to Environmental Politics in the South*, Routledge, London and New York.

Simpson, A. 2015, 'Starting from Year Zero: Environmental Governance in Myanmar', in S. Mukherjee & D. Chakraborty (eds), *Environmental Challenges and Governance: Diverse Perspectives from Asia*, Routledge, London and New York, pp. 152–165.

Simpson, A. 2017,'The Extractive Industries Transparency Initiative: New Openings for Civil Society in Myanmar', in M. Crouch (ed.), *The Business of Transition: Law Reform and Commerce in Myanmar*, Cambridge University Press, Melbourne, pp. 55–80.

Simpson, A. 2018a, 'The Environment in Southeast Asia: Injustice, Conflict and Activism' in M. Beeson & A. Ba (eds), *Contemporary Southeast Asia*, third edition, Palgrave Macmillan, New York, pp. 164–180.

Simpson, A. 2018b, 'Corruption, Investment and Natural Resources', in S. Alam, J. Razzaque & J. H. Bhuiyan (eds), *International Natural Resources Law, Investment and Sustainability*, Routledge, London and New York, pp. 416–434.

Simpson, A. 2018c, 'Environment and Natural Resources', in A. Simpson, N. Farrelly & I. Holliday (eds), *Routledge Handbook of Contemporary Myanmar*. Routledge, London and New York, pp. 417–430.

Simpson, A. 2021a, 'Myanmar: Calling a Coup a Coup', *The Interpreter*, 8 February 2021, viewed 19 April 2021 at <https://www.lowyinstitute.org/the-interpreter/myanmar-calling-coup-coup>.

Simpson, A. 2021b, 'Natural Resources: Wealth and Conflict', in A . Simpson & N. Farrelly (eds), *Myanmar: Politics, Economy and Society*. Routledge, London and New York, pp. 149–168.

Simpson, A. 2021c, 'Myanmar's exile government signs up to ICC prosecutions', *East Asia Forum*, 17 September 2021, viewed 24 September 2021 at <https://www.east-asiaforum.org/2021/09/17/myanmars-exile-government-signs-up-to-icc-prosecutions/>.

Simpson, A. 2021d, 'Two governments claim to run Myanmar. So, who gets the country's seat at the UN?', *The Conversation*, 24 September 2021, viewed 24 September 2021 at <https://theconversation.com/two-governments-claim-to-run-myanmar-so-who-gets-the-countrys-seat-at-the-un-167885>.

Simpson, A. & Farrelly, N. 2020, 'The Rohingya Crisis and Questions of Accountability', *Australian Journal of International Affairs*, vol. 74, no. 5, pp. 486–494.

Simpson, A. & Farrelly, N. 2021a, 'As Killings, Beatings and Disappearances Escalate, What's The End Game in Myanmar?', *The Conversation*, 11 March 2021, viewed 19 April 2021 at <https://theconversation.com/as-killings-beatings-and-disappearances-escalate-whats-the-end-game-in-myanmar-156752>.

Simpson, A. & Farrelly, N. 2021b, 'Interrogating Contemporary Myanmar: The Difficult Transition', in A. Simpson & N. Farrelly (eds), *Myanmar: Politics, Economy and Society*, Routledge, London and New York, pp. 1–12.

132 Simpson and South

Simpson, A. & Farrelly, N. (eds) 2021c, Myanmar: Politics, Economy and Society. Routledge: London and New York.
Simpson, A. & Farrelly, N. 2021d, 'Myanmar's Military Reverts to Its Old Strong-Arm Behaviour — and the Country Takes a Major Step Backwards', The Conversation, 1 February 2021, viewed 19 April 2021 at, <https://theconversation.com/myanmars-military-reverts-to-its-old-strong-arm-behaviour-and-the-country-takes-a-major-step-backwards-154368>.
Simpson, A. & Farrelly, N. 2021e, 'The Rohingya Crisis: Nationalism and its Discontents', in A. Simpson & N. Farrelly (eds), Myanmar: Politics, Economy and Society, Routledge: London and New York, pp. 249–264.
Simpson, A. & N. Farrelly, N. 2021f, 'How A Perfect Storm of Events is Turning Myanmar Into A 'Super-Spreader' COVID State', The Conversation, 30 July 2012, viewed 9 August 2021 at, <https://theconversation.com/how-a-perfect-storm-of-events-is-turning-myanmar-into-a-super-spreader-covid-state-165174/>.
Simpson, A. & N. Farrelly, N. 2021g, 'Treating the Rohingya Like They Belong in Myanmar', The Strategist, Australian Strategic Policy Institute, 17 June 2021, viewed 9 August 2021 at <https://www.aspistrategist.org.au/treating-the-rohingya-like-they-belong-in-myanmar/>.
Simpson, A., Farrelly, N. & Holliday, I. (eds) 2018, Routledge Handbook of Contemporary Myanmar, Routledge: London and New York.
Simpson, A. & Park, S. 2013, 'The Asian Development Bank as a Global Risk Regulator in Myanmar', Third World Quarterly, vol. 34, no. 10, pp. 1858–1871.
Simpson, A. & Smits, M. 2018, 'Transitions to Energy and Climate Security in Southeast Asia? Civil Society Encounters with Illiberalism in Thailand and Myanmar', Society and Natural Resources, vol. 31, no. 5, pp. 580–598.
South, A. 2010, Ethnic Politics in Burma: States of Conflict, second edition, Routledge, London and New York.
South, A. 2018, '"Hybrid Governance" and the Politics of Legitimacy in the Myanmar Peace Process', Journal of Contemporary Asia, vol. 48, no. 1, pp. 50–66.
South, A. & Demartini, L. 2020, 'Towards a Tipping Point? Climate Change, Disaster Risk Reduction and Resilience in Southeast Myanmar', ActionAid Myanmar, Yangon, viewed 12 August 2020 at, <https://myanmar.actionaid.org/publications/2020/towards-tipping-point>.
Thomas, A. 2016, 'Accelerating Threats from Climate Change: Disasters and Displacement in Myanmar', Refugees International, Washington D.C., viewed 1 September 2020 at, <https://www.refugeesinternational.org/reports/2016/myanmar>.
Vijge, M. & Simpson, A. 2020, 'Myanmar's Environmental Governance in Transition: The Case of the Extractive Industries Transparency Initiative' in E. Prasse-Freeman, P. Chachavalpongpun & P. Strefford (eds), Unravelling Myanmar's Transition: Progress, Retrenchment and Ambiguity Amidst Liberalisation, National University of Singapore Press and Kyoto University Press, Singapore and Kyoto, pp. 136–163.
Woods, K. 2015, 'Commercial Agriculture Expansion in Myanmar: Links to Deforestation, Conversion Timber, and Land Conflicts', Forest Trends and UK Aid, DfID, viewed 10 March 2016 at, <http://forest-trends.org/releases/uploads/Conversion_Timber_in_Myanmar.pdf>/.
World Bank 2016, 'Forest Area', World Bank, Washington D.C., viewed 11 June 2016 at, <http://data.worldbank.org/indicator/AG.LND.FRST.ZS>.

8 Innovation and dysfunction

Three decades of climate change governance in the Philippines

Antonio G. M. La Viña and Jameela Joy M. Reyes

Introduction

Any conversation on climate change governance necessitates a historicization of policies that countries have pushed forward to respond to the challenges brought forth by climate change. The Philippines is not an exception.

The Philippines is a disaster risk hotspot with its geographical location, especially its coastal areas, which are among those most vulnerable when disaster strikes. According to the World Risk Report 2018, the Philippines ranks third among all countries with the highest risks worldwide, with an index value of 25.14 percent.[1] This index considers countries' exposure to natural hazards and societal issues, such as poverty and inequality. Moreover, an increasingly warming world, which leads to drastic changes in rainfall patterns and rising sea levels, also contributes to the Philippines' vulnerability to social, physical, and economic destruction. Typhoon Ondoy, for instance, known internationally as Typhoon Ketsana, which hit the Philippines in September 2009, caused approximately PhP 10 billion worth of damage. Typhoon Yolanda, more widely known as Typhoon Haiyan, struck the Philippines in November 2013, affecting more than 16 million Filipinos, and ravaging 44 provinces and hundreds of municipalities. Thousands more remain missing to this day. It is unfortunate and ironic, then, that despite the Philippines contributing approximately 1 percent of the world's total global greenhouse gas emissions, it is a country that has been devastated by, and left vulnerable to, the effects of climate change (La Viña and Guiao 2013).

Bearing this in mind, it is then necessary to note that climate change governance is not novel in the Philippines. A year before the country formally adopted and signed the United Nations Framework Convention on Climate Change in July 1992, then-President Corazon Aquino had already established the Inter-Agency Committee on Climate Change via Administrative Order No. 220, s.1991, which had, for one of its functions, the formulation of policies and response strategies related to climate change. This commitment to address issues relative to climate change was further strengthened when the Philippines ratified the Convention in 1994 and signed the Kyoto Protocol four years after. In the three decades since the Committee's creation, the

DOI: 10.4324/9780429324680-10

Philippines has continued to make strides in climate change governance. However, there is a need to explore this evolution and see whether such approaches have proven to be effective in responding to national and global environmental issues. This is particularly true for the past few years since President Duterte's administration, which started in 2016, and the country's ratification of the Paris Agreement in 2017.

This chapter asks whether or not policy and government frameworks adopted by the Philippines have worked and currently work to address the challenges climate change brings to this country. To answer this question, this chapter will critically look at how climate change governance has evolved in the country and whether the country's changing responses, whether local or international, have effectively answered the challenges of climate change. The measure of effectivity is the government and citizens' knowledge on climate change and the government's initiatives to comply with its Nationally Determined Contributions (NDCs) after the signing of the Paris Agreement. Ultimately, 'effectivity' means achieving the NDC; that is, reducing or stabilizing carbon emissions and increasing resilience and adaptive capacity in the face of climate change.

Governmental response to climate change

The Philippines has always been proactive and, at times, the forerunner in the region in the fight against climate change. This section will lay down the policies and laws which the country developed to address climate change. The earlier mentioned Committee was the first national mechanism to address climate change in the Philippines, which laid the groundwork for the country's current arrangement: a largely multi-agency approach involving as many stakeholders as necessary from the government to non-government organizations to civil societies.

The Presidential Task Force on Climate Change Adaptation and Mitigation was formed in 2007 (PTFCC 2007) through Administrative Order No. 171, s. 2007, later amended in August of the same year (AO 171-A). This Task Force served as the government's primary response to the Fourth Assessment Report on climate change released by the Intergovernmental Panel on Climate Change. The whereas clauses of the law creating the PTFCC cited the Report, which indicated that global temperature would continue to rise. In the short-term, it would increase by about 1.4 to 5.8 °C, leading to an increase in sea level by 18 to 59 centimeters by 2010. The whereas clauses further indicated that the temperature rise was "very likely" caused by human activity and increased emissions. Another reason the Task Force was created was to consider the country's commitments in the UNFCCC and the Kyoto Protocol, which the Philippines ratified in November 2003. Among the Task Force's functions were to conduct a rapid assessment of climate change's impact on the Philippines, especially on the most vulnerable sectors, ensure compliance to air emission standards and combat deforestation and environmental degradation, and undertake strategic approaches to prevent or reduce emissions. The Task Force was also mandated to design concrete risk reduction, mitigation measures and adaptation responses, collaborate with international partners to stabilize emissions, and mainstream climate risk management into development policies, plans, and programs.

In 2009, the Climate Change Act was passed, creating the principal government agency responsible for dealing with climate issues: the Climate Change Commission. The Commission's main role is to be the sole policy-making body of the government, tasked with coordinating, monitoring, and evaluating public programs and action plans on climate change. Since its inception, the Commission has become the main body implementing climate change programs and recommending legislation, policies, and programs for adaptation and mitigation and other related activities. Another primary function of the Commission was to formulate a National Climate Change Action Plan to assess the national impacts of climate change and identify options and appropriate adaptation measures. The current Plan, spanning the years 2011–2028, was produced using a multi-sectoral and multi-stakeholder approach, which covers key climate actions in the strategic priorities, and are defined along thematic outcomes – food security, water sufficiency, ecosystem and environmental stability, human security, climate-smart industries and services, sustainable energy, and knowledge and capacity development.

The Philippine Disaster Risk Reduction and Management Act is the second main piece of legislation in the Philippines that relates to responding to the challenges of an increasingly warming world (DRRM 2010). This Law, which was passed in 2010, principally created a coherent guide to aid the government in building capacity and resilience in the face of natural disasters. The Law emphasized the necessity for contingency planning, mitigation, preparedness, and prevention in response to climate change-induced hazards. Similar to the Climate Change Act, the Law also created an inter-departmental and cross-sectoral body known as the National Disaster Coordinating Council. In 2012, the People's Survival Fund Act was enacted (and amended in 2015) to allow local governments to access funds for adaptation more readily, and sought to "provide long-term finance streams to enable the government to effectively address the problem of climate change" (PSF 2012).

Other complementary climate-related laws in the Philippines that were mentioned in its Intended Nationally Determined Contribution to the Paris Agreement (INDC 2015) include the Renewable Energy Act of 2008 and the Biofuels Act of 2006. The Renewable Energy Act seeks to achieve energy self-reliance through the "adoption of sustainable energy development strategies" to reduce dependence on fossil fuels. In contrast, the Biofuels Act seeks to create measures toward the development and utilization of indigenous renewable and sustainably sourced clean energy sources and mitigate toxic and greenhouse gas emissions.

Internationally, the Philippines has also made strides regarding conversations on climate change. The most prominent of these is the country's signing of the Paris Agreement, which it subsequently ratified (UNFCCC 2015). Monumental in both scale and potential impact, the Paris Agreement, for the first time in the history of climate change negotiations, established a set goal on adaptation and enshrined the written commitment of countries to target net-zero emissions in the second half of the century, limit the global temperature rise to well below 2 °C above pre-industrial levels and to pursue efforts to limit the increase even further to 1.5 °C, enhance the capability of countries to deal with climate change impacts, and make finance flows consistent with a pathway toward low emissions and climate-resilient development. As the 5th Chair of the Climate Vulnerable Forum, an advocacy alliance composed of developing, middle-economy, and small island states, the Philippines has consistently campaigned for a temperature cap of 1.5°C since the COP 20 in Lima, Peru. During the deliberations of the Paris Agreement, the Philippine delegation pushed forward this consistent stance and clamored, in addition to the temperature cap, for "a clear reference to human rights and the inclusion of indigenous people's rights in the negotiating text" (Forest Foundation Philippines and Parabukas 2019, p.42). Its efforts turned into fruition when 112 countries eventually supported the call to the 1.5°C cap.

According to COP 21, the Philippines submitted its Intended NDCs (INDCs), which it communicated to the Convention, and these are premised on pursuing climate mitigation as a function of adaptation. In this regard, one of its main intentions was to undertake emissions reduction of about 70 percent by 2030 relative to its business-as-usual scenario of 2000–2030, which will have to come from the "energy, transport, waste, forestry, and industry sectors" (Forest Foundation Philippines and Parabukas 2019, p.42).

In addition to the INDCs, Parties to the Paris Agreement are also invited to make efforts towards creation of NDCs that reflect their climate actions in order to meet the goals set in the Agreement. According to the Agreement's Article 13.7, Parties should regularly provide a national inventory report of anthropogenic emissions by sources and removals, prepared using good practice methodologies and any information necessary to track progress in implementing and achieving its NDC (UNFCCC 2015). Besides this, and particularly to developing countries like the Philippines, is the invitation to provide information on financial, technology transfer, and capacity-building support needed and received (UNFCCC 2015). Under the Agreement, these Parties are instructed to communicate their NDCs every five years. Having ratified the Agreement in 2017, the Philippines committed to update its intended NDC and submit its NDC before 2020.

On 15 April 2021, the Philippines communicated its NDC to the UNFCCC. Under the document, the country officially committed to

> a projected GHG emissions reduction and avoidance of 75%, of which 2.71% is unconditional and 72.29% is conditional, representing the country's ambition for GHG mitigation for the period 2020 to 2030 for the sectors of agriculture, wastes, industry, transport, and energy
>
> (NDC 2021).

The document further provided that the Philippines

> shall endeavor to peak its emissions by 2030 in the context of accelerating the just transition of its sectors into a green economy and the delivery of green jobs and other benefits of a climate and disaster-resilient and low carbon development to its people, among others
>
> (NDC 2021).

Aksyon Klima Pilipinas (AKP), a national network of civil society organizations working on diverse climate and development-related issues, released a statement post-submission of the NDC, saying that the network recognized the much-awaited submission of the NDC, but they went on to stipulate that the Philippine government,

> in strong partnership with non-government stakeholders, must translate the inscribed words into action, to enforce key policies and measures to strengthen climate change mitigation and adaptation, uphold social justice, and avert or minimize loss and damage, and secure the means of implementation necessary to realize the vision of a climate-resilient, inclusive, and sustainable Philippines
>
> (Aksyon Klima Pilipinas 2021).

Under its adaptation measures, the Philippines committed to undertake adaptation measures across the sectors of agriculture, forestry, coastal and marine ecosystems and biodiversity, health, and human security, among others, in order to pre-empt, reduce, and address residual loss and damage.

The NDC was a result of a series of discussions between government and non-government stakeholders, including civil society, the academe, and the youth, and while the higher GHG emissions reduction target is welcome, the commitments in the NDC are also ambitious, and therefore, the Philippines will need to work doubly harder to meet its targets – including strengthening its resolve to stop the use of coal and banning the issuance of mining agreements.

Reception and reality

The following sections will discuss climate change governance from the Philippines' local government units up to its participation/s in the international sphere, and will look into limitations, roadblocks, and triumphs in the battle against climate change.

Gaps and successes at the local level

Notably, while the Climate Change Act has been generally effective in international climate negotiations, and despite the passage of national laws to respond to climate challenges, the Philippine government has failed to create a set of mechanisms with scaled-up responses to mitigation and adaptation. Despite the creation of a comprehensive policy agenda in the Climate Change Act, which is multi-sectoral and invites stakeholder participation, there have been several implementation difficulties, including the most pressing one of the lack of synergy between and among the different sectors involved, particularly fragmentation (La Viña, Ang & Guiao 2013). However, there have also been a number of successes.

One issue that has cropped up is that, despite being the government's roadmap for its projects and programs relating to climate change, the earlier iterations of the National Climate Change Action Plan showed that it was not fully aligned with the mechanisms already in place, such as the Philippine Development Plan and the Key Result Area-5 programs, activities, and projects (Crepin 2013). Early criticisms of the Climate Change Action Plan show that it did not extensively discuss climate change, particularly concerning adaptation and disaster risk reduction and management. However, there have been efforts to align these programs, as shown in the updated versions of the Climate Change Action Plan and the Philippine Development Plan.

The current Climate Change Action Plan, which outlines the country's agenda for adaptation and mitigation from 2011 to 2028, has seven specific priorities and is thematic: food security, water sufficiency, ecological and environmental stability, human security, climate-friendly industries and services, sustainable energy, and knowledge and capacity development. Coming from the original iteration, the current Climate Change Action Plan further acknowledges the importance of convergence planning among national agencies in the development of the Action Plan since "the strategic priorities are defined along thematic outcomes rather than sectors;" (NCCAP 2011–2028, p.6) thus, its success would require the collaboration of sectoral agencies. The Philippine Development Plan 2017–2022, on the other hand, mentions climate change in several chapters but devotes only a single chapter – chapter 20 – specifically for ensuring ecological integrity and a clean and healthy environment. Despite these efforts to realign, there are still unaddressed gaps. There is also a lack of congruence between the Climate Change Action Plan's priorities and the outcomes that the Philippine Development Plan seeks to achieve, which are not as articulated in the former.

The Climate Change Commission is also not without its organizational issues. Since it is a national agency with a variety of responsibilities, the Commission necessarily has a "limited local presence" (Crepin 2013, p.13). Tasked with being the sole policymaking body for climate programs and action plans, the Commission's vast array of responsibilities means it cannot fully engage with local governments, whether it be about resource distribution or policy questions (Crepin 2013). Confusion also sometimes takes place since the Commission's mandate is broad but not delineated. It is not a collegial body when it should be. It does not have decision-making powers, but it is tasked to coordinate, monitor, and evaluate plans. This arrangement leads to either overlaps or gaps in the functions of the Commission vis-à-vis other agencies that also work towards the integration of climate issues in governance while considering other priorities such as reducing inequality and pushing forward development goals. This is most prevalent when looking at the mandates of the Commission, the National Economic and Development Authority, the government's planning ministry, and the Department of Budget and Management. Thus, very soon after its creation, there was already a need to consolidate the Commission's proper mandate and the extent of its reach. Questions on its authority and capacity, whether technical or financial, to enter into its projects have also surfaced. Inquiries have also been made about the Commission having no ability to hold other government agencies accountable. This means that the latter are not necessarily incentivized to create plans and programs regarding combating climate change. The Climate Change Act is unclear on accountability measures (La Viña, Ang & Guiao 2013) which means that the Commission, after all, may have less power than initially thought.

Another issue is the absence of a clear organizational model that departments can use to deliver climate results. The Departments of Agriculture and Environment and Natural Resources are the only government agencies that have their internal climate units. The Department of Public Works and Highways "does not have a separate climate unit but is expected to create a cross-Departmental cooperation scheme" (Crepin 2013, p.36). These absences, including the lack of ownership from other agencies, should be addressed immediately because they have caused not just confusion but vacuums in the system, and in other cases, stagnation. Ensuring that other government agencies have climate change units is fundamental because climate change is not just about disasters but will largely affect food security (particularly agriculture), trade, and even tourism. Whether these gaps are a result of disinterest or fragmented governance following the lack of coordination and cohesion remains to be seen.

The Climate Change Commission's structure is also a cause for concern. Because its chairperson is the President of the Republic of the Philippines, there have been those who posit that this may lead to a dependence on the President, who may have neither the technical expertise nor the time to head such an organization that needs to act quickly on all matters relating to climate change (La Viña & Guiao 2013). Lastly, when it comes to the Commission's capacity, there is still a lack of mainstream guidelines that other agencies and sectors can look to for guidance, particularly regarding the development of the climate programs.

Capacity-building and knowledge management are also issues cropping up in the Philippine domestic climate change governance. There is no doubt about state capabilities on the national level, with an arsenal of technically knowledgeable staff and experts. On the other hand, the local governments may not have the same access to expert staff who possess specific skills required to create climate action plans and programs. Moreover, there may be other pressing needs that take priority with local government and need a more immediate response; these could range from poverty reduction to job creation to disaster response. The climate change language may also be too technical and therefore inaccessible, although strides have been made to address this concern.

Budget and financing for climate action are also causes for concern. The proposed funding by Congress, for instance, of the Climate Change Commission in 2019 is PhP 63.6 billion, a fraction of its PhP 3.757 trillion total budget, which is not enough for adaptation measures that the Philippines needs to take (Gregorio 2019). It is even more difficult on the local level, especially in areas most often devastated by raging typhoons and floods, situated in the archipelago's remote areas, which are usually poorer. Considering that a huge percentage of local governments' income is derived from the Internal Revenue Allotment, and considering further that the amount transferred to the local government depends on area and population, not on the level of climate vulnerability and potential risk, then it is entirely possible (and is indeed evident when disaster strikes) that the local governments of far-flung areas have a harder time finding funds, not only for disaster prevention but also for recuperation. This is proof that while climate change is a global problem, its impacts are found in specific places, and often the poorest of the poor get the brunt of its harshest effects. There is also a need for the Climate Change Commission, the National Economic and Development Authority, the Department of Budget and Management, and the Department of Finance to identify individual and joint priority policy issues (La Viña, Ang & Guiao 2013) and streamline their mandates regarding climate finance so that there will neither overlaps or gaps in climate action programs, activities, and projects.

However, it is essential to note that there have been successes as well. The People's Survival Fund, as already mentioned, was passed to serve as a funding mechanism for local governments and local community organizations to gain access to a PhP 1 billion replenishable fund to support climate adaptation activities. The Funds were not only provided by the Internal Revenue Allotments to local governments but were also sourced from domestic and international allotments. The Fund's purpose is "to ensure the consideration of climate change impacts such as stronger typhoons, heavier rains, prolonged drought, and other effects" in the formulation of government plans and projects. While helpful, it took a long while before the Fund was fully implemented and before a budget was provided for it. This was because, despite its passage in 2013, it only became fully operational in December 2015. Currently, the Fund has already become a long-term, predictable, and sustainable climate financing source. Concerns on lack of funding should disaster strike are also slightly alleviated because of the Fund's foresight. Of course, any initiatives to scale it up even further will be welcomed.

As a response to earlier hesitations by local governments to complete their Local Climate Change Action Plans, since this often imposes undue administrative burdens on their offices, the Department of the Interior and Local

Government issued a memorandum circular (DILG 2014) in October 2014 listing the guidelines for the formulation of these local climate plans. These guidelines have streamlined the processes that local governments have to undertake to create their local climate action plans and ensure uniformity of documentation and response across local governments.

National-level climate governance

A concern at the national level is that President Duterte, who took the seat in 2016, has, as late as October 2019, pushed for the continuous use of coal while simultaneously calling on the private sector to invest in renewable energy, claiming that both are possible (CNN Philippines 2019a). Notably, coal – and the projected increase in its use – is one of the main reasons why the Philippines is not yet on track to meet its Paris commitments. The increasing exploitation of coal started in 2015, ironically the same year of the Paris Agreement. The government has since then supported the building of around 15 gigawatts of coal-fired power capacity, jeopardizing the Philippines' ability to reach its conditional climate commitments (Climate Action Tracker 2019). Even with the Renewable Energy Law, the Philippines' energy mix still relies heavily on coal, which supplies around 48 percent of the country's electricity. Despite this, coal still cannot cope with peak demand, rationalizing an intentional move towards renewables since the transition can enhance system reliability, reduce emissions, and support the country's 100 percent electrification by 2022 (Climate Action Tracker 2019). It is apparent that the Philippines' climate change governance scheme focuses more on adaptation than mitigation, although strides have been taken towards mitigation as well. Renewable energy is becoming more and more prominent during talks about energy access and efficiency, but there is still a need to concretize plans to lay down the Philippine transition to an energy future that is not reliant on fossil fuels.

The lack of specific mechanisms revolving around agriculture and forestry also shows the shift of focus away from mitigation and, by extension, the lack of concerted efforts to combat climate change and its effects at the local level. Ongoing government projects to turn agricultural and forest land into industrial areas are one example. These programs affect the geographical terrain and lead to the displacement of indigenous peoples and loss of their livelihoods. Mitigation, therefore, has to be highlighted in the Philippine climate policy since the obvious choice for people living in vulnerable and high-risk countries such as the Philippines is already adaptation. Any adaptation plans that the Philippines has will always necessarily be localized. While the principles of common but differentiated responsibilities and historical responsibility support the position that developing countries ought not to be required to commit to mitigation targets if it means having to compromise their economic development, the tension keeps on rising between developing and developed countries as the latter tend to be less than willing to commit to providing extensive financial support. Thus, it may make sense to claim that "it is mitigation commitments that bring in much of the funding to be used to address adaptation needs" (La Viña & Guiao 2013, p.633).

When it comes to disaster risk reduction, experts have mentioned early on the need for a full-time independent disaster agency to focus specifically on handling climate risks. Although the military and civil defense have the workforce, these actors are not necessarily equipped with the skills related to climate change (La Viña & Guiao 2013). In November 2019, the committees on government reorganization and disaster management at the House of Representatives approved a bill creating the Department of Disaster Resilience as a response to the series of earthquakes that shook Mindanao, the southern part of the Philippines, that year, and the uncoordinated efforts to help the thousands of victims of said quakes (Cepeda 2019). The creation of a similar department is also being considered in the Senate (Casayuran 2019).

Aside from these, however, there have been minimal initiatives started by the government to combat climate change since President Duterte took the seat in 2016. There was no mention of climate change in the 2020 State of the Nation Address, which was delivered in July of that year. However, in his 2019 Address, he urged the fast-tracking of the development of renewable energy sources to reduce dependence on traditional energy sources such as coal. On 27 October 2020, Secretary of Energy Alfonso Cusi declared a moratorium on endorsements for greenfield coal power plants (Department of Energy 2020). This was issued in light of the most recent assessment of the Department that the country needed to shift to a more flexible power supply mix. Secretary Cusi also said that he was determined to accelerate the development of the Philippines' indigenous resources and that the department was pushing for the transition from "fossil fuel-based technology utilization to cleaner energy sources" to ensure the country's more sustainable growth.

Moreover, President Duterte lifted the ban on issuing new mining agreements in April 2021 (Ranada 2021). The move to lift the ban was intended to boost the economy in the face of the COVID-19 pandemic, in addition to supporting various government projects. The whereas clauses of the Executive Order provided the rationale for such a decision, to wit: Republic Act No. 10963, or the Tax Reform for Acceleration and Inclusion (TRAIN) Act has doubled the rate of excise tax on minerals, mineral products, and quarry resources from 2 percent to 4 percent, and the Philippines has tapped less than 5 percent of its mineral resources to date. The Executive Order (signed in 2012) barred the signing of new agreements until a new law regarding changing revenue sharing schemes would take effect. This was overturned by the 2021 Executive Order. The new EO granted the Department of Environment and Natural Resources the power to formulate the terms and conditions in the new mineral agreements "that will maximize government revenues and share from production, including the possibility of declaring these areas as mineral reservations to obtain appropriate royalties" (Executive Order No. 130 2021). Many groups have since criticized this decision, saying that the government prioritized money over a healthful ecology (Ranada 2021). Further, Malacañang released a copy of the EO on 15 April, the same day the Philippines submitted its NDC.

International commitments to combat climate change

Internationally, the Philippines' participation to climate action has stagnated a little bit since President Rodrigo Duterte assumed office in 2016, as compared to the Philippines' assertiveness during the Paris Agreement negotiations and in earlier COPs. Nevertheless, the Philippine position on climate change has never wavered since the very beginning, and its interests have remained constant, that is, ensuring that the threat of climate change is avoided given the country's geographic location and economic status, and asserting that global interventions to aid in addressing climate change should not prejudice, but rather forward, the country's sustainable development.

The Paris Agreement, then, was a win for the Philippines, and served as a reminder of the importance of having a synergistic, collaborative group of people who come from all sectors of society, to push forward an agenda that considers the abilities and limitations of the country. The delegation consisted of government officials, technical experts, academics, and members of civil society, who all served important roles in agenda formulation that the government pushed forward. More than all those, the Philippine delegation and participation in Paris proved the importance of taking part in international negotiations that have long-term impacts in countries such as the Philippines, which is among the first countries to be affected by rising sea waters, changes in temperature, and stronger and more frequent typhoons.

During the last day of the negotiations, the Philippines also joined the High Ambition Coalition led by the Marshall Islands to achieve the 1.5°C temperature cap. When asked why the Philippines decided to fight for the 1.5°C cap, Climate Change Commission Secretary Emmanuel de Guzman echoed the concerns of many other developing countries and argued that while the 0.5°C difference may seem unsubstantial to many countries, it would have immense effects on many small islands around the world, including some of those situated in the Philippines – particularly those that are low-lying where the change in temperature will necessarily displace millions of people. Overall, the Philippines' performance in the Paris Agreement is a good analogy of how climate change governance is occurring in the country. Abroad, the Philippines championed principles like ecosystems integrity and climate justice, with a focus on human rights, and performed extremely well internationally. Domestically, however, there is still much work to be done.

It is unfortunate to witness that the strong performance of the Philippines in international negotiations may be experiencing a decline. Everyone, particularly climate workers, applauded the Philippine Senate when it signified its concurrence in the Paris Agreement in 2017; however, there have been very few steps since then to show the country's sincere commitment to achieving its goals. This was made even more evident when, in June 2019, Foreign Affairs Secretary Teodoro Locsin echoed President Duterte's distaste towards climate change conferences and said that he would reject all official participation in these meetings (CNN Philippines 2019b). This pronouncement caused concern to many participants in past conferences, especially given the Philippines' leadership in many of the alliances. During COP 25 in Madrid, only a handful of Filipinos came, led by former senator and current Antique representative Loren Legarda. Those in attendance were not there as negotiators but primarily as observers; a far cry from the Philippines' role in years prior. Those who attended were forced to listen in, even while essential issues, such as loss and damage, were discussed. The issue was based on the "Warsaw International Mechanism for Loss and Damage associated with Climate Change Impacts" established in COP 19 in Poland to address loss and damage resulting from climate impacts, including both extreme, drastic events, and slow onset ones, in developing countries particularly vulnerable to the adverse effects of climate change. Many developing countries, like the Philippines, experience losses and damages during disasters exacerbated by human-induced climate change (cf. Forest Foundation Philippines and Parabukas 2019). Due to differences in priorities, particularly of developed and developing countries, a consensus on the specifics of loss and damage were not arrived at.

Recent developments and moving forward

Between the tail end of November and the middle of December 2020, the Philippines was ravaged by successive typhoons, which severely affected areas in Luzon, the country's biggest island. Among those typhoons was Super Typhoon Rolly, known internationally as Typhoon Goni, a Category 5-equivalent super typhoon. It was one of the strongest tropical cyclones on record, with a strength comparable to that of Typhoon Haiyan in 2013 and Meranti in 2016, both of which equally devastated the country. The other typhoons, which also wreaked destruction and led to loss of both lives and livelihood, were Typhoons Quinta and Ulysses.

The increasing intensity, frequency, and succession of these events is proof of the climate emergency we are faced with today. The Philippines, especially, is considered disaster-prone because of its natural geographic exposure to hazards, its degraded ecosystems, and its people's vulnerability due to poverty, socio-economic marginalization, and a weak governance system, among other factors (Berse et al. 2020). The authors in the previously cited article, therefore, proposed seven key actions to effectively respond to the climate emergency that the country is currently facing, namely: 1) Mainstream nature-based solutions in disaster risk reduction efforts; 2) Institutionalize integrated natural resources and environment management; 3) Invest in safer and durable evacuation centers; 4) Enhance inter-agency collaboration for early warning and response; 5) Promote inter-LGU cooperation in planning, response, and recovery; 6) Strengthen the national focal agency for disaster risk management; and 7) Declare a climate emergency.

In the same article, it was recommended that Congress and local legislative bodies should take the lead on these action plans by adopting resolutions that recognize the existence of the climate emergency and identifying priority mitigation, adaptation, and climate justice policies and measures at both the national and local levels.

There is no doubt that with a looming environmental crisis, more extreme weather conditions will occur. Therefore, there is also a need to strengthen policies regarding deforestation, land use, and illegal logging, all of which contribute to the country's difficulty in protecting itself from the harshest effects of climate change. Engineering solutions to floods, as well, have to be studied very closely, as the government has to ensure that the choices it makes will respond effectively to the specific needs and context of the Philippines and its particular vulnerabilities.

Conclusion

Climate change governance in the Philippines has been evolving in the past 30 years. The country started early with the creation of inter-agency mechanisms in 1991 and with Philippine presidents paying particular attention, especially when the country experienced severe climate emergencies like Typhoons Ketsana in 2009 and Haiyan in 2013. As discussed in this chapter, the Philippine Congress enacted a framework law on climate change and special laws relevant to climate response, but gaps remain. With climate change transforming into a global and national emergency, it is time to revisit and reset the institutional architecture of the Philippine climate response. With the recent release of the Sixth Assessment Report by the Intergovernmental Panel on Climate Change, it is much more important that the Philippines asserts itself in the international arena as a climate leader. Equally important is the need to strengthen its local and national laws in order to be able to prepare for and combat against the harshest effects of climate change.

Among the features of a better climate governance system are strong principles and objectives. In our view, a new law should enshrine climate justice globally and domestically as the overarching goal of climate policy and governance. As mentioned earlier, the Philippines globally was a strong champion for climate justice. Still, as also pointed out in this chapter, it has not been consistent in realizing that norm domestically. Every policy and measure that is developed, adopted, and implemented should be seen through a climate justice prism. This would include ensuring a just transition for the poor and disadvantaged sectors. An integrated mitigation and adaptation approach should also be institutionalized, with agencies like those in charge of energy, agriculture, and forestry encouraged to mainstream climate change in their strategies, programs, and activities. Finally, it might be best to abandon the commission type of agency and instead establish a ministry to lead the climate work, including setting policies, regulating activities, and implementing programs and projects.

Note

1 The index rates the disaster risk of countries owing to five natural hazards: earthquakes, cyclones, floods, droughts, and sea-level rise; and the World Risk Index consists of four components: exposure, susceptibility, coping capacities, and adaptive capacities. For more information, please visit United Nations University-EHS and Bündnis Entwicklung Hift, World Risk Report 2016, pp. 43–51, available at: https://weltrisikobericht.de/wp-content/uploads/2016/08/WorldRiskReport2016.pdf

References

Administrative Order No. 171 2007, *Creating the Presidential Task Force on Climate Change*, Malacañang Palace.

Aksyon Klima Pilipinas 2021, 'Official Statement: On the Submission of PH Nationally Determined Contributions (NDC)', Quezon City.

Berse, K., La Viña, T., Pulhin, J. & Nazal, M. 2020, '[Analysis] 7 Priority Actions to Manage the PH #Climateemergency', *Rappler.com*, viewed at, < https://www.rappler.com/voices/thought-leaders/analysis-7-priority-actions-manage-climate-emergency/>.

Casayuran, M. 2019, 'Creation of Department of Disaster Resilience Pushed', *Manila Bulletin*.

Cepeda, M. 2019, 'House Panels Pass Bill Creating Department of Disaster Resilience', *Rappler.com*, viewed at, <https://www.rappler.com/nation/house-panels-pass-bill-creating-department-disaster-resilience/>.

Climate Action Tracker 2019, *Philippines Current Policy Projections*, viewed at, <https://climateactiontracker.org/countries/philippines/2019-12-02/>.

Climate Change Commission 2012, *National Climate Change Action Plan 2011–2028*, viewed at, <http://climate.emb.gov.ph/wp-content/uploads/2016/06/NCCAP-1.pdf/>.

CNN Philippines 2019a, 'At Opening of Coal Plant, Duterte Calls for Clean Energy', viewed at, <https://www.cnnphilippines.com/news/2019/6/5/Climate-change-conference-Locsin-Duterte.html/>.

CNN Philippines 2019b, 'DFA Chief: No More PH Attendance in Climate Change Talks', viewed at <https://cnnphilippines.com/news/2019/10/16/Duterte-clean-energy-coal-plant.html/>.

Crepin, C. 2013, *Getting a Grip… on Climate Change in the Philippines: Executive Report (English), Public Expenditure Review (PER)*, Washington, D.C.: World Bank Group.

Department of Energy 2020, *DOE Sec. Cusi Declares Moratorium on Endorsements for Greenfield Coal Power Plans*, viewed at, <https://www.doe.gov.ph/press-releases/doe-sec-cusi-declares-moratorium-endorsements-greenfield-coal-power-plants?ckattempt=1/>.

Department of Interior and Local Government 2014, *Memorandum Circular No. 2014–135, Guidelines on the Formulation of Local Climate Change Action Plan (LCCAP)*, viewed at, < https://www.dilg.gov.ph/PDF_File/issuances/memo_circulars/dilg-memocircular-20141022_cd4420b4bd.pdf/>.

Executive Order No. 130 2021, *Amending Section 4 of Executive Order No. 79, s. 2012, Institutionalizing and Implementing Reforms in the Philippine Mining Sector, Providing Policies and Guidelines to Ensure Environmental Protection and Responsible Mining in the Utilization of Mineral Resources*, viewed at, <https://www.officialgazette.gov.ph/downloads/2021/04apr/20210414-EO-130-RRD.pdf/>.

Forest Foundation Philippines and Parabukas 2019, *UNFCCC Negotiations: A Resource Book*, viewed at, <https://static1.squarespace.com/static/5b593f64697a9814301d09f6/t/5d3552b280972800015981ea/1563776502107/UNFCCC+Negotiations+Resource.pdf/>.

Gregorio, X. 2019, 'In Face of Climate Emergency, The Philippines' Funding for its Response Remains Paltry', *CNN Philippines*.

La Viña, A. G. M., Ang, L. G. & Guiao, C.T.T. 2013, 'Governance and Political Analysis of Philippine Climate Change Policy', policy memorandum submitted to the World Bank, Philippines office on 27 June 2013.

La Viña, A. G. M. & Guiao, C. T. T. 2013, 'Climate Change and the Law: Issues and Challenges in the Philippines', *Ateneo Law Journal*, vol. 58, pp. 612–633.

National Economic and Development Authority 2018, *Philippine Development Plan 2017–2022*, viewed at, <https://pdp.neda.gov.ph/wp-content/uploads/2017/01/PDP-2017-2022-10-03-2017.pdf/>.

Ranada, P. 2021, 'Duterte Lifts Ban on New Mining Agreements', *Rappler.com*, viewed at, <https://www.rappler.com/business/duterte-lifts-ban-on-new-mining-agreements/>.

Republic Act No. 9367 2006, *An Act to Direct the Use of Biofuels, Establishing for this Purpose the Biofuel Program, Appropriating Funds Therefor, and For Other Purposes.*

Republic Act No. 9513 2008, *An Act Promoting the Development, Utilization and Commercialization of Renewable Energy Resources and for Other Purposes*, viewed at, <https://www.officialgazette.gov.ph/2008/12/16/republic-act-no-9513/>.

Republic Act No. 9729 2009, *An Act Mainstreaming Climate Change into Government Policy Formulations, Establishing the Framework Strategy and Program on Climate Change, Creating for this Purpose the Climate Change Commission, and for Other Purposes.*

Republic Act No. 10121 2010, *An Act Strengthening the Philippine Disaster Risk Reduction and Management System, Providing for the National Disaster Risk Reduction and Management Framework and Institutionalizing the National Disaster Risk Reduction Management Plan, Appropriating Funds Therefor and for Other Purposes*, viewed at, <https://www.officialgazette.gov.ph/2010/05/27/republic-act-no-10121/>.

Republic Act No. 10174 2012, *An Act Establishing the People's Survival Fund to Provide Long-Term Finance Streams to Enable the Government to Effectively Address the Problem of Climate Change, Amending for the Purpose Republic Act No. 9729, Otherwise Known as the "Climate Change Act of 2009" and for Other Purposes*, viewed at <https://www.officialgazette.gov.ph/2012/08/16/republic-act-no-10174/>.

Republic of the Philippines 2015, *Intended Nationally Determined Contributions.*

Republic of the Philippines 2021, *Nationally Determined Contribution Communicated to the UNFCCC on 15 April 2021.*

United Nations Framework Convention on Climate Change 2015, *Paris Agreement.*

9 Climate change governance in Singapore

Cautious mitigation in a developmental state

Natasha Hamilton-Hart

Acknowledgements

The author thanks David Thompson for his excellent research assistance.

Introduction

Singapore's approach to climate change governance has undergone some shifts since the 1990s but shows many continuity threads over that time. The shifts relate to a significant, albeit modest, escalation of the issue up the policymaking agenda, with senior politicians and official agencies acknowledging the issue as an increasingly serious one. The continuities include extreme caution in making commitments under international climate change governance mechanisms and an unwillingness to adopt policies that would require costly adjustment by either households or industries. The governance of climate change mitigation and adaptation remains elite-led and dominated in all material respects by the People's Action Party (PAP) government, which has held power since Singapore gained internal self-governance in 1959. This chapter argues that Singapore's governance of climate change policy reflects critical elements of its approach to governance more generally: managed consultation with societal groups, centralized coordination across government, and effectively implemented policy commitments. Although the government has recognized the need to make changes to the policy in response to climate change challenges, the political basis of the PAP's rule remains one of securing legitimacy through economic growth. Given that Singapore is a land-scarce, urban economy that has thrived economically based on serving as a trade and service hub, a 'growth first' approach has meant that Singapore has not committed to any absolute reductions in greenhouse gas emissions.

This chapter first describes Singapore's actual greenhouse gas emissions in terms of trends and sectoral composition. The second section shows that absolute increases in emissions are consistent with the way Singapore has framed the climate change issue and defined its responsibilities in the international climate change regime. The third section describes Singapore's mitigation policies, covering a range of initiatives to reduce the rate of growth in

DOI: 10.4324/9780429324680-11

greenhouse gas emissions. The fourth section summarizes Singapore's international commitments other than those relating to its greenhouse gas emissions. The fifth section discusses Singapore's climate change governance in the context of broader features of the Singapore political system.

Emissions trends and profile

Different measures of Singapore's total absolute greenhouse gas emissions show different levels, but the overall trend in emissions is fairly clear: Singapore's emissions have risen significantly over time, as shown in Figure 9.1. The rate of increase has moderated, but the trend of rising absolute emissions contrasts with Singapore's preferred measures of its emissions trajectory, which often show the decline in the emissions intensity of GDP (emissions per dollar of economic growth), or the decline in emissions compared to a steeply rising 'business-as-usual' scenario. The International Energy Agency (IEA) data shown in Figure 9.1 captures only carbon dioxide emissions, whereas the Singapore government's official series (also pictured) shows total emissions in CO_2e (carbon dioxide equivalent gases). The gap is not entirely consistent with the slightly more encompassing national figures. It should be taken as an indicator of data uncertainty, given that, as a non-agricultural country, CO_2 dominates Singapore's emissions profile.

Turning to the sectoral composition of emissions, Figure 9.2 shows that Singapore's direct emissions are dominated by electricity generation, followed by 'other industrial' combustion (primarily the refining industry) and

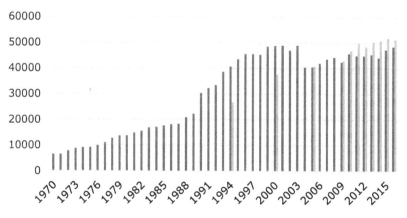

■ Carbon dioxide emissions, kilotonnes

▨ Total greenhouse gas emissions, kilotonnes of CO2 equivalent

Figure 9.1 Singapore emissions trends
Source: Carbon dioxide emissions: IEA energy balance statistics, compiled by Janssens-Maenhout et al. 2017. Total greenhouse gas emissions: National Environment Agency, Singapore.

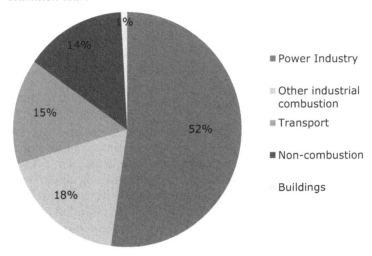

Figure 9.2 Singapore emissions by sector
Source: IEA energy balance statistics, compiled by Janssens-Maenhout et al. 2017.

transport. Although this figure is from the IEA data, the National Climate Change Strategy gives similar primary statistics for 2012: electricity generation at 43 percent, industry at 41 percent, transport at 15 percent, and buildings at one percent. Most "direct emissions from the industrial sector are from the combustion of primary fuels by the refining and petrochemical sector" (NEA 2016, p.36). If secondary emissions from electricity use are considered, industry accounts for 15 percent, buildings 14 percent, and households 6 percent (NCCS, n.d., 'Singapore's emissions profile'). This distribution has been reasonably constant: in 2004, the household and consumer segment accounted for less than 10 percent of the country's total carbon emissions (Hamilton-Hart 2006, p.377).

   The central point of contention about Singapore's emissions reflects a more significant issue with the international climate change regime: how to account for global 'bunker fuels' emissions, which are emissions from the combustion of marine and aviation fuel used in international transport. The UNFCCC excludes the sales of such bunker fuels from national emissions totals. According to the UNFCCC website, the Conference has "continuously addressed" such emissions under the Subsidiary Body for Scientific and Technological Advice (SBSTA). Under the Kyoto Protocol, Annex 1 parties (which Singapore is not) are enjoined to "pursue limitation or reduction of GHGs not controlled by the Montreal Protocol from aviation and marine bunker fuels, working through the International Civil Aviation Organization (ICAO) and the International Maritime Organization (IMO)." However, while parties have an obligation to monitor and report on such fuels separately from their national inventories, they are not subject to national emissions reduction targets.

Singapore has repeatedly, when queried about bunker fuels, referred to the UNFCCC's exclusion of such emissions from national inventories. It does not publicly report on emissions related to its sales of bunker fuels. UNFCCC documentation refers to a memo on bunker fuel emissions sent by the Singapore government to the UNFCCC in 2016 (separate from its formal national communication). Still, the UNFCCC declined to release the memo to the author of this chapter (correspondence, 16 November 2018). Singapore is reported as the world's largest bunker fuel port, with sales of 50.6 million metric tonnes in 2017, according to data from the Maritime and Port Authority of Singapore. A non-official estimate puts Singapore as the largest source in the world for bunker fuel-related emissions, at 42,028 thousand metric tonnes of carbon dioxide in 2014; 14 percent of the world's total bunker fuel-related emissions (Knoema n.d.).

Defining climate change responsibilities

Singapore was initially slow to become involved in the international regime governing climate change. It signed the landmark United Nations Framework Convention on Climate Change (UNFCCC) of 1992 but ratified it only in 1997. It was cautious in its first official communication required under the terms of the treaty: agnostic as to the anthropogenic nature of climate change and blunt on the lack of options available to Singapore to reduce emissions, claiming that because "there is little potential for us to develop alternative sources of energy that are non-fossil fuels," carbon emissions will rise with economic growth (Hamilton-Hart 2006, p.369). The issue did not feature in Singapore's domestic policy discussions until around 2005 when Singapore signed the 1997 Kyoto Protocol to the UNFCCC. This appears to be something of a turning point for Singapore: since then, all official statements relating to climate change have been unequivocal about the reality of the threat of rising greenhouse gas emissions related to human activity.

Singapore made pledges under the UNFCCC regime in Copenhagen in 2009 and Paris in 2016 and has institutionalized climate change policymaking at a high level in the governance system. In public communications, it now frequently describes itself as committed to supporting the international climate change regime. For example, in its Second Biennial Update Report under the UNFCCC, it notes that "as a responsible member of the international community, Singapore is fully committed to playing our part. To underscore our commitment, we ratified the Paris Agreement on 21 September 2016 and were among the first 55 Parties to do so" (NEA 2016, p.6). Two years later, in its Third Update Report issued at the end of 2018, it repeats the language of international responsibility and opens with the blunt statement that climate change impact "has become increasingly palpable around the world" and that the "need for climate action has become more urgent than ever" (NEA 2018, p.6).

Against this clear commitment to the aims of the UN climate change regime, Singapore's strategy is highly cautious and framed to minimize its own responsibilities. It employs several strategies to this end. First, official statements and policymaker speeches emphasize Singapore's small contribution to global emissions. In 2009, Singapore described its emissions as accounting for 0.2 percent of global emissions in 2009 (Chua 2009). By 2016, this changed to Singapore accounting for only 0.12 percent of global emissions (NEA 2016), and in 2018 Singapore's emissions were given as 0.11 percent of the world's total (NEA 2018). Singapore has also stressed that it bears no 'historical debt' for emissions before its independence in 1965 (Chua 2009).

Second, Singapore gives prominent visibility to its 'national circumstances', arguing that it has very limited options to reduce emissions and seeks legitimacy for considering these circumstances in the UN regime. Thus, in its Intended Nationally Determined Contribution statement of 2015, it cites its "limited access to renewable energy" and natural resources, its land area of 716 km^2, high population density (7,540 persons per km2), and a long list of practical impediments to adopting alternative energy sources (Government of Singapore 2015, p.3). Similar language is found on the government's central repository of climate change policy: its National Climate Change Secretariat website.[1] Singapore's national circumstances are frequently summarized in its self-description as a "small, alternative energy disadvantaged city-state", a phrase that is repeatedly used in its communications to the UN regime. Singapore also cites the structure of its economy, resulting from its position as a trade and transport hub and a petrochemicals storage and refining center, as reasons for its emissions. The implication is that others benefit from its economic strategy. For example, a government agency notes that Singapore's "refining and petrochemical plants help create synergies and are part of a business supply network in Southeast Asia, the Western Pacific, South Asia and Australasia" (NEA 2016, p.13).

Singapore seeks legitimacy for giving consideration to such national circumstances in the UN climate regime. Singaporean communications often quote UNFCCC Article 4.8, which states that "Parties shall give full consideration to actions to meet the specific needs and concerns of developing country Parties arising from the adverse effects of climate change and/or the impact of the implementation of response measures." The government claims that three subclauses in the article are of specific relevance to Singapore, namely: 4.8 (a) Small island countries; 4.8 (b) Countries with low-lying coastal areas; and 4.8 (h) Countries whose economies are highly dependent on income generated from the production, processing, and export, and/or on the consumption of fossil fuels and associated energy-intensive products (NEA 2016, p.14). Article 4.10 is also often referred to as it recognizes such countries' circumstances with "serious difficulties in switching to alternatives" (NCCS n.d., 'National circumstances').

This approach is contingent on a third strategy Singapore has adopted to minimize its formal responsibilities: that of identifying itself as a developing country. UNFCCC Article 4.8 is specific to developing countries' needs, and thus Singapore's framing of its duties in line with 'national circumstances' is inherently linked to a developing country identity. Although the UN climate change regime often refers to "developing country Parties", formally the UNFCCC divides parties into Annex 1 (generally, wealthier or 'developed' countries) and non-Annex-1 countries (mostly less wealthy, developing countries, although also including such high-income countries as Kuwait, Saudi Arabia, and the United Arab Emirates). Even at the time that it signed the UNFCCC of 1992, Singapore was already a high-income country. The insistence on maintaining non-Annex 1 status despite some reported pressure is almost certainly a calculated decision to deflect calls for mandatory reductions in greenhouse gas emissions (Hamilton-Hart 2006; Chua 2009). Although this status has at times provoked gentle criticism (e.g., Michaelowa & Michaelowa 2015), it is accepted without issue in parts of the UN climate change system. As late as 2018, for example, a UN newsletter gave a favorable account of Singapore's "commitment with the global climate agenda through South-South cooperation" under the heading 'Developing Countries Showcase National Emission Reductions' (UN Climate Change News 2018).

The fourth strategy that Singapore has used to deflect calls for further formal commitments has been to endorse indicators on which it performs well and carefully choose the emissions targets that it does commit to in order to demonstrate the progress it has already made. This has meant focusing on the 'emissions intensity' of economic growth and pegging targeted future emissions against a generous 'business-as-usual' trajectory. It has also meant vigorous denunciation of measures that present Singapore's emissions in an unflattering light – such as per-capita measures or land-based measures, on the basis that they are unfair (Ong & Lim 2010). The NCCS website rejects per capita efforts, arguing that Singapore's relatively high per capita emissions are "largely due to our small size and dense population". The same website notes Singapore has, in relative terms, very low emissions per dollar of GDP (NCCS n.d., 'Singapore's emissions profile'). Suggestions by some non-governmental groups that a country's emissions be measured by its consumption of greenhouse gas-producing products are similarly rejected. The Environment Minister, for example, was reported as saying that consumption-based measures were "unfair because Singapore is a resource-poor nation that must import almost everything the population needs ... If you look at our utilisation of resources, the way we generate electricity and way we organise our transportation system, we're not perfect yet but we've actually done more than our fair share" (quoted in Syed, 2012). As discussed below, Singapore officials also at times deflect responsibility for the emissions embedded in what it produces for export, on the basis that these exports serve consumer demand elsewhere, but it does not seriously contest the production-based measure of emissions that the UN climate regime is based on. Despite rejecting any binding obligation to reduce emissions, Singapore has, as shown in the next section, made some commitments and adopted a range of policies to mitigate emissions.

Mitigation commitments and policies

At the United Nations Climate Change Conference (CoP 23) held in Bonn in November 2017, Singapore's Environment and Water Resources Minister announced that Singapore would designate 2018 as the Year of Climate Action (Low 2018). That year, Singapore described itself as having "taken ambitious steps" to ensure it was on track to meet emissions targets (MEWR 2018, p.2) and deployed a "comprehensive suite of measures" (MEWR 2018, p.3). The Year of Climate Action was "to encourage greater national consciousness to the urgent need for climate action. Since its launch, we have received more than 300,000 pledges from individuals, businesses, organizations, and educational institutions. In the public sector, all our ministries have pledged to reduce their plastic, water, and electricity consumption" (NEA 2018, p.6). Earlier official reports reference a "comprehensive range of mitigation measures" to reduce greenhouse gas emissions (NEA 2016, p.6). The country's climate change strategy describes its commitments under the UNFCCC as "ambitious" (NCCS n.d., 'Singapore and international efforts'). This section describes the gradual increase in Singapore's obligations regarding greenhouse gas emissions and the raft of policy measures it has introduced. Singapore's policy measures reflect the capacity of its public sector to develop and implement concrete policy initiatives, coordinate across functional areas in a strategic way, and achieve results. In contrast, the targets aimed at, and the official commitments under the UNFCCC, have been very modest.

Singapore's decision to accede to the Kyoto Protocol in 2005 (as a non-Annex 1 Party) did not include any mandatory commitment to reduce emissions. A National Climate Change Strategy was first formulated in 2008, setting out "competency-building" policies and the terms of participation in the international climate change regime (Li & Rajola 2010, p.264). In the lead-up to the UNFCCC CoP (Conference of the Parties) in Copenhagen in 2009, Singapore committed to reducing emissions "by 16% below business-as-usual (BAU) levels in 2020, based on the condition that there should be a legally binding global agreement in which all countries implement their commitments in good faith" (NCCS n.d., 'Singapore and international efforts'). In fact, its voluntary target of reducing the carbon intensity of its economy (emissions per dollar of GDP) by 25 percent compared to 1990 by 2012 had already been met by 2007 (Low 2007). Although the Copenhagen COP did not produce binding commitments to replace the Kyoto Protocol, Singapore's National Climate Change Strategy of 2012 "includes the nation's target to trim its projected greenhouse gas emissions – by as much as 11% by 2020. It will raise this target to 16% if a global, legally binding deal is reached" (Feng 2012). On the basis of the Paris CoP in December 2015, under which Parties committed to submit their own nationally determined emissions reduction schedules, Singapore's commitment thus strengthened somewhat:

With the global agreement at COP-21 in Paris, Singapore is committed to reduce emissions by 16% below 2020 business-as-usual (BAU) level. Having ratified the Paris Agreement on 21 September 2016, Singapore has also formalised our 2030 pledge which builds on our 2020 commitment. As stated in our *Nationally Determined Contribution* (NDC), Singapore aims to reduce its Emissions Intensity by 36% from 2005 levels by 2030, and stabilise its emissions with the aim of peaking around 2030

(NEA 2016, p.50).

These pledges do not commit Singapore to achieve any absolute reductions in emissions. Because Singapore's economy has grown relatively rapidly and will continue to grow, the 'improved' emissions intensity ratio allows for a significant increase in emissions, well above its current trajectory. The emissions intensity figures are set out in Singapore's Intended Nationally Determined Contribution and – in a footnote – it confirms that the target emissions intensity reduction means that "Singapore's emissions are expected to stabilize at around 65 Mt CO_2e [metric tonnes of carbon dioxide equivalent in greenhouse gas emissions] based on current projected growth" (Government of Singapore 2015). This annual emissions level is well above current annual emissions of around 50 $MtCO_2e$. An international monitoring and advocacy group, Climate Action Tracker (2020), rates Singapore's targets as "highly insufficient," noting that if all countries adopted Singapore's approach, projected global warming would reach between 3 and 4 degrees.

Singapore has adopted policy initiatives in four areas to meet its emissions targets. The first is to reduce the carbon intensity of energy used in the country, mainly from a switch in electricity generation from oil to gas, which generates fewer carbon emissions per kilowatt of power. In 2018, the government noted that the switch from fuel oil to natural gas had seen the proportion of Singapore's electricity generated by natural gas rise from 26 percent in 2001 to around 95 percent in 2016 (NEA 2018, p.89). With fuel oil nearly phased out from electricity generation, the relatively low-hanging fruit has been picked.

Further reductions in greenhouse gas emissions from electricity generation will depend on switching to non-fossil fuel generation sources. Since nuclear power generation has been ruled out, this means that policy has focused on renewable energy sources, and it currently views solar energy as "the most promising option for Singapore" (NEA 2018, p.90). This is a notable shift from its position a decade earlier when official statements dismissed solar energy as not viable for Singapore (Hamilton-Hart 2006). To date, the deployment of solar panels is limited, although growing relatively quickly. At the end of 2017, Singapore had 145.8 MWp of solar photovoltaics (PV) installed, compared to 15.3 MWp in 2013, with a target to reach 350MWp by 2020 (NEA 2018, p.90).

The government does not directly subsidize solar energy at the consumer level. It has, however, invested in a variety of research initiatives to improve the cost and efficiency of solar-powered electricity generation. Reflecting a dual concern to promote economic growth in new industries alongside lower greenhouse gas emissions, the lead agency developing the solar industry is the Economic Development Board (EDB). EDB initiatives such as the Solar Capability Scheme (SCS) and Clean Energy Research Testbedding (CERT) schemes have helped develop local Singaporean solar companies in engineering design and installation of solar PV systems, as well as increasing capacity in local research institutes (NEA 2018, p.90). Among these government-funded research institutes, there is the Solar Energy Research Institute of Singapore (SERIS) at the National University of Singapore, established in 2008. As of 2018, its projects as included "energy islands" made up of solar panels floating in the sea, which would supply electricity to nearby industrial zones or living areas. Other projects included a joint initiative between SERIS, the EDB, and the Public Utilities Board to expand "the world's largest" floating photovoltaic (PV) testbed at Tengeh Reservoir (SERIS 2018). SERIS has also worked with the industry to develop research and testing facilities, supporting the establishment of an estimated 100 clean energy companies.

Singapore's ambitions to develop solar energy have grown rapidly. It has moved from emphasizing the solar industry as a source of 'green growth' for the economy to including increased targets for generating its own energy requirements from solar sources. In 2018 official policy aimed to deploy solar systems capable of generating 350 MWp of electricity by 2020, and in 2019 a target of 2GWp by 2030 was announced (NEA 2018, p. 90; Subhani 2020).

The second line of policy to reduce emissions growth brings together a variety of energy efficiency initiatives. Singapore households have long been targeted by many public programs to increase energy efficiency on a voluntary basis. These have continued, now under the National Climate Change Strategy. Officially, mitigation measures in the household sector "are projected to achieve about 0.71–1.07 Mt of CO_2e abatement by 2020, with an estimated 0.57 Mt of CO_2e abatement achieved in 2016" (NEA 2018, p.95). The National Climate Change Strategy details areas such as programs to encourage recycling (from 59 percent to 70 percent of household waste by 2030) and incineration of remaining refuse in waste-to-energy plants. Measures in the waste and water sectors are expected to achieve 0.15 Mt of CO_2e abatement by 2020, with an estimated 0.12 Mt of CO_2e abatement in 2016 (NEA 2018, p.96).

Government targets to increase efficiency in manufacturing and other sectors have continued with a suite of policy measures and incentives. Again, this reflects longstanding policy goals and legislative instruments, notably the Energy Conservation Act (initially enacted in 2012), provide additional mechanisms to require reporting and efficiency measures. A variety of grants and subsidies are available to businesses from the Energy Efficiency Fund under the National Environment Agency. Overall, the government projects to "achieve 1.43 Mt of CO_2e abatement from these energy efficiency measures by 2020, with an estimated 1.27 Mt of CO_2e abatement in 2016" (NEA 2018, p.91). In the public sector, the Public Sector Sustainability Plan 2017–2020 targeted a reduction in electricity consumption of 15 percent by 2020 and water consumption by 5 percent, both over 2013 levels. These targets were expected to "translate to a decrease of 130,000 tCO2e, equivalent to taking 26.650 cars off the road" (MEWR 2018, p.9).

A third policy area for emissions reduction is the greening of Singapore's buildings. Singapore has introduced a certification scheme for more energy-efficient buildings, known as the Green Mark. Its standards "require developers and owners to achieve a 28% energy efficiency improvement from 2005 building codes and include existing buildings if undergoing major retrofitting" (NEA 2018, p.93). The Climate Action Plan (MEWR 2018, p.6) aims for all buildings to be 'green' by 2030. Officially, "mitigation measures in the building sector are projected to achieve 0.87–1.55 Mt of CO_2e abatement by 2020, with an estimated 0.846 Mt of CO_2e abatement in 2016" (NEA 2018, p.93). The basis for measuring these reductions, however, appears to be against older efficiency standards for appliances such as air conditioning units. They fall short of earlier, low-technology design principles for tropical buildings that do not require air conditioning at all. These targets also fall well short of zero-carbon building standards mooted in Europe – and (like the European standards) ignore embedded emissions in construction materials used in Singapore's highly active construction industry, which frequently sees serviceable 'old' buildings demolished to make way for newer (and officially more efficient) buildings. The production of cement is estimated to account for 8 percent of global carbon dioxide emissions, with other building materials such as steel also creating additional emissions (Lehne and Preston 2018).

The fourth policy area for reducing emissions is the transport sector, referring to Singapore's domestic land-based and local area maritime transport, not its international shipping or aviation transport-related emissions, which are excluded from national inventories as per the UN climate regime. For decades, Singapore has had a raft of pricing measures to discourage the use of private motor vehicles to in order to limit road congestion. Further efforts to encourage public transport are now badged under the National Climate Change Strategy, which targets specific steps for taking up of public transport (75 percent of trips made during peak hours by 2030) and limits on the permitted number of new vehicle registrations. Together, "measures in the transport sector are projected to achieve 1.64–1.68 Mt of CO_2e abatement by 2020, with an estimated 0.90 Mt of CO_2e abatement in 2016" (NEA 2018, p.94).

A major new policy initiative targeting reduced emissions across all sectors came in the 2017 Budget Statement, which stipulated that Singapore would adopt a carbon tax. In a departure from shielding the local refining industry, the tax targets upstream generators of carbon emissions, with facilities that emit more than 25,000 tCO_2e of greenhouse gases annually. In practice, this means it will mainly affect refineries, semiconductor, and petrochemical industries (Peloso 2018). The World Bank estimates that the tax will apply directly to "between 30 and 40 energy-intensive companies, accounting for around 80 percent of Singapore's emissions" (World Bank 2018, p.43). Singapore's carbon tax is low and will commence in 2020, based on 2019 emissions. According to the National Climate Change Secretariat website, "To give the industry more time to adjust and implement energy efficiency projects," the projected starting rate was set at $5/tCO_2e, 2019–2023. There was a pledge to review the rate "by 2023", with an announced intention to

> increase the tax to between $10/tCO_2e and $15/tCO_2e by 2030. In doing so, we will take into account international climate change developments, the progress of our emissions mitigation efforts, and our economic competitiveness. Carbon tax revenue will help to fund measures by industry to reduce emissions and provide appropriate measures to ease the transition.

Any introduction of a carbon pricing mechanism is an achievement. However, Singapore's tax rate falls far short of most estimates of a realistic price for carbon that would capture the true impact of carbon emission and lags the prices set by other carbon pricing regimes. Some other countries have set carbon prices much higher, although all fall short of the consensus estimate by the IPCC of US$140 to $590 a tonne from 2030 required to keep the average global temperature increase to less than 1.5 degrees Celsius above pre-industrialization levels (IPCC 2018).

International responsibilities: Global and regional

Singapore views the issue of climate change as one that requires international cooperation, and it positions itself as a responsible member of the international community. As observed by the National Climate Change Secretariat,

> every country needs to play its part to reduce global concentrations of greenhouse gases (GHGs) … We have a deep interest in global efforts to address potential disruptions to natural ecosystems and human societies. Singapore has always been a strong supporter of multilateral approaches to global issues, and we work closely with other countries to tackle the climate challenge.
>
> (NCCS n.d., 'Singapore and international efforts')

Singapore has been fully engaged with the UN climate regime, although its commitments have been minimal for a wealthy country. Singapore has been involved in other actions that reflect some element of international responsibility. Some of this has been in terms of its climate-related research, which extends to implications and trends in its wider region. The Centre for Climate Research Singapore (CCRS) was, for example, established "to develop research expertise in the weather and climate of Singapore and the wider Southeast Asia region" (MEWR 2016). The UN reported Singapore's "commitment with the global climate agenda through South-South cooperation" through supporting "other developing countries to improve their systems to measure, report and verify emission reductions. Singapore also supported other developing countries to build capacity and promoted climate action by sharing successful examples of implemented actions" (UN Climate Change News 2018).

Southeast Asia's premier intergovernmental organization, the Association of Southeast Asian Nations (ASEAN), has been relatively slow to take up the issue. Reflecting its developing country membership (Singapore is the only 'advanced economy' member), the organization has been cautious about mitigation commitments. ASEAN environment ministers in 2010 issued a call for a joint response to climate change. The 2012 Action Plan enjoins members to cooperate on adaptation, research, and information sharing. Under the section on mitigation, it calls for "Exploring the possibility to develop a carbon cap and trade system in the region" and promoting international support for technology transfer to ASEAN. The ASEAN Plan of Action for Energy Cooperation 2016–2025 (published in 2015) somewhat undercuts any sense of responsibility for mitigation, supporting the coal industry, which it expects to grow. The Plan notes that "the ASEAN Coal Sub-sector Network was transformed into the ASEAN Forum on Coal (AFOC) in 1999 with the objective, among others, to promote ASEAN cooperation in the coal sector and to promote intra-ASEAN business opportunities on coal." The ASEAN Blueprint 2025 highlights "Sustainable climate" as one of four areas highlighted under ASEAN cooperation on the environment and mentions a working group of officials on climate change.

As put by Singapore's Deputy Prime Minister,

> ASEAN members and our regional partners are making good progress in advancing the ASEAN Plan of Action for Energy Cooperation till 2025, which will help provide a more conducive environment for investments in renewable energy, energy efficiency, and cleantech. Under this Plan of Action, ASEAN members aim to increase the component of renewable energy in the ASEAN Energy Mix to 23% by 2025 and reduce energy intensity in ASEAN by 20% in 2020 over 2005 levels
>
> (Teo Chee Hean 2018).

Singapore, as Chair of ASEAN in 2018, convened an Expanded Special ASEAN Ministerial Meeting on Climate Action, which comprises ASEAN and China, Japan and Korea as dialogue partners. It is hard to measure the effects of this type of international meeting as no concrete actions typically ensue. However, it is possible to argue that Singapore has helped to keep climate change on the regional policy agenda.

Climate change politics, Singapore style

Singapore's approach to meeting the challenge of climate change is consistent with its broader national system of governance. The previous sections have described an approach that is extremely cautious about adopting any binding commitments to reduce emissions while also seeking to maintain an image as a responsible international actor. Although its obligations are limited, Singapore has moderated its earlier trajectory of rising emissions. This achievement reflects the capacity of the government to design and implement policy changes in a coherent and effective way. Its refusal to adopt policies that would reduce emissions in absolute terms reflects a longstanding commitment to prioritize economic growth. This section describes the major elements of the broader political economy that underpins both the achievements and the limitations of Singapore's climate governance. These elements can be summarized as:

- Government dominance;
- Developmental state partnership with business;
- Limited independent influence of environmental advocates;
- Instrumental and conditional pursuit of environmental goals; and
- Effective coordination of policymaking and implementation.

Climate change policy in Singapore reflects the priorities of Singapore's political leadership, which has consistently defended its prerogative to set the policy agenda. Singapore's politics remains dominated by the ruling People's Action Party, which has controlled the government since 1959 and tolerated very little political opposition. Despite political liberalization since the 1990s, it remains categorized in terms such as "soft authoritarian" or "electorally legitimized authoritarian regime" (Barr 2014). Somewhat greater space for civil society advocacy groups since around 2000 means that some analysts describe the regime as having shifted from being a "closed hegemonic party state" to a "competitive authoritarian regime" (Ortmann 2015). Such characterizations capture PAP dominance but underplay the importance of elections and certain types of consultation. Elections are competitive and serve as a barometer of government legitimacy. As such, the PAP leadership has been sensitive to declines in the party's share of the vote and frequently offers calibrated concessions designed to shore up its popularity with voters generally, but it has not conceded to demands that might threaten its political dominance (Chua 2017).

Government dominance of the policy agenda does not preclude significant partnerships with both local and foreign businesses. In the early decades of independence, the Singapore government could be fairly described as having a very high level of autonomy from local business, which it tended to hold at arm's length (Rodan 1989). However, from the mid-1980s, the government moved self-consciously to incorporate business voices in high-level discussions of economic strategy (Low 2001). A dense network of personal linkages, as well as more formal collaborative mechanisms, have linked business and government since then (Hamilton-Hart 2000). A sustained commitment to state-facilitated economic growth and industrial development in partnership with business means that Singapore can be counted alongside other East Asian developmental states such as South Korea and Taiwan (Yeung 2017). The Singapore state, like the other developmental states of East Asia, faces increasingly complex challenges from shifts in the global economy, the growing independence of local firms, and the turn to innovation-led growth, all of which have in some ways undermined its ability to steer the economy (Hamilton-Hart & Yeung 2021). However, even if achievements have been thwarted in some cases, there has been no retreat from state activism and deliberate attempts to influence the terms of Singapore's engagement with the global economy. Although state-dominated, this overall developmental orientation includes dialogue and collaboration with business players.

The Committee on the Future Economy, convened in January 2016, is emblematic of both systematic state activism and partnership with industry. Its 2017 report provides a comprehensive overview of dozens of initiatives for investment partnerships, research, skills development, and other activities aimed at securing continued high-quality growth in Singapore (CFE 2017). Members of the committee (chaired by two ministers) included representatives from major businesses and the committee references consulting with 9,000 stakeholders. This major economic strategy document puts Singapore's climate change strategy in context: climate change and environmental sustainability more generally merit only two paragraphs in a 144-page report. It thus notes that the government should ensure that "we pay attention to the quality of the environment" and references international audiences: "Our efforts to build a City in a Garden have been well received both at home and abroad. Given the threat of climate change, Singapore should play its part in contributing to global efforts to improve environmental sustainability. Our reputation as a City in a Garden has placed Singapore on the tourism map. It has enhanced the liveability for all who call Singapore home." The only elaboration comes in a recommendation for Singapore to "lead the way as a model city in sustainability, through more aggressive investment and deployment of new solutions in the energy sector", including the development of solar electricity technologies and investment in infrastructure (CFE 2017, p.91, 99).

Notably, environmental sustainability is valued instrumentally as a means to maintain "a high-quality living environment" (CFE 2017, p.14, 63). The government-managed twinning of instrumental environmentalism (the environment is valued as a means to attract investment or sustain growth) and developmentalism is consistent with Singapore's record of 'environmental authoritarianism' more generally (Han 2017) and state developmentalism with specific reference to low-carbon 'green growth' as an economic strategy (Dent 2018). Where there is a trade-off between economic growth and the environment, growth takes priority.

Environmental advocacy groups play a limited role in Singapore's climate change strategy. 'Constructive' engagement with societal organizations such as the Singapore Environment Council has been welcomed by the government, particularly in socializing messages around energy conservation, waste reduction, and hygiene. Representatives of environmental groups are often invited to dialogues with agencies such as the National Environment Agency. Government-nominated MPs, intended to serve as 'independent' voices in parliament, have from time-to-time included leaders of major environmental NGOs. Local media, both the online media that is independent of the government and the major print newspapers, report on the views of local and foreign environmental advocates regarding climate change and related issues. Such advocates, however, operate in sharply circumscribed ways. Government agencies may listen to various voices but pick and choose whose views they consider constructive.

Consultation – at least in public – seems to occur more often after the fact than before. In the case of the carbon tax, for example, the policy was announced in February 2017 – a month before a consultation paper went out seeking input on specific questions set by the National Climate Change Secretariat (NCCS 2017). Submissions were not made public but were summarized in a later document, with each query or suggestion paired against the government's response (NCCS n.d., Annex A. responses to feedback and suggestions on the carbon tax'). Demonstrating the priority given to economic growth and business concerns, the only major deviation from the initial carbon tax proposal was a *lowering* of the tax from an earlier proposed $10 to $20 per tonne of carbon equivalent to $5. Given the prominent role of oil majors in Singapore and the sustained record of companies such as Exxon-Mobil in lobbying against any action on climate change, it may count as something of an achievement that a statement from ExxonMobil on the Singapore tax observed that "The risk of climate change is clear and warrants action. A uniform price of carbon applied consistently across the economy is a sensible approach" (quoted in Vasagar 2017).

At times, the local print media has relayed criticisms of the government's climate change record, but generally alongside official rebuttals of the criticisms. A 2009 news report, for example, reported that Singapore's "insistence on being classified as a 'Non-Annex 1' country under the Kyoto Protocol has come under pressure in recent times" (Cheam 2009). Online media outlets and environmental advocacy groups have gone much further in criticisms of Singapore's record of high per-capita emissions, unambitious mitigation targets, and refusal to accept responsibilities commensurate with its income level (for example, Green Future Solutions 2015; Lee 2017; The Online Citizen 2009a). The print media largely, however, reinforces the government's defense of its record. When academic research underlined Singapore's high negative environmental impact (Bradshaw, Giam & Sodhi 2010), the local print media gave coverage to official rebuttals (Vaughan 2010; Ong & Lim 2010; Feng 2012).

Official justifications for Singapore's record form part of a cohesive system of governance designed to plan, implement, and rationalize policy in a coordinated way. The coordination across government and with industry partners is in many ways reflective of a high degree of state capacity for coherent, long-term policymaking. Unlike in some more plural systems, climate change policy has not been relegated to environmental agencies. There is a policy-making structure that allows for a whole-of-government response. The major decision and planning mechanism on climate change policy is the Long-Term Emissions and Mitigation Working Group (LWG), which "studies how Singapore can stabilize its long-term emissions. LWG examines options for emission reduction and identifies the capabilities, infrastructure, and policies needed for long-term mitigation" (National Climate Change Secretariat, 'Inter-ministerial committee on climate change'). The LWG is co-chaired by two very senior civil servants and includes heads of all major economic development and infrastructure agencies as well as environmental ones. The LWG sits under the Inter-Ministerial Committee on Climate Change, established in 2007. As of 2019, this Inter-Ministerial Committee was chaired by former Deputy Prime Minister Teo Chee Hean, then Senior Minister and Co-ordinating Minister for National Security.

Conclusion

Climate change governance in Singapore is reflective of the wider Singapore system of governance. Broadly, this means it is coherent, state-dominated, sensitive to the concerns of business, and predicated on a growth-first mindset that elevates economic performance above other considerations. Singapore may be more 'ordinary' than it was in terms of its record in avoiding policy disarray and achieving exceptional performance (Barr 2016), but it nonetheless still possesses a remarkable capacity for coordinated policy planning in complex issue areas. Set against the country's strategic, long-term planning capacity and ambitious programs for achieving continued economic growth, the climate change strategy is substantively very modest. How much Singapore's 'exceptional circumstances' should weigh into any assessment is inherently contestable. Clearly, it is not a developing country and should not be treated as such in the international climate change regime. It is, however, far from being the only rich country that enjoys non-Annex 1 status in the UNFCCC.

What should we make of Singapore's claims of being simply unable to realize absolute reductions in emissions due to its small size and reliance on carbon-intensive industries, including its role as a major port and air hub? In many ways, Singapore's claims of being unable to decarbonize in the short term or even medium term without damaging economic growth are no different from those made by many other countries. It is the same claim that countries as different as New Zealand, China, and the United States frequently make. The sobering conclusion is that climate sustainability requires urgent and large cuts to absolute emissions that are not consistent with the pursuit of economic growth under the current growth model. Singapore's growth model is thus not sustainable, but in this respect it is not exceptional.

Note

1 National Climate Change Secretariat, at https://www.nccs.gov.sg/climate-change-a nd-singapore/national-circumstances/overview.

References

Barr, M. 2014, 'The Bonsai Under the Banyan Tree: Democracy and Democratisation in Singapore', *Democratization*, vol. 21, pp. 29–48.

Barr, M. 2016, 'Ordinary Singapore: The Decline of Singapore Exceptionalism', *Journal of Contemporary Asia*, vol. 46, no. 1, pp. 1–17.

Bradshaw C. J. A., Giam X. & Sodhi N. S. 2010, 'Evaluating the Relative Environmental Impact of Countries', *PLoS ONE*, vol. 5, no. 5, p.e10440, doi:10.1371/journal.pone.0010440.

Cheam, J. 2009, 'Black Marks on Green Blueprint', *Straits Times*, 7 May 2009.

Cheam, J. 2012, 'A Role Model in Fighting Climate Change', *Straits Times*, 30 October 2012. Chua B. H. 2017, *Liberalism Disavowed, Communitarianism and State Capitalism in Singapore*, NUS Press, Singapore.

Chua M. H. 2009, 'Emissions Target a Deft Balancing Act for S'pore', *Straits Times*, 4 December 2009.

Climate Action Tracker 2020, 'Singapore', viewed at, <https://climateactiontracker. org/countries/singapore>.

Committee on the Future Economy (CFE) 2017, 'Report of the Committee on the Future Economy', Ministry of Trade and Industry, Singapore, viewed at, <https:// www.mti.gov.sg/Resources/publications/Report-of-the-Committee-on-the-Future-Economy>.

Dent, C. M. 2018, 'East Asia's New Developmentalism: State Capacity, Climate Change and Low-Carbon Development', *Third World Quarterly*, vol. 39, pp. 1191–1210.

Feng, Z. 2012, 'Why S'pore Green Targets Are "Smaller"', *Straits Times*, 15 June 2012.

Government of Singapore 2015, 'Singapore's Intended Nationally Determined Contribution (INDC) and Accompanying Information', Submitted to UNFCCC for COP21 July 2015, viewed at, <http://www4.unfccc.int/submissions/INDC/Published %20Documents/Singapore/1/Singapore%20INDC.pdf>.

Green Future Solutions 2015, 'The Paris Agreement: What It Means for Singapore and What More Can We Do?', 16 December 2015, viewed at <http://www.green future.sg/category/insights/page/2/>.

Hamilton-Hart, N. 2000, 'The Singapore State Revisited', *The Pacific Review*, vol. 13, pp. 195–216.

Hamilton-Hart, N. 2006, 'Singapore's Climate Change Policy: The Limits of Learning', *Contemporary Southeast Asia*, vol. 28, pp. 363–384.

Hamilton-Hart, N. & Yeung, H. 2021, 'Institutions Under Pressure: East Asian States, Global Markets and National Firms', *Review of International Political Economy*, vol. 28, pp. 11–35.

Han, H. 2017, 'Singapore, a Garden City: Authoritarian Environmentalism in a Developmental State', *Journal of Environment and Development*, vol. 26, pp. 3–24.

IPCC 2018, 'Chapter 2: Mitigation Pathways Compatible With 1.5°C in the Context of Sustainable Development', Draft report, 4 June 2018, viewed at, <https://report. ipcc.ch/sr15/pdf/sr15_chapter2.pdf>.

Janssens-Maenhout, G., Crippa, M., Guizzardi, D., Muntean, M., Schaaf, E., Olivier, J. G. J., Peters, J. A. H. W. & Schure, K. M. 2017, *Fossil CO2 and GHG Emissions of All World Countries*, EUR 28766 EN, Publications Office of the European Union, Luxembourg.

Knoema n.d., 'CO2 Emissions from Bunker Fuels', viewed at, <https://knoema. com/atlas/topics/Environment/CO2-Emissions-from-Fossil-fuel/CO2-emissions-from-bunker-fuels>.

Lee, J. 2017, 'WWF S'pore Says If Everyone Consumes Like S'poreans, We'll Need 4 Planet Earths to Survive', *Mothership Singapore*, 3 August. 2017, viewed at, <https://m othership.sg/2017/08/wwf-spore-says-if-everyone-consumes-like-sporeans-well-need-4-p lanet-earths-to-survive/>.

Lehne, J. & Preston, F. 2018, *Making Concrete Change: Innovation in Low-Carbon Cement and Concrete*, Chatham House, London.

Li, B. & Rajola, V. 2010, 'Climate Change and Disaster Risks: The Singapore Response', in R. Shaw, J. Pulhin & J. J. Pereira (eds), *Climate Change Adaptation and Disaster Risk Reduction: An Asian Perspective*, Emerald Publishing, Bingley.

Lim, T. 2018, 'Budget 2018 Is An Immoral Budget', *The Online Citizen*, 5 March 2018, viewed at, <https://www.theonlinecitizen.com/2018/03/05/lim-tean-budget-2018-is-an-immoral-budget/>.

Low, A. 2007, 'Singapore's Green Drive: Is Enough Being Done?', *Straits Times*, 12 May 2007.

Low, L. 2001, 'The Singapore Developmental State in the New Economy and Polity', *The Pacific Review*, vol. 14, no. 3, pp. 411–441.

Low, M. 2018, '2018 as Singapore's Year of Climate Action', Policy Brief 21, 29 January 2018, Energy Studies Institute, National University of Singapore.

Michaelowa, A. & Michaelowa, K. 2015, 'Do Rapidly Developing Countries Take Up New Responsibilities for Climate Change Mitigation?', *Climatic Change*, vol. 133, no. 3, pp. 499–510.

Ministry of the Environment and Water Resources (MEWR) 2008, 'Singapore's National Climate Change Strategy', viewed at, <https://www.elaw.org/system/files/ Singapore_Full_Version.pdf>.

Ministry of the Environment and Water Resources (MEWR) 2016, 'Climate Action Plan 2016: A Climate-Resilient Singapore for a Sustainable Future', viewed at, <https://www.nccs.gov.sg/docs/default-source/publications/a-climate-resilient-singapore-for-a-sustainable-future.pdf/>.

Ministry of the Environment and Water Resources (MEWR) 2018, 'Climate Action Plan 2018: Take Action Today for a Sustainable Future', viewed at, <https://www. mewr.gov.sg/docs/default-source/default-document-library/a5finalmerw.pdf>.

NCCS (National Climate Change Secretariat) 2017, 'Climate Change Strategy and Carbon Pricing: Consultation Paper', 20 March 2017, viewed at, <https://www.nccs. gov.sg/docs/default-source/default-document-library/climate-change-strategy-and-ca rbon-pricing.pdf>.

NCCS (National Climate Change Secretariat) n.d., 'Annex A. Responses to Feedback and Suggestions on the Carbon Tax', viewed at, <https://www.nccs.gov.sg/docs/defa ult-source/default-document-library/annex-a.pdf>.

NCCS (National Climate Change Secretariat) n.d., 'Inter-Ministerial Committee on Climate Change', viewed at <https://www.nccs.gov.sg/who-we-are/inter-ministeria l-committee-on-climate-change/>.

NCCS (National Climate Change Secretariat) n.d., 'National Circumstances', viewed at <https://www.nccs.gov.sg/singapores-climate-action/overview/national-circumstances/>.

NCCS (National Climate Change Secretariat) n.d., 'Singapore's Emissions Profile' viewed at, <https://www.nccs.gov.sg/singapores-climate-action/singapore-emissions-profile/>

NCCS (National Climate Change Secretariat) n.d., 'Singapore and International Efforts', viewed at, <https://www.nccs.gov.sg/singapores-climate-action/singapore-and-international-efforts/>.

NEA (National Environment Agency, Singapore) 2016, *Singapore's Third National Communication and Second Biennial Update Report*, National Environment Agency, Singapore.

NEA (National Environment Agency, Singapore) 2018, *Singapore's Fourth National Communication and Third Biennial Update Report*, National Environment Agency, Singapore.

Ong, P. and Lim C. H. 2010, 'Green Study Disregards Singapore's Circumstances', *Straits Times*, 19 May 2010.

Ortmann, S. 2015, 'Political Change and Civil Society Coalitions in Singapore', *Government and Opposition*, vol. 50, pp. 119–139.

Peloso, M. 2018, 'Solar Systems Key to S'pore's Renewable Energy Development', *Business Times*, 6 September 2018.

Rodan, G. 1989, *The Political Economy of Singapore's Industrialization: National State and International Capital*, Palgrave Macmillan, New York.

Schulz, N. 2010, 'Delving into the Carbon Footprints of Singapore – Comparing Direct and Indirect Greenhouse Gas Emissions of a Small and Open Economic System', *Energy Policy*, vol. 38, pp. 4848–4855.

SERIS (Solar Energy Research Institute of Singapore) 2018, 'SERIS Celebrates a Decade of Research Excellence and Innovation', press release, 5 April 2018, viewed at, <https://www.seris.nus.edu.sg/doc/events-and-news/050418-SERIS-Press-release.pdf/>.

Subhani, O. 2020, 'S'pore's Clean Power Growth May Beat Expectations: Report', *Straits Times*, 1 January 2020.

Syed, S. 2012, 'Singapore's 'Supertrees' Spark Green Thoughts', *BBC News*, 18 June 2012, viewed at, <https://www.bbc.com/news/business-18015741/>.

Teo Chee Hean 2018, 'Address by Deputy Prime Minister Teo Chee Hean at the Expanded-Special Asean Ministerial Meeting on Climate Action (E-SAMCA)', 10 July 2018.

The Online Citizen 2009a, 'ECO Singapore Urges the Government To Be More Proactive', 19 December 2009, viewed at, <https://www.theonlinecitizen.com/2009/12/19/government-must-act-against-climate-change/>.

The Online Citizen 2009b, 'Lee Hsien Loong Tells the World That Singapore Is a Developing Country', 18 December 2009, viewed at, <https://www.theonlinecitizen.com/2009/12/18/national-statement-of-singapore-at-cop15/>.

UN Climate Change News 2018, 'Developing Countries Showcase National Emission Reductions', 9 May 2018viewed at, <https://unfccc.int/news/developing-countries-showcase-national-emission-reductions>.

Vasagar, J. 2017, 'Singapore Carbon Tax Set to Squeeze Oil Groups', *Financial Times*, 21 February 2017.

Vaughan, V. 2010, 'Is Singapore the Worst Environmental Offender?', *Straits Times*, 14 May 2010.

World Bank 2018, *State and Trends of Carbon Pricing 2018*, Washington, D.C.

Yeung, H. 2017, 'State-led Development Reconsidered: The Political Economy of State Transformation in East Asia since the 1990s', *Cambridge Journal of Regions, Economy and Society*, vol. 10, pp. 83–98.

10 Climate change governance and (il)liberalism in Thailand

Activism, justice, and the state

Adam Simpson and Mattijs Smits

Introduction

Climate change will affect all of Southeast Asia. In Thailand, as in many countries, these impacts may well be catastrophic (Marks 2011; Simpson 2018). In this chapter, we situate the effects, responses, and governance of climate change in Thailand in the context of the literature on environmental authoritarianism (Beeson 2010) and concomitant activist responses. The debate around environmental authoritarianism primarily focuses on whether less liberal regimes can contribute to better environmental outcomes (Sonnenfeld & Taylor, 2018). There is, however, an additional component in this debate. The primary rationale for addressing climate change is to ensure better outcomes for human society in the near and distant future. There is no other *raison d'être* for broad-based climate change action. The earth and most of its lifeforms would likely continue in a world where human civilization has been destroyed by climate change. Since climate action is, therefore, primarily based on the concept of intergenerational justice, policies should similarly be driven by values of climate justice for those already living (Robinson & Shine 2018). Therefore, policy outcomes for climate change mitigation and adaptation should be achieved through actions that reduce inequalities and inequities, thereby avoiding the worst impacts of climate change on human society.

Thailand provides a fascinating case study in this regard. Its history is characterized by persistent illiberalism and authoritarianism (Pavin Chachavalpongpun 2021) but also by periods of liberalism and dynamic environmental activism, particularly compared with much of Southeast Asia (Elinoff and Lamb 2020; Simpson 2014; Unger & Patcharee Siroros 2011). Thai movements have undertaken campaigns against specific fossil fuel projects such as the Bo Nok Coal-Fired Power Station, which was successful, and the Trans-Thai Malaysian Gas Pipeline, which was not (Simpson & Smits 2017). Thai environmental movements have been at the forefront of the fight for climate justice (Simpson & Smits 2018), but the magnitude of the climate crisis, equitable climate change mitigation, and adaptation also require governments responsive to civil society activism.

DOI: 10.4324/9780429324680-12

Despite some significant environmental victories, the ability of environmental movements in Thailand to substantially influence policy and political outcomes has always tended to reflect accommodation by existing political power structures. These power structures, often allied to the monarchy and linked to structural inequality (Hewison 2014a), run deeply through Thai society and stretch back to its earliest history. In turn, this political history is deeply intertwined with energy and (later) climate politics, discourses, and infrastructure development. For example, some of the largest hydropower dams in the country carry names of royal family members, showing the relations between energy development, state-led modernization projects, and the ruling elite (Smits 2015). The energy sector is a crucial part of Thai efforts to mitigate climate change, since in 2016 the sector contributed 77 percent of Thailand's total greenhouse gas emissions (Pemika Misila et al. 2020).

Thai national politics has been volatile since the transition to constitutional monarchy in 1932, with periods of democratic rule interspersed with regular coups and military rule. A period of stability in the late 1990s and early 2000s ended with a coup in 2006 and then another in 2014, with which the country is still grappling. Soon after the 2014 coup, its leader, General Prayut Chan-o-cha, established a military-dominated national assembly, which duly elected him as prime minister (Veerayooth Kanchoochat & Hewison 2016).

An election in March 2019 under a military-authored constitution was profoundly flawed (Human Rights Watch 2019) and resulted in General Prayut removing his uniform to become a 'civilian' prime minister. Since the coup, the government has provided mixed messages regarding its commitment to climate change mitigation and renewable energy. For instance, General Prayut has opposed electricity from renewable or alternative energy sources on the basis that it results in higher power bills while also committing Thailand to reduce greenhouse gas emissions by 20 percent compared with the business-as-usual (BAU) scenario by 2030, according to Thailand's Nationally Determined Contribution (NDC) (Bhuridej 2020, p.3).

In 2020 however, despite restrictions on political gatherings of more than five people and as the country grappled with the COVID-19 pandemic, persistent mass democracy protests erupted across Bangkok, with predominantly young Thais demanding reform of the monarchy and the constitution. The 2020 protests focused primarily on political transformation rather than climate change, but previous protests by young Thais have petitioned the government to declare a 'climate emergency' and cease using fossil fuels (Patpicha Tanakasempipat 2019). A resurgence of demonstrations in early 2021 indicated that the conflict was likely to be protracted, although the military and monarchy have a history of outlasting protesters (Baker 2016). Regardless of the political outcome, it is expected that Thailand's economy will continue making some progress towards decarbonization. Still, it may not be enough to avoid the catastrophic impacts of climate change (International Energy Agency 2021). If left to the current military-backed government, it may further entrench the military within society, both politically and economically, at the expense of equity, efficiency, and justice.

For a long time, Thailand and other Southeast Asian countries only marginally contributed to global climate change. This impact is now increasing: in 2018, the region had over 8 percent of the global population, over 6 percent of global GDP, and around 5 percent of global energy demand (International Energy Agency 2019). To capitalize on recent technological advances while addressing this impact, there is now evidence of a policy shift within the government towards renewable energy, but the traditional bureaucratic channels and tendering processes are being bypassed. It looks more like the process is primarily designed to benefit the military rather than promote equity and efficiency (Greacen 2021). This demonstrates that democratic checks and balances, including civil society activism and participation in economic decarbonization, are crucial, not only to ensure that it occurs at all but also to ensure a focus on climate justice so that it does not worsen inequalities (Bulkeley, Edwards & Fuller 2014; Simpson and Smits 2018). It appears that the military is aware of this power within civil society since it has introduced a new illiberal NGO law in 2021, which threatens to undermine Thailand's status as both the regional hub for international NGOs and a home for a dynamic domestic civil society sector (Kavi Chongkittavorn 2021).

There is limited literature on the climate change governance activities of civil society organizations in Thailand (Haris, Mustafa & Ariffin 2020). In this chapter, we examine the potential for civil society influence under illiberal regimes, which are common throughout Southeast Asia. We then review Thailand's climate and energy policies and the impact of environmental organizations and movements, with a case study on the Thai Climate Justice Network (TCJ). The TCJ demonstrates how the ability of environmental organizations to organize, campaign, and influence society on climate justice has fluctuated with the nature of the contemporary political regime.

Climate change governance under illiberal regimes

The competition and contestation around energy and climate change governance is usually the result of the respective interests that drive the policy processes (Clapp, Newell & Brent 2018; Simpson 2007). A push for energy security is often instigated by national political economy interests and pursued by both liberal and illiberal political regimes (Simpson 2013b). In contrast, climate change concerns and policies have been primarily driven by liberal regimes at both the national and international level with "authoritarian discourses, state violence, and state-sanctioned private violence … increasingly evident in efforts to keep fossil fuels flowing" (McCarthy 2019), although some authoritarian states have gradually adopted climate policies and rhetoric (Myers 2020). In less developed or middle-income countries, such as Thailand, these contrasts appear starker due to the extreme variations in political regime type, which is reflected in the different opportunities available to civil society actors to influence policy.

The rise of climate change as a global issue and the emergence of authoritarian China as the world's largest carbon emitter has ignited debates regarding the efficacy of illiberalism in tackling both energy and climate security and environmental issues more broadly. This development has given rise to illiberal environmental governance conceived as "environmental authoritarianism" by Beeson (2010), which Gilley (2012, p.288) synonymizes with "authoritarian environmentalism". Beeson identifies two dimensions of this concept: an unresponsive state that prohibits or ignores democratic or civil society activism, and the state compunction to obey environmental policies at the expense of individual freedoms.

Beeson (2010) and Shearman and Smith (2007) argue that authoritarian regimes in East and Southeast Asia, due to rapid reaction times and the mobilization of state resources, may prove more capable of responding to complex environmental problems than some of the region's democracies. China is the oft-quoted source of this argument, primarily due to its immense recent solar and wind power investments. As Gilley (2012) argues, however, this approach is better at producing policy outputs than outcomes, although even the policies themselves can be fragmentary. Local officials in China "regularly fabricate their energy use reports" and central authorities are often more concerned with pleasing party bosses rather than producing effective outcomes (Gilley 2012, p.297). Opaque policy processes, in addition to limits on civil society activism and free speech, make it extremely difficult for the public monitoring of targets; in a democratic country these might be undertaken by civil society or a free media, which would be likely to result in more coherent policies and better implementation and outcomes (ibid. p.298).

Opacity of governance, a lack of free media, and civil society restrictions also mean that injustices are ignored, misrepresented, or receive little domestic attention (Human Rights Watch 2021). In general, civil society organizations can only influence environmental outcomes when they have political space to advocate (Pacheco-Vega & Murdie 2020) but their influence is also highly dependent on their strategies, tactics and operation, with community-level projects providing a key route to effect change under extreme illiberalism (Simpson & Smits 2018).

The literature in this area is limited and somewhat inconclusive with major journals giving the issue only partial attention (Hayes et al. 2021; Meyer & Chang 2021). Fredriksson and Wollscheid (2007) suggest that democratic regimes result in better environmental policies than the alternative, while Neumayer (2002) claims that they do not necessarily result in better outcomes. Povitkina (2018) finds in a study of 144 countries that carbon emissions are only lower in democracies within low-corruption contexts; if corruption is high there is no noticeable difference between authoritarian and democratic regimes. Shahar (2015) argues that freely operating environmental movements and associated democratic modes of governance offer a better route for energy and climate security.

Stevenson and Dryzek (2014) argue that deliberative forms of environmental policymaking are crucial for addressing a problem as complex and as integrated into energy and economic concerns as climate change. However, genuine public participation and deliberative decision-making are extremely limited throughout Southeast Asia (Beeson 2010). Hobson (2012, p.976) argues that climate change activism can either promote or impede democratization, but there is little evidence that illiberalism provides better outcomes.

On balance, it is difficult to determine whether a democratic regime will definitively result in better climate policies and outcomes (Pickering, Bäckstrand & Schlosberg 2020). States remain the best placed actors to facilitate socio-ecological transformation due to their powers to regulate, tax, and redistribute (Eckersley 2021). Nevertheless, the ability of environmental activists to highlight climate risks and associated injustices in countries such as Thailand provides a key pathway for societal awareness and resultant climate action that ameliorates, or at least does not worsen, existing injustices.

Climate change impacts in Thailand

While Thailand faces a plethora of long-term environmental issues, the impacts of global climate change are likely to exacerbate existing environmental issues significantly. Climate change is expected to result in more temperature extremes, affecting regional weather and climate patterns in Southeast Asia. Much of Thailand is low-lying, coastal, or otherwise susceptible to weather extremes such as cyclones that are likely to increase in frequency and intensity by climate change as water temperatures increase. Although too much water is often a significant contributor to insecurity, drought is also becoming a problem due to increased monsoonal variability. The most devastating potential impacts of global climate change will affect different geographical areas through many climate-related disasters, including droughts, landslides, floods, and tropical cyclones (Webersik 2010, p.85).

The 2011 floods inundated most of central Thailand and were the worst in 50 years. This critical event is an example of the political economy of climate change impacts and associated social inequalities. Climate change has resulted in substantially increased pre-monsoon rainfall in the Chao Phraya River Basin in recent decades and a significant sea-level rise at the river outlet, which increased the severity of the 2011 floods. Other non-climate-related environmental impacts, which are common across Bangkok and Thailand as a whole, such as land subsidence, deforestation, urbanization, and removal of natural attenuation basins (like wetlands), also contribute to flooding (World Bank 2010, p.25).

Although many parts of Thailand experienced the 2011 floods, not all communities or people were affected or treated equally. The management of floods and other disasters in Thailand has been organized by elites and their bureaucracies to be deployed in ways that serve their interests and not those of more politically marginalized groups. This was particularly evident in the

2011 floods where privileged areas of the industrial sector and the associated Thai elites' assets were protected while other, less fortunate areas with less political and economic connections were sacrificed (Lebel, Manuta & Garden 2011; Marks 2015; Salamanca & Rigg 2016).

This preferential treatment is replicated in flood resilience policies for Bangkok, prioritizing economic assets and structural flood protection, with little attention on adaptation measures and the social impacts in less privileged societal groups (Laeni, van den Brink & Arts 2019). Prioritizing dominant players in policies that mitigate or adapt to climate change have the effect of reinforcing structural inequalities within society and undermining the pursuit of climate justice (Archer & Dodman 2015; Bulkeley, Edwards & Fuller 2014). Therefore, Thailand's institutional structure and inequitable political economy hinder its capacity to adequately address climate change (Marks 2011), and limits the involvement of civil society in decision-making processes where they could advocate on behalf of vulnerable or marginalized communities. While Thailand has historically offered more political space for civil society than many of its Southeast Asian neighbors (Simpson 2018), this opportunity has been limited since the 2014 coup.

Climate change governance in Thailand

Despite decades of government instability in Thailand, with democratic governments regularly removed by military coups, the country has been a reliable international participant in the various United Nations climate change agreements. The government ratified the 1992 United Nations Framework Convention on Climate Change (UNFCCC) in December 1994, and its two key milestones: the 1997 Kyoto Protocol in August 2002 and the 2015 Paris Agreement in September 2016 (UNFCCC 2020).

As a UNFCCC Non-Annex 1 country, however, the requirements for emissions reductions for Thailand have been minimal, and greenhouse gas emissions steadily increased from around 81 Mt in 1990 to almost 250 Mt in 2013, with the proportion of natural gas emissions gradually increasing to 84 Mt, while oil contributed 95 Mt and coal 69 Mt (International Energy Agency 2020). Nevertheless, total carbon emissions have not increased recently, with a slight but steady decrease observable since 2016.

Under the Kyoto Protocol, over 150 Clean Development Mechanism (CDM) projects were registered in Thailand. More recently, the Thai Greenhouse Gas Management Organization (TGO), an independent public organization set up by the Thai government, tried to develop new and voluntary market-based instruments to mitigate climate change (Smits 2017).

The central government has also taken up climate change adaptation and mitigation policy. The Office of Natural Resources and Environmental Policy and Planning (ONEP), under the Ministry of Natural Resources and Environment, developed Thailand's first Climate Change Master Plan in the period 2010–2012 during relatively liberal democratic governments (Simpson & Smits 2019).

Since the 2014 coup, there has been some progress on climate change policy, with Thailand submitting its Intended Nationally Determined Contribution (INDC) to the UNFCCC in October 2015, stating that it intends to reduce its greenhouse gas emissions by 20 percent from the projected business-as-usual (BAU) level by 2030 – a reduction of 111 Mt-CO_2e to 555 Mt-CO_2e (Raweewan Bhuridej 2015). Cabinet approved an updated Climate Change Master Plan 2015–2050 in July 2016 (ONEP 2015). In May 2017, Cabinet endorsed Thailand's NDC Roadmap on Mitigation 2021–2030, including sectoral action plans and progress reports to the ONEP every six months (Teerapong Laopongpith 2019).

Energy policy and activism

The recent decrease in Thailand's carbon emissions is partly due to three decades of largely progressive policy innovation in energy and the environment, particularly compared with the rest of Southeast Asia (Greacen & Greacen 2004). However, it is also partly due to importing energy resources from its more authoritarian neighbors – notably Myanmar and Laos – while exporting the associated environmental problems (Simpson 2015, 2018; Smits 2015). Since energy accounts for around three-quarters of Thailand's climate emissions, these energy policy shifts have been crucial to reducing emissions (Bhuridej 2020, p.3). This policy innovation was the result of both relatively amenable governments and the region's most dynamic environmental movements, which experienced notable early successes in the 1980s with the blocking of Nam Choan Hydroelectric Dam in Kanchanaburi Province in 1988 (Forsyth 2001, p.5; Rigg 1991, p.46) and a ban on logging in 1989.

Following the dam's blocking, the state-owned Electricity Generating Authority of Thailand (EGAT) began focusing on cross-border energy projects to import energy from the then more authoritarian neighboring countries, Myanmar and Laos. Projects included the Yadana, Yetagun, and Zawtika Gas Pipelines, and a range of proposed dams on the Thanlwin (Salween) River in Myanmar, as well as the Nam Theun 2 Dam, Xayaburi, and Don Sahong Dams in Laos. These projects reduced the domestic emission intensity of electricity production but had adverse impacts on local communities' social and environmental well-being in Myanmar and Laos (Piya Pangsapa & Smith 2008; Simpson 2007, 2013a). From the early 2000s, when gas from the Myanmar Yadana pipeline started flowing, gas often contributed around 75 percent of Thailand's electricity, with approximately one-third imported from Myanmar (International Energy Agency 2020; Simpson 2014).

Despite the continued dominance of EGAT and associated state energy utilities in Thailand's electricity market, it has one of the most progressive renewable energy policies in the region, with reforms dating back to 1992 establishing markets with feed-in tariffs for independent power producers, small power producers, and very small power producers (initially 1 MW).

There are now many small entrepreneurs active in this sector (Sopitsuda Tongsopit & Greacen 2013).

Although it was essentially the government's neoliberal tendencies that launched the energy reforms, civil society and NGOs have also been influential in its development. Nevertheless, over the last decade, there have been some backward steps regarding energy governance. Much of the effective energy and electricity governance structures that had been established over the 1990s and 2000s were undermined with the success of the renewable energy sector creating fertile ground for well-connected corporations to extract rents. With high rents added onto costs, inevitable price increases were associated with renewable energy in general, causing potentially long-lasting damage to community support for the renewable energy sector as a whole.

This outcome seems consistent with General Prayut's lauding of fossil fuels as an energy source. For instance, he instructed the Energy Ministry in mid-2015 to boost 'public understanding' about the high cost of producing electricity from renewable energy sources, which he argued would lead to higher power bills (*Bangkok Post*, 14 August 2015). The Prime Minister also used his absolute authority under Section 44 of the Interim Constitution to exempt all kinds of power plants, gas processing plants, and other utility plants from regulations under the Town and City Planning Act (*Prachatai*, 22 January 2016).

In May 2015, the National Energy Policy Committee approved Thailand's first Power Development Plan 2015–2036 (EPPO 2015). The first public hearing for the formulation of the PDP 2015 was held in August 2014 (Chavalit Pichalai 2015), three months after the coup, which had banned protests and restricted political freedoms (Hewison 2014b). Journalists and activists had been arrested for voicing opposition to the military government resulting in a less than conducive environment for dissenting voices.

Activist groups critiqued the process of developing the PDP 2015, and its focus on large hydropower and coal, with new coal-fired power stations slated for the south of the country. The Network of People Affected by the Power Development Plan 2015, supported by the Thai Climate Justice Working Group, wrote an open letter to the Prime Minister and the Minister of Energy requesting the cancellation of PDP 2015 and the establishment of a more transparent and democratic process (Network of People 2015).

The first Alternative Energy Development Plan 2015–2036 (AEDP 2015) was developed under the PDP 2015 and administered by the Department of Alternative Energy Development and Efficiency within the Ministry of Energy (DEDE 2015). This provides a framework for boosting renewable energy use in the country, although the beginnings of this energy transition had already begun, even with limited government support (Simpson & Smits 2017).

Renewable energy comprised around 17 percent of the total final energy consumption in 2015, but much of this was drawn from longstanding large hydroelectric dams (IRENA 2017). Potential increases in the percentage of renewable energy under most plans tend to rely on the importation of large hydropower from Laos, which displaces not only local populations but also causes negative impacts on downstream livelihoods and environments (Green & Baird 2020). Renewable energy was anticipated to increase to 28 percent in 2036 under the business-as-usual (BAU) model. The International Renewable Energy Agency (IRENA) estimated that with relatively modest changes to policy settings, these estimates could be increased to 37 percent (IRENA 2017).

From the mid-2010s, an increase in the use of biofuels and, to a lesser extent, wind and solar PV resulted in increasing shares within total electricity production, with a concomitant reduction in the proportion of large hydro and coal power. By 2019, natural gas provided around 65 percent of the electricity in Thailand, with solar PV and wind contributing around 2.6 percent and 1.9 percent, respectively (see Table 10.1).

While solar and wind together still only contributed less than 5 percent of the total, their installed capacity increased significantly after 2015, with solar doubling and wind increasing fivefold in four years (see Table 10.2).

Nevertheless, the government continued to provide conflicting signals to the energy sector. In March 2018, for example, the Minister of Energy announced that the government would stop purchasing electricity from renewable energy projects for five years due to increased electricity costs (Sopitsuda Tongsopit 2018). As indicated above, however, electricity cost increases during this time

Table 10.1 Main sources of electricity (GWh)

	2019	*Percentage of Total*
Natural gas	127,442	65.1
Coal	34,390	17.6
Bioenergy	18,508	9.4
Hydropower	6,434	3.3
Solar	5,182	2.6
Wind	3,655	1.9

Source: International Energy Agency (2020)

Table 10.2 Renewable energy installed capacity (MW)

	2015	*2019*	*Percentage increase*
Hydropower	3,639	3,667	0.7
Bioenergy	3,231	4,258	31.8
Wind	234	1,507	544
Solar	1,425	2,987	109.6

Source: IRENA (2020)

can be attributable to rents extracted by businesses with close connections to the military government, rather than issues relating to the technologies themselves.

Despite this vacillation over renewable energy, the government unveiled a revised PDP 2018–2037 (MoE 2018a), which the National Energy Policy Council approved in January 2019, two months before the March 2019 elections – the first since the 2014 coup, and the first held under the 2017 constitution drafted by the military. The new plan provided expanded ambitions for both renewable energy and the role of private operators, with the contribution from coal sharply decreasing from earlier estimates (Chatrudee Theparat & Yuthana Praiwan 2019). The plan reduces the proportion of power generated by the state-run EGAT from 35 percent in the previous plan to 24 percent, with small solar power operators particularly encouraged. Total power capacity by 2037 is expected to be 77,211 MW, with 5,857 MW imported, mainly via hydropower dams in Laos and Cambodia. The contribution of natural gas was expected to fall to 53 percent and coal to 12 percent, with non-fossil fuels increasing to around 35 percent (Hong 2019).

Despite the current small contribution of solar to total electricity, Thailand has installed the most solar capacity in Southeast Asia and was also the first to institute the equivalent of a feed-in tariff (Sopitsuda Tongsopit & Greacen 2013). Under the 2019 rooftop solar program, the Energy Ministry expects rooftop solar to reach 10,000 MW by 2037 (Pugnatorius 2020; Solar Magazine 2019). This contributes towards the goal in the updated Alternative Energy Development Plan 2018 (MoE 2018b) of installed solar power increasing to 15,574 MW by 2037, a fivefold increase from 2019 and around 20 percent of the total. The previously time-consuming licensing process for rooftop solar was streamlined into a notification process with owners able to sell excess electricity to the grid.

While the above-mentioned developments as laid out in the various Thai PDPs have always received a substantive amount of criticism from civil society actors, there were at least technocratic principles underpinning them (Greacen 2021). In 2021, however, the army started to explore alternative avenues by announcing its plan to develop up to 30,000 MW of solar farms on its own land without involving the Ministry of Energy, let alone NGOs. Given that the Thai energy systems already have excessive amounts of overcapacity (up to 59 percent in 2020), this unprecedented move is likely to further enrich the military at the expense of efficiency and equity. Instead of mitigating climate change, it is more likely to further entrench military power and do so behind closed doors, far removed from any civil society interference (Greacen 2021).

These changes in government policy towards a more favorable view of solar energy accompanied a new government discourse around 'Thailand 4.0' (Archanun Kohpaiboon 2020). Under this discourse, Prayut and the government began to see a range of new technologies and their delivery through mechanisms such as 'Smart Cities' as key to both continued economic growth and political survival. However, the sometimes anarchic nature of hi-tech development is not well suited to the top-down decision-making and the picking of winners, which is a characteristic of the military regime and its semi-civilian successor. In addition, ad hoc and contradictory policy positions – such as those relating to renewable energy – and corruption and political instability have led to difficulties in attracting hi-tech foreign investments to the country.

Climate activism: The Thai Climate Justice Network

Having sketched the historical and more recent developments in national climate and energy politics in Thailand, we now zoom in on a case of a specific civil society organization in this field: the Thai Climate Justice Network (TCJ). We do this to show how particular organizations navigate the changing landscape of climate and energy policy in Thailand. In particular, the pendulum of authoritarianism versus more liberal regimes the country has seen over time.

TCJ was a network of around ten Thai civil society groups set up in 2008 to work on climate mitigation in the agriculture, energy, and forestry sector. The network organized events, campaigns, and other activities on diverse topics related to climate change through its Facebook presence and website. As a network organization, activities were also linked to the interests and objectives of the different member organizations. It generally adopted a very critical stance towards climate change policy in Thailand.

An early essential activity of the TCJ was to critically follow and comment on the development of the Climate Change Master Plan, developed by ONEP in 2010. TCJ saw this as a key policy document and, therefore, a pivotal issue to campaign on for people to understand the whole picture of climate change. They tried to join the public meetings and influence the Master Plan. ONEP, meanwhile, tried to keep TCJ out of these negotiations. The Director suggested they were open to public dialogue with civil society organizations who were interested in implementation or policy formulation; still, he was concerned about NGOs like TCJ because he "did not want them to destroy the process" (personal communication with ONEP Director, 20 April 2015).

The above example shows that it is difficult for organizations like TCJ to influence the policymaking process in Thailand. First, they are often not welcome to engage in policy dialogues if they are seen as being 'too critical'. Secondly, it is not easy to target the most relevant policies. In this case, the Climate Change Master Plan proved to be relatively unimportant in the political landscape in Thailand. TCJ eventually characterized it as a 'paper tiger' because, according to them, ONEP did not have the power to influence certain key sectors, such as agriculture and energy.

Another focal point of TCJ was the use of market-based climate policy instruments, such as Payment for Ecosystem Services and carbon markets. They critiqued some of the many CDM projects in Thailand, raising concerns about rice husk gasification causing excessive local pollution, and funding for the controversial Mae Mo lignite power plant (UNFCCC 2016). In 2013, they wrote an open letter about Thailand's REDD+ readiness plan (REDD Monitor 2013), where they criticized the limited opportunities for civil society organizations to provide input on this plan and for failing to include the plight of forest communities. They also argued that REDD+ had the potential to aggravate conflicts.

This example shows that there is still plenty of scope within illiberal governments to make use of global governance instruments that function under liberal (market-based) governance. The irony is that there is often no requirement of adhering to liberal principles within these instruments, such as freedom of consent and public participation, which is a persistent critique of mechanisms such as the CDM and REDD+ (Pearse and Böhm 2014; Newell, 2012).

Although TCJ had some successes, the 2014 coup was a major setback and a major shift towards a less liberal regime. It affected the work of TCJ in two main ways. First, in the immediate aftermath of the coup, people were totally absorbed with the political situation and its immediate consequences. The general interest in, and ability to discuss, environmental issues in large groups decreased. TCJ had planned to launch an extensive campaign, but this was difficult and attracted little attention at the time. The second consequence was that the coup meant less space for civil society organizations to work. In the first period after the coup, the government imposed martial law which prevented political gatherings of more than five people. Under the new constitution, Prayuth replaced martial law with Article 44, a law granting the leader absolute power (Corben 2016). The concentration of power was also felt by the TCJ, who argued that the military government was pushing forward projects without taking public participation into account (Simpson & Smits 2018). These cumulative impacts eventually led TCJ to cease all public activity.

Conclusion

This chapter analysed energy and climate governance in Thailand through the lens of environmental authoritarianism. While this debate mainly centers on whether liberal or illiberal regimes contribute to better environmental or climate outcomes, we also investigated the role played by civil society and whether such regimes contribute to climate justice. The Thai political context generally alternates between more and less liberal regimes, and we mapped the parallel developments and fates of climate policy, energy policy, and activism in the country.

At a general level, the political developments over the last few decades in Thailand demonstrate that we cannot simply state that illiberalism or environmental authoritarianism necessarily does, or does not, result in climate action. This is partly because the changes in regimes in Thailand tend to be frequent, so there is some continuity of governance and bureaucracies within the state, whatever the regime in power. Nevertheless, the political changes mask more profound underlying social and political inequalities and processes in the country, and the effects are not immediately visible. Thus, while climate and energy governance in the context of Thailand has been relatively progressive on some fronts, achieving just and equitable outcomes has proved complicated, and civil society is often barred from participating in or critiquing the direction of climate and energy policy.

For instance, the government's relatively liberal renewable energy policies are nevertheless rife with technocratic and elitist processes that, particularly during military rule, serve to reinforce the political and economic dominance of the military at the expense of transparency, accountability, and societal equity. Given that the rationale for climate action is intergenerational justice, it seems clear that climate mitigation or adaptation which exacerbates contemporary inequalities and injustices is little better than no action at all. The rise and fall of the Thai Climate Justice Network demonstrates the barriers to civil society activism in Thailand, particularly under more illiberal regimes. The group was eventually silenced in the aftermath of the 2014 military coup. It is difficult not to conclude that climate justice outcomes would be better served when these civil society organizations have the political space for policy critique and advocacy.

These developments in Thailand, where climate change governance is linked to activism, justice and the state, hold broader relevance for other countries in the region. This is not only because Thailand is one of the most significant contributors of greenhouse gas emissions in Southeast Asia but also because of the importance of understanding climate change governance under less liberal regimes, which are plentiful throughout the region. To the north and west of Thailand, Laos and Cambodia are critical examples of states with longer histories of stable illiberal governments, whereas to the east Myanmar has returned to a much less liberal regime following a coup in 2021. In these and other countries in the region and the world, it is essential to look beyond the immediate political situations and investigate both the immediate and longer-term political economies of states to determine the consequences for climate and energy governance and justice. With the impacts of climate change becoming more severe and the regional share of global greenhouse gas emissions growing, adaptation *and* mitigation will become increasingly important in Southeast Asia. While it is clear that climate action is possible under more illiberal regimes, without the participation and support of civil society, these countries may either fail to meet their targets or achieve these targets while exacerbating injustices within the country, which undermines the entire rationale for action.

References

Archanun Kohpaiboon 2020, 'Thailand 4.0 and Its Challenges', *East Asia Forum*, 17 April 2020, viewed 26 November 2020 at, <https://www.eastasiaforum.org/2020/04/17/thailand-4-0-and-its-challenges/>.

Archer, D. & Dodman, D. 2015, 'Making Capacity Building Critical: Power and Justice in Building Urban Climate Resilience in Indonesia and Thailand', *Urban Climate*, vol. 14, no. 1, pp. 68–78.

Asia-Pacific Forestry Commission 2001, 'Forests Out of Bounds: Impact and Effectiveness of Logging Bans in Natural Forests in Asia-Pacific', Food and Agriculture Organization of the United Nations, Bangkok, viewed 20 November 2020 at, <http s://coin.fao.org/coin-static/cms/media/9/13171037773780/2001_08_high.pdf>/.

Baird, I. G. & Quastel, N. 2015, 'Rescaling And Reordering Nature–Society Relations: The Nam Theun 2 Hydropower Dam and Laos–Thailand Electricity Networks', *Annals of the Association of American Geographers*, vol. 105, no. 6, pp. 1221–1239.

Baker, C. 2016, 'The 2014 Thai Coup and Some Roots of Authoritarianism', *Journal of Contemporary Asia*, vol. 46, no. 3, pp. 388–404.

Beeson, M. 2010, 'The Coming of Environmental Authoritarianism', *Environmental Politics*, vol. 19, no. 2, pp. 276–294.

Bhuridej, R. 2020, 'Thailand's Updated Nationally Determined Contribution (NDC): Letter to UNFCCC Secretariat', Office of Natural Resources and Environmental Policy and Planning (ONEP), Bangkok, 20 October 2020, viewed 6 November 2020 at, <https://www4.unfccc.int/sites/ndcstaging/PublishedDocuments/Thailand%20First/ Thailand%20Updated%20NDC.pdf>.

Bulkeley, H., Edwards, G. A. S. & Fuller, S. 2014,'Contesting Climate Justice in the City: Examining Politics and Practice in Urban Climate Change Experiments', *Global Environmental Change*, vol. 25, pp. 31–40.

Chatrudee Theparat & Yuthana Praiwan 2019, 'National Power Plan Expands Private Output', *Bangkok Post*, 25 January 2019, viewed 26 November 2020 at, <https://www. bangkokpost.com/business/1617382/national-power-plan-expands-private-output/>.

Chavalit Pichalai 2015, 'Thailand's Power Development Plan 2015 (PDP 2015)' (PowerPoint Presentation), Director General, Thai Energy Policy and Planning Office, Ministry of Energy, Bangkok.

Clapp, J., Newell, P. & Brent, Z. W. 2018, 'The Global Political Economy of Climate Change, Agriculture and Food Systems', *The Journal of Peasant Studies*, vol. 45, no. 1, pp. 80–88.

Corben, R. 2016, 'Analysts: Thai Military Looks to Extend Government Oversight', *VoA*, 10 August 2016, viewed 27 November 2020 at, <http://www.voanews.com/a/ legal-experts-thai-military-looking-to-extend-government-oversight/3458313.html>.

DEDE 2015, 'Alternative Energy Development Plan: AEDP2015', Department of Renewable Energy Development and Energy Efficiency (DEDE), Ministry of Energy, Bangkok, viewed 16 September 2020 at, <http://www.eppo.go.th/images/ POLICY/ENG/AEDP2015ENG.pdf>.

Eckersley, R. 2021, 'Greening States and Societies: From Transitions to Great Transformations', *Environmental Politics*, vol. 30, no. 1–2, pp. 245–265.

Elinoff, E. & Lamb, V. 2020, 'Environmental Politics in Thailand: Pasts, Presents, and Futures', in P. Chachavalpongpun (ed.), *Routledge Handbook of Thailand*, Routledge, London and New York.

EPPO 2015, 'Thailand Power Development Plan 2015–2036 (PDP2015)', Energy Policy and Planning Office (EPPO), Ministry of Energy, Bangkok, viewed 16 September 2020 at, <https://www.egat.co.th/en/images/about-egat/PDP2015_Eng.pdf>.

Forsyth, T. 2001,'Environmental Social Movements in Thailand: How Important is Class?', *Asian Journal of Social Sciences*, vol. 29, no, 1, pp. 35–51.

Fredriksson, P. G. & Wollscheid, J. R. 2007, 'Democratic Institutions Versus Autocratic Regimes: The Case of Environmental Policy', *Public Choice*, vol. 130, no. 3–4, pp. 381–393.

Gilley, B. 2012, 'Authoritarian Environmentalism and China's Response to Climate Change', *Environmental Politics*, vol. 21, no. 2, pp. 287–307.

Greacen, C. S. 2021, 'Firms Line Up for Slice of Solar Pie', *Bangkok Post*, doi:4 March 2021, viewed 9 May 2021 at, <https://www.bangkokpost.com/opinion/op inion/2077963/firms-line-up-for-slice-of-solar-pie>.

Greacen, C. S. & Greacen, C. 2004, 'Thailand's Electricity Reforms: Privatization of Benefits and Socialization of Costs and Risks', *Pacific Affairs*, vol. 77, no. 3, pp. 517–541.

Green, W. N. & Baird, I. G. 2020, 'The Contentious Politics of Hydropower Dam Impact Assessments in the Mekong River Basin', *Political Geography*, 83: pp. 1–12.

Hammond, G. P. & Pearson, P. J. G. 2013, 'Challenges of the Transition to a Low Carbon, More Electric Future: From Here To 2050', *Energy Policy*, vol. 52, pp. 1–9.

Haris, S. M., Mustafa, F. B. & Ariffin, R. N. R. 2020, 'Systematic Literature Review of Climate Change Governance Activities of Environmental Nongovernmental Organizations in Southeast Asia', *Environmental Management*, vol. 66, no. 5, pp. 816–825.

Hayes, G., Jinnah, S., Kashwan, P., Konisky, D. M., MacGregor, S., Meyer, J. M. & Zito, A. R. 2021, 'Trajectories in Environmental Politics', *Environmental Politics*, vol. 30, no. 1–2, pp. 4–16.

Hewison, K. 2014a, 'Considerations on Inequality and Politics in Thailand', *Democratization*, vol. 21, no. 4, pp. 846–866.

Hewison, K. 2014b, 'Thailand: The Lessons of Protest', *Asian Studies: Journal of Critical Perspectives on Asia*, vol. 50, no. 1, pp. 1–15.

Hobson, C. 2012, 'Addressing Climate Change and Promoting Democracy Abroad: Compatible Agendas?', *Democratization*, vol. 19, no. 5, pp. 974–992.

Hong, C.-S. 2019, 'Thailand's Renewable Energy Transitions: A Pathway to Realize Thailand 4.0', *The Diplomat*, 9 March 2019, viewed 16 September 2020 at, <https://thediplomat.com/2019/03/thailands-renewable-energy-transitions-a-pathway-to-realize-thailand-4-0/>.

Human Rights Watch 2019, 'Thailand - Structural Flaws Subvert Election: Stacked Senate, Media Restrictions, Repressive Laws Undermine Right to Vote', Human Rights Watch, New York, 19 March 2019, viewed 17 May 2021 at, <https://www.hrw.org/news/2019/03/19/thailand-structural-flaws-subvert-election/>.

Human Rights Watch 2021, 'China: Crimes Against Humanity in Xinjiang', Human Rights Watch, New York, 19 April 2021, viewed 17 May 2021 at, <https://www.hrw.org/news/2021/04/19/china-crimes-against-humanity-xinjiang/>.

International Energy Agency 2019, 'Southeast Asia Energy Outlook 2019', International Energy Agency (IEA), Paris.

International Energy Agency 2020, 'Thailand', International Energy Agency (IEA), Paris, viewed 15 September 2020 at, <https://www.iea.org/countries/thailand/>.

International Energy Agency 2021, 'Net Zero by 2050: A Roadmap for the Global Energy Sector', International Energy Agency (IEA), Paris, viewed 19 May 2021 at, <https://www.iea.org/reports/net-zero-by-2050/>.

IRENA 2017, 'Renewable Energy Outlook: Thailand', The International Renewable Energy Agency (IRENA), Abu Dhabi, viewed 8 September 2020 at, <https://www.irena.org/publications/2017/Nov/Renewable-Energy-Outlook-Thailand/>.

IRENA 2020, 'Renewable Energy Statistics 2020 ', The International Renewable Energy Agency (IRENA), Abu Dhabi, viewed 8 September 2020 at, <https://www.irena.org/publications/2020/Jul/Renewable-energy-statistics-2020/>.

Kabiri, N. 2016, 'Public Participation, Land Use and Climate Change Governance in Thailand', *Land Use Policy*, vol. 52, pp. 511–517.

Kavi Chongkittavorn 2021, 'Examining the Implications of Thailand's New Draft Law on NGOs', *The Irrawaddy*, 12 May 2021, viewed 17 May 2021 at <https://www.irra waddy.com/opinion/guest-column/examining-the-implications-of-thailands-new-dra ft-law-on-ngos.html/>.

Laeni, N., van den Brink, M. & J. Arts 2019, 'Is Bangkok Becoming More Resilient to Flooding? A Framing Analysis of Bangkok's Flood Resilience Policy Combining Insights from Both Insiders and Outsiders', *Cities*, vol. 90, pp. 157–167.

Lebel, L., Manuta, J. B. & Garden, P. 2011, 'Institutional Traps and Vulnerability to Changes in Climate and Flood Regimes in Thailand', *Regional Environmental Change*, vol. 11, no. 1, pp. 45–58.

Looney, R. E. 2017, 'Introduction', in R. E. Looney (ed.), *Handbook of Transitions to Energy and Climate Security*, Routledge, London and New York.

Marks, D. 2011, 'Climate Change and Thailand: Impact and Response', *Contemporary Southeast Asia*, vol. 33, no. 2, pp. 229–258.

Marks, D. 2015, 'The Urban Political Ecology of the 2011 Floods in Bangkok: The Creation of Uneven Vulnerabilities', *Pacific Affairs*, vol. 88, no. 3, pp. 623–651.

McCarthy, J. 2019, 'Authoritarianism, Populism, and the Environment: Comparative Experiences, Insights, and Perspectives', *Annals of the American Association of Geographers*, vol. 109, no. 2, pp. 310–313.

McDonald, M. 2013, 'Discourses of Climate Security', *Political Geography*, vol. 33, pp. 42–51.

Meyer, J. M. & Chang, J. 2021, 'Continuities and Changes; Voices and Silences: A Critical Analysis of the First Three Decades of Scholarship in Environmental Politics', *Environmental Politics*, vol. 30, no. 1–2, pp. 17–40.

MoE 2018a, 'Thailand's Power Development Plan (PDP) 2018 Rev. 1 (2018-2037)', Ministry of Energy, Bangkok, Thailand.

MoE 2018b, 'Alternative Energy Development Plan 2018-2037 (AEDP 2018-2037)', Ministry of Energy, Bangkok, Thailand.

Myers, S. L. 2020, 'China's Pledge to be Carbon Neutral By 2060: What It Means', *New York Times*, 23 September 2020, viewed 9 May 2020 at, <https://www.nytimes. com/2020/09/23/world/asia/china-climate-change.html/>.

Network of People 2015, 'Open Letter: People's Demand to Cancel PDP2015 and Start a New Transparent Process', Network of people affected by Power Development Plan 2015 (PDP 2015), 7 September 2015, viewed 17 February 2016 at, <http://www.thaiclimatejustice.org/knowledge/view/126/>.

Neumayer, E. 2002, 'Do Democracies Exhibit Stronger International Environmental Commitment? A Cross-Country Analysis', *Journal of Peace Research*, vol. 39, no. 2, pp. 139–164.

Newell, P. 2012, 'The Political Economy of Carbon Markets: The CDM and Other Stories', *Climate Policy*, vol. 12, no. 1, pp. 135–139.

ONEP 2015, 'Climate Change Master Plan B.E. 2558–2593 (2015–2050)', Office of Natural Resources and Environmental Policy and Planning (ONEP), Ministry of Natural Resources and Environment, Bangkok, viewed 27 November 2020 at, <http s://climate.onep.go.th/wp-content/uploads/2019/07/CCMP_english.pdf/>.

Pacheco-Vega, R. & Murdie, A. 2020, 'When Do Environmental NGOs Work? A Test of The Conditional Effectiveness of Environmental Advocacy', *Environmental Politics*, vol. 30, no. 1–2, pp. 180–201.

Parichart Promchote, Wang, S.-Y. S. & Johnson, P. G. 2016, 'The 2011 Great Flood in Thailand: Climate Diagnostics and Implications from Climate Change', *Journal of Climate*, vol. 29, no. 1, pp. 367–379.

Patpicha Tanakasempipat 2019, 'Young Climate Strikers' Drop Dead' at Thai Environment Ministry', *Reuters*, 20 September 2019, Bangkok, viewed 20 November 2020 at, <https://www.reuters.com/article/us-climate-change-strike-thailand-idUSKBN1W509Y/>.

Pavin Chachavalpongpun (ed.) 2021, *Coup, King, Crisis: A Critical Interregnum in Thailand*, Council on Southeast Asia Studies, Yale University, New Haven.

Pearse, R. & Böhm, S. 2014, 'Ten Reasons Why Carbon Markets Will Not Bring About Radical Emissions Reduction', *Carbon Management*, vol. 5, no. 4, pp. 325–337.

Pemika Misila, Pornphimol Winyuchakrit & Bundit Limmeechokchai 2020, 'Thailand's Long-Term GHG Emission Reduction in 2050: The Achievement of Renewable Energy and Energy Efficiency Beyond The NDC', *Heliyon*, vol. 6, no. 12.

Pickering, J., Bäckstrand, K. & Schlosberg, D. 2020, 'Between Environmental and Ecological Democracy: Theory and Practice at the Democracy-Environment Nexus', *Journal of Environmental Policy & Planning*, vol. 22, no. 1, pp. 1–15.

Piya Pangsapa & Smith, M. J. 2008, 'Political Economy of Southeast Asian Borderlands: Migration, Environment, and Developing Country Firms', *Journal of Contemporary Asia*, vol. 38, no. 4, pp. 485–514.

Povitkina, M. 2018, 'The Limits of Democracy in Tackling Climate Change', *Environmental Politics*, vol. 27, no. 3, pp. 411–432.

Pugnatorius 2020, 'Seven Opportunities in Thailand's Solar Energy: Outlook 2021', Pugnatorius Ltd, Bangkok, viewed 27 November 2020 at, <https://pugnatorius.com/solar/>.

Raweewan Bhuridej 2015, 'Submission by Thailand to UNFCCC: Intended Nationally Determined Contribution and Relevant Information', Secretary General, Office of Natural Resources and Environmental Policy and Planning, Bangkok, 1 October 2015, viewed 18 February 2016 at, <http://www4.unfccc.int/submissions/INDC/Published%20Documents/Thailand/1/Thailand_INDC.pdf>.

REDD Monitor 2013, 'Thai Climate Justice Working Group Slams Thailand's Readiness Preparation Proposal: "The Participatory Process Was Problematic and the Content of the Draft Was Defective"', REDD Monitor, viewed 27 November 2020 at, <http://www.redd-monitor.org/2013/03/19/thai-climate-justice-working-group-slams-thailands-readiness-preparation-proposal/>.

Rigg, J. 1991, 'Thailand's Nam Choan Dam Project: A Case Study in the "Greening" of South-East Asia', *Global Ecology and Biogeography Letters*, vol. 1, no. 2, pp. 42–54.

Robinson, M. & Shine, T. 2018, 'Achieving a Climate Justice Pathway to 1.5 °C', *Nature Climate Change*, vol. 8, pp. 564–569.

Salamanca, A. & Rigg, J. 2016, 'Adaptation to Climate Change in Southeast Asia: Developing a Relational Approach', in P. Hirsch (ed.), *Routledge Handbook of the Environment in Southeast Asia*, Routledge, London and New York.

Shahar, D. C. 2015, 'Rejecting Eco-Authoritarianism, Again', *Environmental Values*, vol. 24, no. 3, pp. 345–366.

Shearman, D. & Smith, J. W. 2007, *The Climate Change Challenge and the Failure of Democracy*, Praeger, Connecticut.

Simpson, A. 2007, 'The Environment-Energy Security Nexus: Critical Analysis of an Energy "Love Triangle" in Southeast Asia', *Third World Quarterly*, vol. 28, no. 3, pp. 539–554.

Simpson, A. 2013a, 'Challenging Hydropower Development in Myanmar (Burma): Cross-Border Activism Under a Regime in Transition', *The Pacific Review*, vol. 26, no. 2, pp. 129–152.

Simpson, A. 2013b, 'Challenging Inequality and Injustice: A Critical Approach to Energy Security', in R. Floyd & R. Matthew (eds), *Environmental Security: Approaches and Issues*, Routledge, London and New York, pp. 248–263.

Simpson, A. 2014, *Energy, Governance and Security in Thailand and Myanmar (Burma): A Critical Approach to Environmental Politics in the South*, Routledge, London and New York.

Simpson, A. 2015, 'Democracy and Environmental Governance in Thailand', in S. Mukherjee & D. Chakraborty (eds), *Environmental Challenges and Governance: Diverse Perspectives from Asia*, Routledge, London and New York, pp. 183–200.

Simpson, A. 2018, 'The Environment in Southeast Asia: Injustice, Conflict and Activism', in M. Beeson & A. Ba (eds), *Contemporary Southeast Asia*, third edition, Palgrave Macmillan, New York, pp. 164–180.

Simpson, A. & Smits, M. 2017, 'Transitions to Energy and Climate Security in Thailand', in R. E. Looney (ed.), *Handbook of Transitions to Energy and Climate Security*, Routledge, London and New York, pp. 296–311.

Simpson, A. & Smits, M. 2018, 'Transitions to Energy and Climate Security in Southeast Asia? Civil Society Encounters with Illiberalism in Thailand and Myanmar', *Society and Natural Resources*, vol. 31, no. 5, pp. 580–598.

Simpson, A. & Smits, M. 2019, 'Illiberalism and Energy Transitions in Myanmar and Thailand', *Georgetown Journal of Asian Affairs*, vol. 4, no. 2, pp. 45–57.

Smits, M. 2015, *Southeast Asian Energy Transitions: Between Modernity and Sustainability*, Routledge, London and New York.

Smits, M. 2017, 'The New (Fragmented) Geography of Carbon Market Mechanisms: Governance Challenges from Thailand and Vietnam', *Global Environmental Politics*, vol. 17, no. 3.

Solar Magazine 2019, 'Thailand Solar Energy Profile', *Solar Magazine*, 5 March 2019, viewed 26 November 2020 at, <https://solarmagazine.com/solar-profiles/thailand/>.

Sonnenfeld, D. & Taylor, P. L. 2018, 'Liberalism, Illiberalism, and the Environment', *Society & Natural Resources*, vol. 31, no. 5, pp. 515–524,

Sopitsuda Tongsopit 2018, 'Give Renewable Energy A Chance', *Bangkok Post*, 30 March 2018, viewed 27 November 2020 at, <https://www.bangkokpost.com/opinion/opinion/1437543/give-renewable-energy-a-chance/>.

Sopitsuda Tongsopit & Greacen, C. 2013, 'An Assessment of Thailand's Feed-In Tariff Program', *Renewable Energy*, vol. 60, pp. 439–445.

Stevenson, H. & Dryzek, J. S. 2014, *Democratizing Global Climate Governance*, Cambridge University Press, Cambridge.

Teerapong Laopongpith 2019, 'Thailand's Climate Change Policies and Actions', Paper delivered at 'The Workshop for Accelerating Climate Actions in Bangkok', Office of Natural Resources and Environmental Policy and Planning (ONEP), Bangkok, 24 May 2019, viewed 22 October 2020 at, <http://www.bangkok.go.th/upload/user/00000231/web_link/jica/s%20Climate%20Change%20Policies%20and%20Actions.pdf/>.

UNFCCC 2016, 'Project 8664: Energy Efficiency Improvement of Mae Moh Power Plant Through Retrofitting Turbines', United Nations Framework Convention on Climate Change (UNFCCC) Secretariat, Bonn, viewed 27 November 2020 at, <http://cdm.unfccc.int/Projects/DB/TUEV-RHEIN1355199535.37/view/>.

UNFCCC 2020, 'United Nations: Climate Change', United Nations Framework Convention on Climate Change (UNFCCC) Secretariat, Bonn, viewed 20 November 2020 at, <https://unfccc.int/>.

Unger, D. H. & Patcharee Siroros 2011,'Trying to Make Decisions Stick: Natural Resource Policy Making in Thailand', *Journal of Contemporary Asia*, vol. 41, no. 2, pp. 206–228.

Veerayooth Kanchoochat & Hewison, K. 2016, 'Understanding Thailand's Politics', *Journal of Contemporary Asia*, vol. 46, no. 3, pp. 371–387.

Webersik, C. 2010, *Climate Change and Security: A Gathering Storm of Global Challenges*, Praeger, Santa Barbara.

World Bank 2010, 'Climate Risks and Adaptation in Asian Coastal Megacities', World Bank, Washington D.C., viewed 15 September 2020 at, <http://documents1.worldbank.org/curated/en/866821468339644916/pdf/571100WP0REPLA1egacities01019110web.pdf>.

World Bank 2020, 'Forest Area', World Bank, Washington D.C., viewed 20 November 2020 at, <https://data.worldbank.org/indicator/AG.LND.FRST.ZS?locations=TH/>.

11 Governing climate across ontological frictions in Timor-Leste

Alexander Cullen

Introduction

Timor-Leste is the youngest and most impoverished country in Southeast Asia and faces some of the most difficult climate change challenges in the region. While responsible for one of the lowest greenhouse gas (GHG) emissions per capita, its challenging topographical character suggests it is one of the world's ten most climate vulnerable nations (GCF 2019). In tension with this issue is a formal economy highly dependent on entrenched but dissipating Petro-capitalism to fund basic government services. Assessments of climate change impacts on the largely rural population remain somewhat nascent, but articulate an emergent severity pressing upon already stressed livelihood and environmental needs (Barrowman & Kumar 2018). Yet sufficient resources for climate change mitigation or adaptation are limited by the immediate needs of health, education, and other pressing development indicators for which Timor-Leste remains low. While this known gap presents difficulties in registering, analysing, and administering state responses, it also offers opportunities for local knowledges to meaningfully contribute to climate change governance and understanding in Timor-Leste.

Localized customary knowledge for managing and adapting to climate phenomena is rich. The climate and the relationship to weather variances are integral components of rural livelihoods in Timor-Leste. These relationships are not just mediated through material landscape and livelihood change, but also charged with spiritual potency and entangled in a specifically Timorese form of cosmopolitics (Palmer 2020). In reference to Isabelle Stengers, cosmopolitics can be thought of as the ways that human entanglements with non-human actors may be foregrounded (Stengers 2010). It is through such considerations, political agency and voice are ceded to non-humans (or nature) to create and govern 'common worlds'. Therefore, in complex cosmologies of Timor-Leste, different phenomena such as 'climate change' and forms of life are not only interconnected, but they are also activated by political agency yielded to the more-than-human.

This is now generating increasing attention towards customary knowledge and localized resource management as a potentially viable pathway for

DOI: 10.4324/9780429324680-13

meaningful future climate responses. However, mobilization or formal support remains comparatively low at the Timorese state level. When it or other NGO planning programs do recognize customary climate knowledge, they rarely afford credence to the ontological dimensions of climate politics or multi-natural experiences of Timorese environmental change. Given the weakness of formal governance interventions in a socially and topographically challenged country with few resources, ignoring such issues risks exacerbating already severe climate impacts and their unequal effect on the country's most marginal.

An approach to climate governance that foregrounds indigenous knowledges and management radically challenges conventional scientific approaches, institutions, and state power; as well as the global governance mechanisms that sustain, and are sustained by them (Hope 2021). But as will be argued, in Timor-Leste customary institutions (and in particular their socio-climate knowledge) are critical to everyday governance, not least because of the paucity of state administration and its resource limitations outside of the capital Dili. In this regard, taking local ontologies of weather change as seriously as 'scientific understandings' is necessary for Timorese climate responses to be consequential. For this to occur while also avoiding social conflict or exacerbating deepening environmental loss, it is pertinently vital.

This chapter will show that customary climate governance remains necessary for meaningful outcomes in Timor-Leste, and a lack of attention locally risks entrenching current livelihood struggles and localized marginality. To do this, it will first outline the geography and weather experiences of Timor-Leste relevant to contemporary and future climate impacts emergent in current modelling. Attempting to meet these impacts through adaptation and mitigation is complicated by the diverse micro-geographies and gaps in data. The issues of these gaps and the limits in state capacity are highlighted by summarizing the formal governance mechanisms and international targets Timor-Leste has sought to employ. However, in comparison to other development needs they are shown to lack prioritization, or the resilience of customary institutions in mediating community-environment management in the every day. The potency and considerable knowledge of customary climate ontologies is unpacked before demonstrating occurrences of socio-climatic failure when this is ignored.

In everyday governance, there often exists shared customary and Western epistemologies that reflect similar recognition of climate phenomena but are articulated through different knowledge systems. Yet, only superficial recognition of the complex epistemological work of custom and its entanglement with socio-natural change in Timorese 'weatherscapes' remains. A continued lack of recognition risks exacerbating state failures to mitigate, adapt, or manage increasingly frequent climate change-related impacts as well as other cosmo-political consequences, such as community conflict or reluctance in support.

Timor-Leste is particularly vulnerable to climate events and current assessments predict the likelihood of wide-ranging impacts socially, environmentally, and economically. Many of these will not be unique to the Timorese experience, however its post-conflict context generates additional complexities that trouble the 'conventional wisdom' of state-directed governance (O'Lear 2016) or growing private interest in global climate finance (Bracking 2019). The embryonic and centralized nature of its bureaucratic structures, revitalized customary institutions, varied micro-geographies, and conflict history present unique governance challenges. As a Least Developed Country (LDC), governance remains focused on primary development concerns, relegating the anxieties of climate change adaptation and mitigation mechanisms to second-tier priorities.

While formalized institutions of the state across different scales are increasing in robustness, in rural Timor-Leste, local customary institutions and cosmologies remain more important to local everyday experiences, environments, and lifeworlds. Attention to the importance of these and the local knowledges that bring these into being largely remains ignored with inadequate credence to the experiences and politics within local ontologies. This has ramifications, not only in terms of disaster and livelihood adaptations to climate changes, but also reverberates politically across and through social landscapes. Furthermore, the experiences of climate change are being rendered apolitical by its current discourse, which limits effective responses, particularly in terms of local knowledge mobilization and community governance ownership. Beyond this, the notion of apolitical climate experiences succors the government creep into new sites of previously informal customary rule, limiting its engagement with atmospheric and cosmological imbalance.

Understanding the role of geography

The critical onto-politics of Timor-Leste climate governance emerges with such vigor because its varied topographical, social, and environmental geography ensures climate vulnerabilities and impacts are profound, wicked, and many. These and a myriad of longstanding post-conflict challenges complicate governance aimed at effectively mitigating or adapting to the array of impacts projected in the future. These include extreme weather events such as drought and flooding, soil degradation and erosion, rise in zoonotic diseases, frequent loss of critical infrastructure, displacement, depletion of fish stocks, and food insecurity. Furthermore, the country's specific physical geographical constraints, that dynamically shift across small distances, increases not just its risk to climate change but also the difficulties of response. Variability complicates integrated climate responses and mitigation, troubling uniform policy and project prescriptions drafted in the capital Dili. While terrestrially small by global norms at 14,874 km^2, the country nonetheless experiences substantial variations of localized micro-geographies that in turn generate a diverse range of climate vulnerabilities, social landscapes, and cosmopolitical challenges.

Much of its land compromises the eastern half of Timor Island, as well as the enclave Oecussi in the western half and Atauro Island north of Dili. Atauro and Oecussi's physical separation necessitates weekly ferry connections that are frequently disrupted by storms and mechanical failure (Cullen & Marx 2015), further diminishing possibilities for timely networked responses needed there. The coast is fringed by a rich coral marine life substantially important to fisher livelihoods and wider coastal ecologies. A high mountainous spine of peaks and valleys runs west to east, dividing the dry north from the wetter south. This spine ensures that there is little gentle relief, with much of the land on slopes greater than 40 percent. What little flat land exists is mainly on the southern coastal plain and utilized for agriculture. Much of the land is covered in shallow rocky soils which are alkaline, not particularly fertile, do not store water well, and are easily eroded, particularly when hit by intense downpours (Monk, Fretes & Reksodiharjo-Lilley 1997). Timorese soil's fragility and thinness are partly due to being derived from limestone and metamorphosed marine clay. High tropical humidity also results in organic matter decomposing quickly, reducing nutrient input into the soil (World Bank 2009).

Weather patterns vary considerably across the country. A multiplicity of microclimates endures due to the wide diversity of environments that exist in rapid slope and altitude changes. Two primary weather cycles persist on either side of the mountain ridge, with some monomodal north-coast towns amongst the driest in Southeast Asia (600mm/year) (Trainor 2010). The south of the country experiences bimodal rainfall pattern producing a longer (7–9 month) wet season with two rainfall peaks starting in December and again in May (Keefer 2000 in Barnett, Jones & Dessai 2003). In these areas, rainfall can average 1,500–4,000 mm/year (Trainor et al. 2008). Timor's climate is significantly susceptible to ENSO influences (Fox 2001), with El Niño bringing prolonged drought and La Niña extended periods of rain. The variability of the climate produces various challenges to successful agricultural production, of which 70 percent of the rapidly growing population of 1,170,00 (DNE RDTL 2015) rely on for their livelihoods. By 2030 the population may have reached 1.6 million. Traditional subsistence farming, often using swidden techniques, remains the most common form of agriculture for the estimated 140,000+ rural households (Oxfam 2008). These forms of livelihood are extremely vulnerable to climate variability and change, with many experiencing an annual 'hungry season'. Deforestation remains a key concern across the country and is estimated to be occurring at approximately 1.7% percent.

Only 10.4% percent of the total land area is considered arable (World Bank 2015), and thus farmers regularly cultivate steeply sloped areas (USAID 2004). Historically, Timor's groups have preferred to augment shifting agriculture on limestone terraces or the mixed marine-based soils of ridges and mountain slopes, with more intensive farms on alluvial plains and beside riverbanks (Fox 2001). The most widely produced crops are maize, rice, and cassava. However, rice is the highly preferred staple and is primarily grown in relatively flat areas on the south side of the island where water availability is less constrained. Many supplement their income with sales of surplus harvest and cash crops such as coffee, which is the nation's second largest export. This is gradually becoming less sustainable due to changing climatic conditions and fluctuating global market linkages that temper re-investment (Kanamaru 2020).

These agricultural systems are principally managed through traditional customary institutions that are of primary importance to everyday governance compared to relatively weak state administrative institutions. Largely suppressed under Indonesian occupation, customary rural institutions re-emerged in the institutional vacuum of post-independence. They remain the central form of social and political organization, identity, ontological framework, and mediator of community–environment relations (Fitzpatrick, McWilliam & Barnes 2013). For many, customary law remains far more integral to the governance of everyday livelihoods, ecologies, and the lifeworlds they bring into being. These elements produce unique challenges in terms of governing climate in Timor-Leste, especially concerning alleviating, contesting, and comprehensively mapping its predicted impacts, of which there are many.

Writing the impacts of Timorese climate change

Research into experiences and predictions of climate change impacts for Timor-Leste remain limited and grounded largely in global and regional scientific approaches to understanding climate variance. Nonetheless, these provide a pertinent starting point for navigating climate change issues and ontological difference in a post-conflict country facing ongoing complex development challenges. Due to colonial and occupational violence, Timor-Leste's socioeconomic and ecological systems were in fragile states at independence. They still remain so two decades later, with current simulations problematizing an exacerbation of this fragility.

Climate models for Timor-Leste predict future increases in the intensity of extreme short-term precipitation events (Kirono, Chiew & Kent 2010), uncertainty over the frequency of droughts and floods, reduced groundwater, and a more vigorous hydrological cycle (Barnett, Jones & Dessai 2003; Wallace et al. 2012). The climate trend is for shorter, more erratic wet seasons and longer dry seasons. The likely results are more prolonged droughts, more intense flooding, and increased landslide/erosion risks (TLNDMG, Australian Bureau of Meteorology & CSIRO 2011). These will be further exacerbated by more extreme ENSO events, thereby increasing existent experiences of agroecological vulnerability and infrastructure destruction that plagues the country every year. Drought is already more severe in El Niño years, with over 50 percent of the population experiencing severe food insecurity in 2015–2016 (FAO 2018), and increasing rainfall extremes during La Niña (TLNDMG, Australian Bureau of Meteorology & CSIRO 2011). The frequency and intensity of tropical cyclones is also expected to rise significantly.

Annually, landslides and landslips cut off major connecting roads while bridges crossing Timor's shallow but widely braided rivers are washed out by short but intense rain events leading to flash floods. Costly newly built road infrastructure is routinely damaged, sometimes only shortly after completion. This, in turn, dramatically slows transportation between Dili and the districts, compromising market chain connections, food security, and general logistic networks. Electricity posts traversing highly sloped areas are also washed out or slip, causing temporary disconnection from the national grid for certain villages. It is estimated that the costs of damage to infrastructure by extreme weather events are annual economic losses of USD$250 million (GCF 2019). This is substantial considering the national budget for 2020 was proposed at USD$1.6 billion. The landslides, floods, erosion, and droughts responsible for these losses will increase in frequency and intensity with changing temperatures.

Modest estimates suggest a temperature rise of 1.5°C by 2040 (Molyneux et al. 2012), likely to increase to more than 2.5°C by 2070 (ADB 2013). The increase in temperature is expected to facilitate a wider spread of zoonotic disease, including recently eliminated diseases such as malaria, to higher altitudes where most of the population lives. Concurrently an increase in ocean acidification and coral bleaching events will lead to severe marine stock loss (Rosegrant et al. 2016), on which most coastal communities depend. Further increases in sea level rise and storm surges will continue, which is currently experienced at 5.5mm/year, inundating the low-lying coastal plain and its infrastructure. By 2095, sea-level rise could be as low as 18cm or as high as 79cm (GCF 2019). Local groundwater will be more vulnerable to seawater intrusion (Geoscience Australia 2012), contributing to more significant water scarcity.

These elements will exacerbate current food insecurity issues experienced by agricultural communities who already expect an annual hungry season. The impacts of climate change on crop yields have been well-documented, most concluding that under current and future climate conditions, low growth and production is expected (ADB 2013) while being further constrained by high predicted population growth (Molyneux et al. 2012). These elements and the likelihood they will exacerbate political vulnerability present a particularly complicated set of challenges for a post-conflict country that has already experienced severe political unrest, as seen during the 2006 crisis (Van der Auweraert 2012; Scambary 2009).

Extant climate change mitigation and adaptation efforts

Like other Small Island Developing States (SIDS), the scale and profundity of climate impacts are in stark contrast to the globally disproportionate low levels of carbon emissions per capita at 0.4 GHG tCO_2eq/cap. This is less than 8 percent of the global average and only 0.02 percent of a large emitter such as Australia (Ritchie 2019). Nonetheless, the economy remains dominated by oil/gas extractive processes from the Timor Sea, accounting for 85 percent of the government budget. Other economic arenas are as yet unviable or undeveloped. Future state development plans are concentrated in further expansion and exploitation of remaining oil and gas reserves. This is manifest in state preparation of the Tasi-Mane development, a large infrastructure investment of pipelines, harbor, airport, highways, and processing facilities stretching across the south-coast. In-country processing is still under negotiation but has not stemmed infrastructure development, nor the large displacement of many villagers from the land needed for this. The state and other development actors recognize the importance of climate emissions reduction, although it is relevant to note that the most significant aid donors for climate change programs in Timor-Leste (Australia, USA, Canada) are also the world's largest emitters.

Unlike the rest of Southeast Asia, Timor-Leste has a minimal number of climate change projects. Those that do exist are primarily donor-led or small-scale initiatives run by non-governmental organizations (NGOs) (Chandra, Dargusch & McNamara 2016) and inadequate in addressing the complex mitigation and adaptation needs. This includes poor disaster and emergency responsiveness, low government capacity, coordination, and rapid engagement (Sagar et al. 2018). Partly this is due to a lack of comprehensive data and understanding on socio-climatic relations in Timor-Leste. However, residual post-conflict development challenges remain across health, education, infrastructure, economy, and other socio-political arenas. These command immediate attention socially and politically, and therefore primary allocation of the minimal available resources and financing.

Nevertheless, the government of Timor-Leste has made progress in terms of acting and ratifying international obligations concerning climate action through global governance mechanisms. The country ratified the UNFCCC on 11 April 2006 and then the Kyoto Protocol in 2008. It was one of the few countries formally offering land to Kiribati people should theirs become uninhabitable due to rising sea levels (RTCC 2012). It submitted its National Adaptation Program of Action on climate change (NAPA) in 2011 (RDTL 2011) and Intended Nationally Determined Contributions (INDCs) in 2014, and these both included climate vulnerability assessments.

Bolstering this formalized turn to improved climate recognition, the UN resolution on implementing the Sustainable Development Goals (SDGs) was adopted and passed by parliament in 2015. This purported expanded action on climate change but attempts to draw a road map for meeting the 2030 SDG goals proved difficult in practice. Support across INGOs, civil society groups, and the formal economy in Timor-Leste for the SDGs is strong, but they are poorly understood and remain difficult to articulate. This has limited the implementation of processes or policies that would contribute towards them, including those of climate governance (Courvisanos & Jain 2019). Resultingly, the SDGs to which Timor has committed and its NDCs lack alignment and fail to take the other into account.

Since ratifying the Paris Agreement in 2016, Timor-Leste's INDCs as per agreement have shifted beyond intentions to NDCs with hard targets. These are shorter deadlines to provoke better and adaptive management, with the first reached in 2020. As governance mechanisms, NDCs are meant to effectively outline GHG emission targets and actions, including mitigation and adaptation that ratified countries will undertake (under specifically communicated periods). For Timor-Leste, the focus is almost entirely adaptation as previously specified in its NAPA, rather than mitigation. As such it doesn't have an emissions reduction target, but rather has made commitments through various sectors such as agricultural improvements that require green finance. This is rationalized by its minimal share of global emissions at 0.003 percent total and possessing one of the lowest carbon contributions per capita despite being highly vulnerable to climate change.

This program of work emphasizing agricultural improvements has continued under Timor's second national communication for the UNFCCC as published in 2020. Mitigation where it does exist centres on possible forest regeneration and its securitization through protected areas. But there has also been ad-hoc administrative controls attentive to reducing rural firewood reliance, and limiting customary practices of swidden farming (RDTL 2020). This is of concern because the removal of the Indonesian kerosene subsidy at independence created a high demand on fuelwood that continues to this day (Sandlund et al. 2001).

Through guidance from the NAPA, numerous projects have been funded by international donors (and more recently the Green Climate Fund) focusing on agriculture, infrastructure, disaster risk reduction, and mangrove coastal management. Such funding is comparatively small compared to others in the region at just of USD\$22 million thus far. There have also been problems assessing effectiveness due to inadequate implementation of overarching monitoring and evaluation frameworks. Nonetheless, the possibilities of new climate aid and donor financing are increasingly seen by the government as a means to meet financial or resource shortfalls. Most recently, this has meant attention directed to program accreditation for the Green Climate Fund. This has at times meant programs addressing critical issues of climate-proofing infrastructure, slope stabilization, and water security are amalgamated into ad-hoc governance processes and approaches. State climate adaptation governance has suffered from poor intra-departmental coordination and communication while tending to be top-down, with little local involvement reflecting the ongoing disconnect between the national and local (Mercer et al. 2014). The lack of credible scientific data sets in Timor-Leste continues to hinder effective planning and analysis as information has either not been recorded or has been lost or destroyed (Norton & Waterman 2008). In response, the government has created a dedicated National Directorate for Climate Change situated within the Ministry of Commerce, Industry, and Environment and steered by a Secretary of State for the Environment.

This issue of missing credible climate data as a necessary conduit for planning and adaptation governance has begun to draw attention to other sites of practice and value. Increasingly, it is voiced that there is existent rich customary forms of knowledge concerning local climate experiences, their mediation, and the weather-worlds they bring into being. Furthermore, it is now widely recognized the importance local customary knowledge and practices can play in reducing risk and improving disaster preparedness in the Pacific (Hiwasaki et al. 2014). The current lack of administrative capacity and related knowledge gaps as enacted by the state should provide robust rationalization for meaningful customary engagement and knowledge co-production, especially in the rural margins. However, this has been largely diaphanous in practice, or when actually mobilized in program delivery, done so with little consideration for customary and political dimensions, risking damaging socio-environmental consequences including conflict.

The power of custom in the Timorese state and weatherscape

Customary institutions and their constituted knowledge remain the most resilient and pertinent forms for governing the everyday in rural Timor-Leste. Since independence, the performance of state administration and its infrastructure has been concentrated most acutely in Dili, while their felt power diminishes quickly outside of peri-urban areas. Here, the promises of development have been slower to intercede and instead of the state, it is often customary institutions mediating localized conflict, usufruct rights, justice, spiritual wellbeing, and livelihoods (Palmer & de Carvalho 2008). The failures of state and development programs are usually related to an unwillingness for meaningfully recognition of this. In this respect, the incorporation and support of customary institutions offers far more than complementary local knowledges to scientific assessment. It structures and provides the most salient form of governance, articulating how things are entangled with each other, and its centrality to people's lives.

For many Timorese, land, identity, and cosmological health are reconciled through a complex set of socio-political relations tied to the ancestral origin of place (Bovensiepen 2011). Through ceremonial performance, customary farming, and attention to "feeding the land" (McWilliam & Traube 2011), it remains spiritually potent and populated with a myriad of more-than-human beings that arbitrate in the production of material, atmospheric, and spiritual well-being. That which is sacred or spiritually powerful is known as *lulic*, and may coalesce or concentrate in certain areas such as village wood groves (McWilliam 2001). *Lulic* requires sagacious management lest a risk of manifesting physical or cosmological misfortune, environmental disaster, sickness, and even death. For instance, unauthorized trespassing, neglect, or cutting of *lulic* trees can incur repercussions such as volatile rain events that ruin harvests or cause landslides. Care and attunement (c.f. Stengers 2010) to the changes in the cosmopolitical landscape is therefore essential to ensure its productivity and guarantee good cropping weather. Such landscape changes may be intricately signposted by numerous biological and cosmological events, including the appearance of certain species or apparitions that manifest on the land or in the atmosphere over the agricultural calendar. These knowledge claims are sometimes at odds with normative developmental expectations for resource management and claims of expertise.

Timorese customary understandings, world-making, and ontologies of climate are complex and differ in many ways to the Western scientific articulations that structure formalized forms of Timorese climate research and governance (c.f. Hiwasaki et al. 2014). They also differ in how they make sense of socio-political and cosmological community relations to their lived weatherscapes. This affects the deployment of appropriate and measured governance responses that best minimize climate impacts on livelihoods while also limiting state-customary governance synergies and climate knowledge co-production.

Given the deficiencies of state capacities, certain suggestions have been made for better incorporating and utilizing customary resource management mechanisms as climate response. This is made particularly pertinent by the meagre state financial or human resources available for environmental management or protected areas. *Lulic* forests and the customary obligations to their preservation have been tentatively highlighted as a practice that should generate greater formal encouragement. *Tara bandu* is perhaps more frequently touted as the most important and viable customary form of local resource management (Meitzner Yoder 2007). It is often described as a ritual prohibition of certain local resources or environmental activities for a prescribed amount of time. It may be enacted across various situations beyond forest landscapes for mediating water access (Palmer 2015) and fishing (Alonso-Población et al. 2018), by hanging or displaying the prohibited items. Transgression instigates the application of a predetermined penalty. *Tara bandu* not only regulates people's relationship to the environment, but also with other people, animals, and ancestral power. *Tara bandu* has been somewhat promoted by the UNFCCC as a possible practice for climate response in Timor-Leste which can be incorporated into law and practice (Chandra, Dargusch & McNamara 2016).

Linking local knowledge to international climate governance

Burgeoning reforestation projects have thus far sought payment for environmental services or endured government-implemented replanting schemes that often obliged community participation. These have enjoyed mixed success (Bond, Millar & Ramos 2020; Miyazawa 2013), yet local knowledge could at least complement or inform their design for stronger outcomes. The government remains interested in exploring the REDD+ program as well as other forms of green financing. However, past studies examining the potential for carbon sequestration have assessed this as low (Lasco & Cardinoza 2007), with potential for vegetative mass or carbon heavy ecosystem sinks as low. In addition, much of the green financing sought thus far has been for larger agricultural improvements that often marginalize customary swidden farming systems and the knowledge they depend upon.

While important as progress, these particular developments risk engendering simplified environmental aims rather than critical concern for the complex dimensions of ritualized livelihood practices. Given the deficiencies of data recorded/available on climate impacts and experiences, this is even more pertinent. There are dangers of limiting effective learning from traditional knowledge of weather management and its likely erosion without ongoing attenuation. This produces greater unknowns in terms of community response to environmental change. Furthermore, climate change simplifications to intense weatherscapes of inclemency become understood materially as environmental spatial fixes to be amended by protected areas or suppressing agricultural activities. In coastal communities, the conceptual separation of land and sea by conventional climate adaptation governance is in contrast to traditional environmental governance and thus an obstacle to building resilience (Burns 2019).

Beyond this, the government and the NGO sector have begun mainstreaming climate into community engagement programs. They seek improvement in local resource management where viable through new discourses of state environmental rationality and discipline (Cullen 2020). Environmental education is seen as a central pillar to positive climate outcomes and often program summaries stress a community's low levels of education and/or histories of natural resource exploitation as funding justification (c.f. Ismail et al. 2019). These discourses posit localized tree cover and biodiversity loss (particularly through timber felling) as central concerns in halting erosion as well as national deforestation. Furthermore, global climate change and the impacts it brings are repeatedly articulated as the immediate result of localized deforestation, or, more explicitly, singularized logging. Donor-led adaptation programs thereby conceptualize climate change vulnerability as a biophysical issue rather than a consequence of the dynamic interactions between political, institutional, economic, and social (Barrowman & Kumar 2018) or cosmological structures (Burns 2019).

Outside of Dili 'climate development' has largely manifested as informal 'socializations' with village administrators or community hall meetings organized by under-resourced local coordinators. They themselves (and the communities they interact with) recognize the authority and power that extends from claims of formalized expertise emanating from educated Dili. However, such actors often only poorly grasp the finer scientific rationale of climate science. Thus, engagement frequently metastases into explanations rendered unto simple equivalences: climate change (*klimata mudansa*) are experiences of bad or unexpected weather.

In this manner, the logic of climate change is reduced to intense drought and flood phenomena that frequently trouble agricultural production during ENSO events. These events are understood as a direct causality of the cutting of local trees, and therefore one should not log lest climate change and its dire weather be invoked. Through various programs or engagement, the government has encouraged this direct link between localized logging transgressions and experiences of weather disasters. While such logic might prove misconstrued, it nonetheless mirrors simplistic Western media representations of climate change causality. Furthermore, it works to invoke the ontological repercussions of complex customary relations to landscape change and capture them within the state's resource management remit.

Seeing the failures to take customary climate governance seriously

These concerns are illustrated by the challenges faced during the last major La Niña event. During this period, constant rain ruined crop harvests as the wet season refused to yield for planting, distressing already precarious food security. In a particular case, the unseasonal weather was identified as 'climate change' and therefore naturally linked to some specific actions of local timber cutting. A villager clearing land close to a *lulic* grove was soon blamed as specifically causing climate change, and demands were made of him for reparations. When these were not forthcoming, threats of violence were made to his person before he eventually had to leave the village for safety reasons. Elsewhere, inadequate state institutional responses and land degradation exacerbated by expropriation have eroded land use resilience. The resulting crop failures, land erosion, and livelihood insecurity were carefully framed by certain village administrators as merely the outcome of the 'climate change'. The depoliticization of the disaster experience was deliberate so as to avoid displeasing state administrators and increasing their competitiveness for sparsely limited government disaster relief supplies.

Finally, in a separate area, the failure of rains to abate was clearly identified as 'climate change' – but as a result, made separate from the scope of customary response and community agency. Requests to the *Lian Nain* (village elder) to perform potent customary rituals to halt the rain were dismissed by him as unviable, as he clarified that the rain was impossible to stop because it was climate change. In constructing environmental concerns through a scientific framework (and thereby necessitating scientific solutions), the state may be positioning itself as the preeminent management authority. Still, it also removes local agency and responsibility for the environment that is entangled in custom. Such environmental issues risk becoming solely the domain of the state, limiting community participation that has been integral for ecosystem and socio-environmental management.

It is evident from Dili-based environmental organizations and local government workers, that they observe helpful and perhaps even obvious opportunities for synergies across competing understandings of landscape. In this manner, superstitious, traditional beliefs in atmospheric ramifications for forest violations reflect a teleological mirror of the contemporary scientific world. The crude explanation and framing of climate change in community environmental 'education' renders a messy and multifaceted issue problematically singularized and causally simple. This is at once 'frictious' (Tsing 2005) with the customary complexity of knowing and being as socio-environmental relations and the mediation of worldly potency that is entangled in localized climate. Its profundity requires serious and engaged consideration of the myriad of factors connected to land and weatherscape in Timor-Leste. Failure to do so risks unintended outcomes, including violence, but also a dilution of rich customary knowledge. Furthermore, it forecloses on the possibilities for expansive understanding and dynamic response capacity to climate-related phenomena that emerge when credence is afforded to local customary knowledge and systems.

Conclusion

Timor-Leste faces some of the most difficult climate governance challenges due to its susceptibility to climate impacts, but also its low capacity for meaningful administrative response. Custom holds important local knowledge for managing climate-related risk but the political ontologies of these weatherscapes are often marginalized. The lack of willingness on the part of formal administrative systems and their associate technologies to embrace customary knowledges and the authority that facilitates its material practices is related to aspirations of Timorese citizen modernity. Customary knowledges and structures do not fit snugly within contemporary administrative ones. Couching climate as crises to be managed within state development ideation expands the reach of government rule, territorializing the rural customary margins and reshaping farmers as new environmental citizens (Cullen 2020). As the primary aim, it matters not whether local farmers or government interlocutors fully grasp an understanding of (scientific) climate change contributions, effects, and impacts. The consequences of this are the marginalization of localized knowledge contributions and its forms of action. The making of recent weather phenomena as climate change works to make 'the rain' the domain of government and science and disempowered of customary agency. Local responses become more muted.

Divergent stakeholders recognize the benefit of mitigating and adapting eroding slopes to the heightened risks of weather events. However, the production of climate change knowledges at the interface of custom and state produces problematic and erroneous understandings. Local community members are being told that they will be addressing climate change and its events by not cutting trees when they themselves are not responsible for such events. Meanwhile, the government produces the majority of national carbon emissions through ongoing (although retreating) oil production. Diminished respect by formalized institutions for complex customary knowledges and governance reflects a poor understanding of how communities will respond to epistemological challenges, landscape change, and related cosmopolitical potency; a potency that structures and renews the richness of customary power, everyday life, and identity. Such potency and the localized power it sustains are eroded by shallow forms of 'Western' climate knowledges. These remained couched within a modernist development approach that territorializes resources, trees, rivers, slopes, fields, and atmospheres of localized entanglement. Thus, once under 'flawed' customary management, they are now the dominion of the state. Meanwhile, modern and developed Timorese citizenship is advanced as those that 'don't cut down the forest'.

The consequences of this local disempowerment are not only a loss of local mitigation and adaptation knowledge, but also a retraction and reluctance of local agency in climate response. To some, the idea of climate change has remade the management of rain the domain of government and science. Furthermore, climate change is reimagined as a singular disaster rather than embedded in a web of complex political and ontological relations that stretch temporally and spatially. The idea of an uncertain but looming climate crisis compelling state technical intervention, ignores customary and indigenous peoples' lived experiences through ongoing crises of (neo)-colonial violence and environmental devastation (Whyte 2018). If climate justice (not just governance) is to concurrently dismantle coloniality, thorough ethnographic understanding of climate change through a moral meteorology of indigenous ontologies is necessary (Burman 2017). In Timor-Leste, this is particularly acute given the endured genocide, land expropriation, and customary suppression during Indonesian occupation. In effect, climate change events and their impacts are unmoored from their social, historical, environmental, and economic context, including legacies of colonial occupation or state expropriation of land. Communities have begun to be more reluctant to identify government policy and its failure to cause livelihood loss rather than climate change weather impacts.

Global climate knowledge is complex, powerful, and essential in assessing and strategizing timely responses to climate change impacts. But its reification has worked also to erase local understandings and, overall, indigenous knowledges have been left off global climate science (Yeh 2015). Meanwhile, the IPCC remains a preeminent source for climate knowledges and associated authority within the country. What Timor-Leste's current governance situation demonstrates is that its climate change risks are many, and it is already enduring these in light of one of the lowest GHG emissions per capita. These will likely increase in coming decades and despite the emergence of some adaptation programs, the state lacks adequate support and resources to meaningfully engage with these. More international support and climate financing may prove vital to managing greater weather risks as more work is done in assessing these. However, marginalizing the complex customary epistemologies and residual socio-relations to land leads to a diaphanous conceptual understanding of customary climate management and broader local conflict.

References

ADB 2013, *The Economics of Climate Change in the Pacific*, Asian Development Bank, Mandaluyong City.

Alonso-Población, E. *et al.* 2018, 'Narrative Assemblages for Power-Balanced Coastal and Marine Governance. Tara Bandu as a Tool for Community-Based Fisheries Co-Management in Timor-Leste', *Maritime Studies*, vol. 17, no. 1, pp. 55–67.

Barnett, J., Jones, R. N. & Dessai, S. 2003, *Climate Change in Timor Leste*, University of Melbourne & CSIRO, Melbourne.

Barrowman, H.M. & Kumar, M. 2018, 'Conceptions of Vulnerability in Adaptation Projects: A Critical Examination of the Role of Development Aid Agencies in Timor-Leste', *Regional Environmental Change*, vol. 18, pp. 2355–2367.

Bond, J., Millar, J. & Ramos, J. 2020, 'Livelihood Benefits and Challenges of Community Reforestation in Timor Leste: Implications for Smallholder Carbon Forestry Schemes', *Forests, Trees and Livelihoods*, vol. 29, no. 3, pp. 187–204.

Bovensiepen, J. 2011, 'Opening and Closing the Land: Land and Power in the Idaté Highlands', in A. McWilliam & E. G. Traube (eds), *Land and Life in Timor-Leste: Ethnographic Essays*, ANU, Canberra, pp. 47–60.

Bracking, S. 2019, 'Financialisation, Climate Finance, and the Calculative Challenges of Managing Environmental Change', *Antipode*, vol. 51, no. 3, pp. 709–729.

Burman, A. 2017, 'The Political Ontology of Climate Change: Moral Meteorology, Climate Justice, and the Coloniality of Reality In the Bolivian Andes', *Journal of Political Ecology*, vol. 24, pp. 921–938.

Burns, V. 2019, 'Oceanic Embodiments: Living ENSO Events in Coastal Timor-Leste', *Political Geography*, vol. 70, pp. 102–116.

Chandra, A., Dargusch, P. & McNamara, K. E. 2016, 'How Might Adaptation to Climate Change by Smallholder Farming Communities Contribute to Climate Change Mitigation Outcomes? A Case Study from Timor-Leste, Southeast Asia', *Sustainability Science*, vol. 11, no. 3, pp. 477–492.

Courvisanos, J. & Jain, A., 2019, *Commitment of Stakeholders to the SDGs in Timor-Leste*, Deakin and Federation University, Melbourne.

Cullen, A. 2012, 'A Political Ecology of Land Tenure in Timor Leste: Environmental Contestation and Livelihood Impacts in the Nino Konis Santana National Park', in M. Leach *et al.* (eds.), *Peskiza foun kona ba / Novas investigações sobre / New Research on / Penelitian Baru mengenai Timor-Leste*, Swinburne Press, Hawthorn, pp. 58–165.

Cullen, A. 2020, 'Transitional Environmentality – Understanding Uncertainty at the Junctures of Eco-Logical Production in Timor-Leste', *Environment and Planning E: Nature and Space*, vol. 3, no. 2, pp. 423–441.

Cullen, A. & Marx, S. 2015, *A Political Economy of Public Transportation in Timor-Leste*, The Asia Foundation, Dili.

DNE RDTL 2015, 'Population and Housing Census 2015 – Preliminary Results', Government of Timor-Leste, viewed at, <http://www.statistics.gov.tl/wp-content/up loads/2015/10/1-Preliminary-Results-4-Printing-Company-19102015.pdf>.

FAO 2018, 'Terminal Report: Enhancing Food Security and Nutrition and Reducing Disaster Risk Through the Promotion of Conservation Agriculture (in Timor-Leste)', Dili.

Fitzpatrick, D., McWilliam, A. & Barnes, S. 2013, *Property and Social Resilience in Times of Conflict: Land, Custom and Law in East Timor*, Ashgate Publishing, Farnham.

Fox, J. J. 2001, 'Diversity and Differential Development in East Timor: Potential Problems and Future Possibilities', in H. Hill & J. M. Saldanha (eds.), *East Timor – Development Challenges for the World's Newest Nation*, Asia Pacific Press, Singapore.

GCF 2019, 'Democratic Republic of Timor-Leste Country Programme Green Climate Fund', viewed at, <https://www.greenclimate.fund/sites/default/files/document/timor-leste-country-programme.pdf>.

Geoscience Australia 2012, 'Vulnerability Assessment of Climate Change Impacts on Groundwater Resources in Timor-Leste', pp. 1–85.

Hope, J. 2021, 'The Anti-Politics of Sustainable Development: Environmental Critique from Assemblage Thinking in Bolivia', *Transactions of the Institute of British Geographers*, vol. 46, no. 1, pp. 208–222.

Hiwasaki, L. *et al.* 2014, 'Process for Integrating Local and Indigenous Knowledge with Science for Hydro-Meteorological Disaster Risk Reduction and Climate Change Adaptation in Coastal and Small Island Communities', *International Journal of Disaster Risk Reduction*, vol. 10, PA, pp. 15–27.

Ismail, C.J. *et al.* 2019, 'Comparative Study on Agriculture and Forestry Climate Change Adaptation Projects in Mongolia, the Philippines, and Timor Leste', In P. Castro *et al.* (eds.), *Climate Change-Resilient Agriculture and Agroforestry*, Springer, Switzerland, pp. 413–430.

Kanamaru, T. 2020, 'Production Management as an Ordering of Multiple Qualities: Negotiating the Quality of Coffee in Timor-Leste', *Journal of Cultural Economy*, vol. 13, no. 2, pp. 139–152.

Kirono, D. G. C., Chiew, F. H. S. & Kent, D. M. 2010, 'Identification of Best Predictors for Forecasting Seasonal Rainfall and Runoff in Australia', *Hydrological Processes*, vol. 59.

Lasco, R. D. & Cardinoza, M. M. 2007, 'Baseline Carbon Stocks Assessment and Projection of Future Carbon Benefits of a Carbon Sequestration Project in East Timor', *Mitigation and Adaptation Strategies for Global Change*, vol. 12, no. 2, pp. 243–257.

McWilliam, A. 2001, 'Prospects for the Sacred Grove: Valuing Lulic Forests on Timor', *The Asia Pacific Journal of Anthropology*, vol. 2, no. 2, pp. 89–113.

McWilliam, A. & Traube, E.G. 2011, 'Land and Life in Timor-Leste: Introduction', in A. McWilliam & E. G. Traube (eds.), *Land and Life in Timor-Leste*, ANU Press, Canberra, pp. 1–21.

Meitzner Yoder, L.S. 2007, 'Hybridising Justice: State-Customary Interactions over Forest Crime and Punishment in Oecusse, East Timor', *The Asia Pacific Journal of Anthropology*, vol. 8, no. 1, pp. 43–57.

Mercer, J. *et al.* 2014, 'Nation-Building Policies in Timor-Leste: Disaster Risk Reduction, Including Climate Change Adaptation', *Disasters*, vol. 38, no. 4, pp. 690–718.

Miyazawa, N. 2013, 'Customary Law and Community-Based Natural Resource Management in Post-Conflict Timor-Leste', in J. Unruh & R. C. Williams (eds.), *Land and Post-Conflict Peacebuilding*, Routledge, London, pp. 511–532.

Molyneux, N. *et al.* 2012, 'Climate Change and Population Growth in Timor Leste: Implications for Food Security', *AMBIO: A Journal of the Human Environment*, vol. 41, no. 8, pp. 823–840.

Monk, K. A., Fretes, Y. D. & Reksodiharjo-Lilley, G. 1997, *The Ecology of Nusa Tenggara and Maluku*, Perplus Editions (HK) Ltd, Singapore.

Norton, J. & Waterman, P. 2008, *Reducing the Risk of Disasters and Climate Variability in the Pacific Islands: Timor-Leste Country Assessment*, World Bank, Washington D.C.

O'Lear, S. 2016, 'Climate Science and Slow Violence: A View from Political Geography and STS on Mobilizing Technoscientific Ontologies of Climate Change', *Political Geography*, vol. 52, pp. 4–13.

Oxfam 2008, 'Timor-Leste Food Security Baseline Survey Report', Oxfam Australia, viewed at, <http://aid.dfat.gov.au/Publications/Documents/et-food-security-survey.pdf>.

Palmer, L. 2010, 'Enlivening Development: Water Management in Post-Conflict Baucau City, Timor-Leste', *Singapore Journal of Tropical Geography*, vol. 31, pp. 357–370.

Palmer, L., 2015, *Water Politics and Spiritual Ecology: Custom, Governance and Development*, Routledge, London and New York.

Palmer, L. 2020, 'The Cosmopolitics of Flow and Healing in North-Central Timor-Leste', *The Australian Journal of Anthropology*, vol. 31, no. 2, pp. 224–239.

Palmer, L. & de Carvalho, D. D. A. 2008, 'Nation Building and Resource Management: The Politics of "Nature" In Timor Leste', *Geoforum*, vol. 39, no. 3, pp. 1321–1332.

RDTL 2011, 'Timor-Leste National Adaptation Programme of Action to Climate Change', pp. 1–108.

RDTL 2020, *Timor-Leste: Second National Communication Under the UNFCC*, Dili.

Ritchie, H. 2019, 'Where in the World Do People Emit the Most CO2?' *Our World in Data*, viewed at, <https://ourworldindata.org/per-capita-co2>.

Rosegrant, M. W. *et al.* 2016, 'Economic Impacts of Climate Change and Climate Change Adaptation Strategies in Vanuatu and Timor-Leste', *Marine Policy*, vol. 67, C, pp. 179–188.

RTCC 2012, 'East Timor Could Become Home to Kiribati's Climate Refugees', *Climatechangenews.com*, viewed 2 February 2021 at, <https://www.climatechangenews.com/2012/09/20/east-timor-could-become-home-to-kiribati's-climate-refugees/>.

Sagar, V.C. *et al.* 2018, *Integrating a Disaster Response Architecture in Timor-Leste: Opportunities and Challenges*, S. Rajaratnam School of International Studies, Singapore.

Sandlund, O. T. *et al.* 2001, 'Assessing Environmental Needs and Priorities in East Timor. Final Report', NINA-NIKU, Foundation for Nature Research and Cultural Heritage Research, Trondheim, Norway.

Scambary, J. 2009, 'Anatomy of a Conflict: The 2006–2007 Communal Violence in East Timor', *Conflict, Security & Development*, vol. 9, no. 2, pp. 265–288.

Shepherd, C. & Palmer, L. 2015, 'The Modern Origins of Traditional Agriculture: Colonial Policy, Swidden Development and Environmental Degradation in Eastern Timor', *Bijdragen tot de Taal-, Land- en Volkenkunde / Journal of the Humanities and Social Sciences of Southeast Asia*, vol. 171, no. 23, pp. 281–311.

Stengers, I. 2010, *Cosmopolitics*, University of Minnesota Press, Minneapolis.

TLNDMG, Australian Bureau of Meteorology & CSIRO 2011, *Current and Future Climate of Timor Leste*, Australian Government.

Trainor, C. R. 2010, *Timor's Fauna Influence of Scale, History and Land-use on Faunal Patterning*, CDU, Darwin.

Trainor, C. R. *et al.* 2008, 'Birds, Birding and Conservation in Timor-Leste', *Birding Asia*, vol. 9, pp. 16–45.

Tsing, A. 2005, *Friction*, Princeton University Press, New Jersey.

USAID 2004, 'FAA 118 / 119 Report Conservation of Tropical Forests and Biological Diversity in East Timor'.

Van der Auweraert, P. 2012, *Dealing with the 2006 Internal Displacement Crisis in Timor-Leste*, ICTJ, New York.

Wallace, L. *et al.* 2012, *Vulnerability Assessment of Timor-Leste's Groundwater Resources to Climate Change*, Geoscience Australia, Canberra.

Whyte, K. 2018, 'Indigenous Science (Fiction) for the Anthropocene: Ancestral Dystopias and Fantasies of Climate Change Crises', *Environment and Planning E: Nature and Space*, vol. 1, no. 2, pp. 224–242.

World Bank 2015, 'Arable Land (% of Land Area)', *data.worldbank.org*, viewed 17 December 2015 at, <http://data.worldbank.org/indicator/AG.LND.ARBL.ZS/ countries>.

World Bank 2009, *Timor-Leste: Country Environmental Analysis, Sustainable Development. East Asia and Pacific Region*, Washington D.C.

Yeh, E. T., 2015, 'How Can Experience of Local Residents Be "Knowledge"? Challenges In Interdisciplinary Climate Change Research', *Area*, vol. 48, no. 1, pp. 34–40.

12 Climate change governance in Viet Nam

Party leadership, decentralization, and transitions

Koos Neefjes

Introduction

This chapter analyses the recent history of climate change policy and governance in Viet Nam, including the political economy of climate change adaptation and energy transition, focusing on the Mekong Delta region.

Climate change policymaking and governance occur within the political system led by the Communist Party of Viet Nam (CPV) and a strong central government, whereas administrative decentralization to provincial authorities has been taking place in the past decades. Climate change became a critical concern in about 2007, and Viet Nam's government issued the first major policy in 2008. This was followed by CPV resolutions, central government policies and provincial actions plans, a related governance structure, and with international finance following in their wake.

Vietnamese leaders and media initially stressed the country's vulnerability to the effects of climate change, which was a justification for demanding international climate finance. Viet Nam was also becoming a major greenhouse gas (GHG) emitter because of rising fossil fuel consumption, and gradually the awareness grew that emissions reduction could be economically attractive. How climate change vulnerability and the need to reduce emissions is playing out depends on the interests and influence of the main stakeholders, which include the ministries, state-owned enterprises (SoEs), provincial authorities, non-governmental organizations (NGOs), academia, and the international aid community.

Shifts in the interests and comparative power of such stakeholders has had significant implications for public investment in climate change adaption as well as regulation of the private sector, including policies on renewable energy generation. I discuss such changes specifically for the Mekong Delta region. This region is highly vulnerable to the effects of climate change and produces significant amounts of food for international markets. The Mekong Delta region has potential for wind and solar power generation, but until recently it was slated to host 14 coal-thermal power plants by 2030, potentially leading to globally significant long-term GHG emissions.

The next section addresses the role of the CPV and the recent decentralization process. In the third section, I give an overview of climate-related

DOI: 10.4324/9780429324680-14

policies and statements in international climate negotiations, and I discuss the role of international aid agencies in shaping Viet Nam's climate policies, as it is seeking international support to adapt to climate change and reduce GHG emissions. Subsequently the governance structure is summarized, which is followed by a section on climate change adaptation and GHG mitigation in the Mekong Delta region.

I conclude that the CPV consolidated emerging views on climate change and renewable energy in its resolutions. Some ministries have increased their influence over others, and administrative decentralization has led to increased influence of provincial authorities on public investment in climate change responses. However, the changing interests and increasing power of certain stakeholder groups do not necessarily lead to optimal outcomes for climate resilience or reduced GHG emissions.

Communist Party leadership and administrative decentralization

The CPV leads the country, as set out in the constitution (SR Viet Nam 2013, Article 4). It forms the primary political space in Viet Nam. It has affiliated 'mass organizations', such as the Fatherland Front, Farmers' Union, and Women's Union, that are operating in parallel with the government administration at all levels. Almost all significant positions in the central and local authorities are occupied by CPV members. CPV members select CPV leaders. The government (the executive) is checked by representative bodies, including the National Assembly and the People's Councils at the provincial and lower levels (SR Viet Nam 2013, Article 6). Most representatives on these bodies are members of the CPV and may have functions in the government or mass organizations. Thus, many leaders have dual or even triple roles in the Party, government, a mass organization, the National Assembly, or a People's Council. CPV resolutions have provided political directions on climate change.

There has been a process of political and administrative change in the past three decades that reinforces the provincial government, which has implications for climate change governance. Vasavakul (2019) writes that the CPV's cadre rotation mechanism has accommodated elites, in which

> central leaders up for promotion are rotated to the provincial level while successful and well-connected provincial leaders are promoted to national-level positions [... and the] combination of sustained economic growth, expansion of political space, accountability, and tolerance of small-scale public protests has been another factor in strengthening regime–society legitimization.
>
> (p. 64)

McGrath (2018) analysed changes in relations between the central state and provinces over the decade to 2017. Decentralization aimed to transfer authority to provinces and manage budgets, programs, and public services more effectively. The Law on the Organization of Local Government (2015) and the Budget Law (2016) increased local influence as the country moved towards a market-oriented economy. Local authorities became responsible for managing the construction, operation, and management of public works, roads, water supply, waste and wastewater, and drainage. Tax revenues were shared between the central and local levels, increasing control over the budget by provinces. Equitization of SoEs, the rapid growth of some large domestic private enterprises (with links to political power), and considerable foreign direct investment took place in the decade to 2017. This affected state and province tax revenue, as well as relations between national and local elites and between the business sector and the state. Provinces with high revenue increased their autonomy, while poorer provinces remained in an 'asking-giving' relationship with the central government, which can include informal conditions. McGrath (2018) sees this as serving local and national leaders' mutual interests and a "disincentive for provinces to assume responsibilities and control resources". From 2016, international aid loans to projects in provinces must be partially repaid by the provinces. This makes the poorer provinces reluctant to accept investment in, e.g., climate change-relevant flood protection works, unless they could keep more revenue or get higher central government transfers. And provinces accepting the responsibility to repay loans also gain influence in investment decisions.

Interest groups must operate within the CPV-State political space to achieve their goals. Insiders include SoEs and professional associations, and outsiders include private businesses and NGOs – though some of their personnel are also CPV members. Political economy analysis looks at the power or influence of interest groups and links between them. Political ecology analysis is similar but is focused on power relations concerning resources such as land, water, biodiversity, and air, which are essential in the climate change context (Neefjes 2000). I use these concepts below, to analyse climate change policy evolution and the emergence of climate change as a determinant in public and private investment, planning, and regulation. This builds on Neefjes and Dang Thi Thu Hoai (2017), who analysed the relative power and links between stakeholders for the case of energy transition in Viet Nam. They show that stakeholders in energy transition include energy SoEs and provincial authorities, whose interests and influence have been changing.

McGrath (2018) believes that decentralization has impacted positively on the effectiveness of public investment and service delivery. However, he does not address competition between provinces over the central-level budget to construct, e.g., provincial industrial zones and airports, instead of achieving economies of scale from such infrastructure's regional functions. Provincial cooperation for regional action instead of competition could improve attractiveness for investments, e.g., in renewable energy. It could also enable adaptation actions, e.g., through better cooperation on water management during river floods. Better regional cooperation is an objective of the Mekong Delta integrated regional master planning effort, which I discuss below in 'Governance of climate change in the Mekong Delta region'.

Viet Nam's climate change policies

The primacy of climate change vulnerability

Viet Nam ratified the United Nations Framework Convention on Climate Change (UNFCCC) of 1992 and the Kyoto Protocol of 1998. It made institutional arrangements to implement its commitments under these treaties, and, with international support, it submitted its "initial communication" to the UNFCCC in 2003 (SR Viet Nam 2003). Engagement with the global process in the 1990s and early 2000s was limited to some experts in climatology and hydrology, as well as energy, industry, waste, forestry, and agriculture because GHG emissions in these sectors must be monitored. National policy and action were limited, but weather, sea levels, and river flows were monitored by the authorities, and some international cooperation took place on research of vulnerabilities to climate change (Adger & Kelly 2000; IMHEN & NCAP 2008).

Viet Nam has a history of climate-related natural disasters and has developed response mechanisms to river floods, tropical storms and storm surges in the coastal zone, and landslides in the mountains (Trần Thục et al. 2015). In the early years of the 21st century, NGOs working on disaster risk management, including the Viet Nam Red Cross (VNRC), started developing climate change adaptation projects (VNRC & NRC 2006). The Global Environment Facility (GEF) funded a variety of technical assistance projects, aiming to reduce GHG emissions by improving energy efficiency. But there were no major Vietnamese public or private investments in climate change resilience or GHG emissions reduction. Climate change policy and action were limited and partially driven by international agencies.

In 2006 and 2007, the world in general, and Viet Nam specifically, experienced a sudden rise in attention to and awareness of climate change. Major international publications reached the media and policy circles in the run-up to the 13th Conference of Parties (CoP 13) to the UNFCCC in Bali, Indonesia, in 2007. This included the 'fourth assessment' by the Intergovernmental Panel on Climate Change (IPCC 2007), a report on the economics of global climate change (Stern 2007), a World Bank model study on the effects of sea-level rise (Dasgupta et al. 2007), and the Human Development Report (HDR) 2007/08 on climate change by the United Nations Development Programme (UNDP 2007). UNDP-Viet Nam and a Vietnamese researcher had contributed case studies to the HDR (Chaudhry & Ruysschaert 2007; Nguyen Huu Ninh 2007). The full HDR 2007/08 was translated into Vietnamese and attracted a great deal of attention from policymakers, the media, and scientists. Al Gore and the IPCC were jointly awarded the Nobel Peace Prize for their efforts to build up knowledge about climate change. The above-mentioned Vietnamese researcher, Nguyen Huu

Ninh, was also a contributing author of the IPCC's fourth assessment. As a co-recipient of the Nobel Peace Prize, he was frequently invited to talk shows on national TV. The HDR 2007/08, the IPCC report, and Dasgupta et al. (2007) all showed the vulnerability of the Mekong Delta to sea-level rise and the potential effects on human displacement – this still is a recurrent headline.

Climate change negotiations and international finance

International aid agencies, diplomats, and NGOs entered dialogues with leaders of the Ministry of Natural Resources and Environment (MONRE), the Ministry of Agriculture and Rural Development (MARD), and others in the run-up to CoP 13 in 2007. Viet Nam only had a small number of officials with sufficient knowledge and language skills to engage with the international process, but this changed rapidly in subsequent years, as the government enlarged its delegation to the negotiations and international agencies trained officials. The view was that Viet Nam is particularly vulnerable to climate change effects, and developed countries must provide support because their GHG emissions caused climate change. Viet Nam argued at CoP 13 and 14 that it is among the countries most seriously affected by climate change (SR Viet Nam 2007), and "ranks among the top 5 most impacted countries by the rise of the sea level [...] and therefore needs support from [...] OECD countries [who should] set up a special support programme for the 5 most-hit countries" (SR Viet Nam 2008b).

This vulnerability ranking was based on the above-mentioned model study (Dasgupta et al. 2007), which considered 84 coastal countries affected by sea-level rise; so not all climate change effects and developing countries. The proposal was highly unrealistic in the international context, but it expressed Viet Nam's priorities and its perceived right to international finance. Media reports continued to stress Viet Nam's climate change vulnerability around CoP 15 in 2009, which the Prime Minister attended.

Viet Nam has been successful in attracting international climate finance and technical assistance compared to other developing countries. Around 2007–2008, international aid agencies started to mainstream climate change in their programs. Denmark and others supported MONRE in the formulation of Viet Nam's first national policy on climate change, the National Target Program to Respond to Climate Change (NTP-RCC) (SR Viet Nam 2008a). Denmark financed its implementation, as it had high expectations of Viet Nam's constructive contribution to CoP 15 in Copenhagen. MONRE was the principal partner, based on its leading role in hydrometeorology and related international representation to the UNFCCC, the World Meteorological Organization (WMO), and the IPCC. The NTP-RCC includes tasks of MONRE, and it hails the start of mainstreaming climate change responses into sector and province policies. It also shows the importance of MARD; because agriculture is weather dependent, MARD leads on disaster and water management, and forestry is important for climate change mitigation. MARD was quick to issue its action plan on climate change, accompanied by claims on international finance.

Also in 2008, the aid agencies of Japan and France initiated the Support Program to Respond to Climate Change (SPRCC). This concerns policy loans to the government that are disbursed after achieving 'policy actions' agreed upon between the government and donors. Other agencies joined the SPRCC later, notably the World Bank. The SPRCC is co-led by MONRE and the Ministry of Planning and Investment (MPI) and has been Viet Nam's primary source of international climate finance in the period 2009–2020. SPRCC finance goes to the general budget but is partially earmarked to climate change investments managed by MARD, other sector ministries, and provinces. The Climate Public Expenditure and Investment Review' demonstrated for the period 2010–2013 that about 90 percent of public resources for climate-relevant actions had been allocated to adaptation, mainly in water management and transport (MPI-WB-UNDP 2015). This was confirmed in a draft update for the period 2010–2020, also showing that roughly 1/3 of public climate change expenditure had been from Official Development Assistance (ODA), mainly in the form of loans (UNDP-Viet Nam 2021).

Viet Nam's vulnerability to climate change and access to international climate finance were primary drivers for climate change policymaking. International stakeholders played a role in this, using soft conditionalities, but their influence should not be overstated. The leading ministries had stakes in the financial resources, technical cooperation, and in policy formulation and implementation, whereas their views on policy did not seem vastly different from those of international agencies and experts.

Gradually increasing commitments to GHG emissions reduction

Along with most other developing countries, Viet Nam saw GHG emissions reduction as the historical responsibility of developed countries. Mitigation would therefore require international public financing and national policies expressed this expectation, including the National Climate Change Strategy (NCCS) of 2011, and the Viet Nam Green Growth Strategy (VGGS) of 2012 (SR Viet Nam 2011, 2012c). Both were followed by national, sectoral, and provincial action plans, some of which included unrealistically ambitious investment projects in the hope of securing public funding (McGrath 2018). The National REDD+ Action Plan (NRAP) was issued with an expectation to attract international finance too (SR Viet Nam 2012b).[1] The formulation, financing, and implementation of the NTP-RCC, SPRCC, NCCS, VGGS, and NRAP were all steered by the National Climate Change Committee (see the section 'Climate change governance structure and processes').

MONRE was created in 2003 and was initially uninfluential. However, based on its meteorology mandate, it has, within a few years, accessed new finance, received a lot of media attention, taken the lead in high-profile international relations and formulation of domestic policies, and has been instrumental in setting the political climate change agenda. Notably, work started in 2011 on what became CPV Resolution No. 24 on climate change

(CPV 2013). This was done in parallel with NCCS and VGGS formulation and led by senior officials-cum-CPV members of MONRE. This resolution is not detailed, but the NCCS and VGGS are consistent with it (even though issued before). In 2020, this is still the primary political document on climate change in Viet Nam, guiding all subsequent law formulation and climate change policies. And Viet Nam included a chapter on climate change in the Law on Environmental Protection, also led by MONRE (SR Viet Nam 2014; SR Viet Nam 2020b).

From 2012 onwards, the realization took hold that some GHG emissions mitigation actions could be economically attractive. Different technologies were assessed using 'marginal emission abatement cost curves' (MACCs) to inform the VGGS of 2012. This was done with international assistance and repeated for Viet Nam's Intended Nationally Determined Contribution (INDC) that was submitted to the UNFCCC (SR Viet Nam 2015). The Prime Minister announced the country's two GHG emissions mitigation targets at CoP 21 in Paris: one that will be achieved with domestic means and one contingent on international support. Afterwards, MONRE led the drafting of the Plan for Implementation of the Paris Agreement, which affects sector and provincial plans and investments (SR Viet Nam 2016b). It also initiated a review of the INDC and subsequently drafted the updated NDC, which was submitted to the UNFCCC (SR Viet Nam 2020a).

GHG emissions reduction and the energy sector

The NDC states that public resources will focus on climate change adaptation, and GHG emissions mitigation targets must be achieved mainly with private capital. Figure 12.1 presents the base-year data (2014), projections of business-as-usual (BAU) emissions, and the two mitigation targets for 2030. This shows that mitigation ambition is low, while the energy sector is responsible for the largest share in emissions. If Viet Nam achieved its 27 percent target, it would emit about 6.2 tonnes CO_2eq/capita in 2030 (population 110 million), which would be higher than the 5.4 tonnes CO_2eq/capita by the European Union (EU) in 2030 (as per the EU's NDC).

Viet Nam could further reduce GHG emissions, especially in the energy sector, which would deliver economic, social, and environmental benefits (UNDP-Viet Nam 2018). However, the energy sector was not much engaged with climate change matters. The Ministry of Industry and Trade (MOIT) guides all energy production and trade. Denmark supported a national energy efficiency program from late 2008 onwards (in parallel with the NTP-RCC), and this program has reached its third phase, which suggests that MOIT favors it (SR Viet Nam 2019a). However, until about 2017, MOIT held back on non-hydro renewable energy, which resulted in high and rising emissions in the BAU as well as low mitigation targets (Figure 12.1).

Nevertheless, gradual changes happened in the energy sector. This was because of influence of the international community, domestic NGOs, and

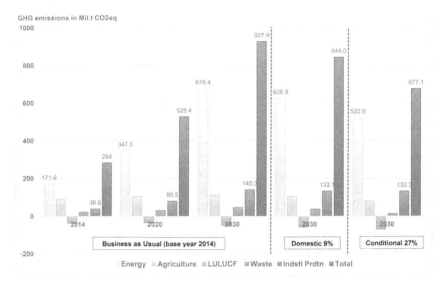

Figure 12.1 GHG emissions in 2014, BAU projections for 2020 and 2030, and NDC
targets 2030
Source: Author's graph, with data from SR Viet Nam 2020a.

scientists; the pressure of rapid growth in power demand; and because of costs reduction of non-hydro renewable energy. In the period 2014 to 2020, the EU allocated over €300m in grant finance towards achieving Sustainable Development Goal 7 on access to energy, renewable energy, and energy efficiency. This is a significant contribution towards Viet Nam's 'conditional' mitigation target (Figure 12.1). The Renewable Energy Development Strategy was issued, with EU encouragement, and the power development plan was revised (PDP7-revised), adjusting the planned total fossil fuel-based power generation downwards (SR Viet Nam 2016a).

Support policies for renewable energy were issued, including feed-in-tariffs (FiTs) on solar Photovoltaics (PV) and wind power (SR Viet Nam 2017a, 2018, 2020d). The FiTs had expiry dates for guaranteeing a good power purchase price for 20 years. As a result, by 31 December 2020, more than 100,000 rooftop solar PV systems were installed across the country with a total capacity of 9 giga-Watts (GW), plus solar power plants with a total capacity of 10 GW. Viet Nam dramatically overachieved its targets and became one of the leading countries on solar PV in just two years. But because power transmission capacity is insufficient to absorb all power during the sunniest hours, this was followed by a curtailment schedule for 86 solar power plants, which will thus not be able to sell all electricity produced (EVN 2021). This demonstrates that policy changes can lead to rapid renewable energy deployment, making ambitious emissions mitigation possible. The power development plan for the period to 2030 (PDP8) is in draft at the time of writing. It is expected to include more solar and wind power generation, and increased power transmission and storage capacity compared to PDP7-revised, but GHG emissions might increase until at least 2045 because of continued expansion of coal and gas-thermal power.

Climate change governance structure and processes

Policies and governance structures

Governance structures and processes developed along with the climate change policies outlined above. The Steering Committee for the NTP-RCC was established in 2008 and was replaced by the National Committee on Climate Change (NCCC) (SR Viet Nam 2012a). The NCCC is headed by the Prime Minister, with the minister of MONRE as Vice Chairman and several sector ministers as members (SR Viet Nam 2012a). The NCCC's aims include leading, coordinating, and monitoring national strategies and programs on climate change, green growth, energy efficiency, REDD+ and other climate change issues, and the negotiation of treaties on climate change. The NCCC's standing office, located in MONRE, is responsible for developing the NCCC's work program and monitoring the implementation of climate change responses. Ministries have assigned climate change focal points, with some specialized committees on issues such as REDD+ (in MARD) and green growth (in MPI). Several provincial and city authorities established Climate Change Coordination Offices.

The Central Committee for Flood and Storm Control (CCFSC) is a similar (and older) structure on disaster risk management, in accordance with the Law on Natural Disaster Prevention and Control. Its focus is on preparing for and responding to floods, typhoons, landslides, and droughts. There are provincial committees too, with representatives from many departments and the Women's Union. The Prime Minister also leads the CCFSC, and the MARD minister is the Standing Vice-Chair. Disaster management relates closely to climate change adaptation, and thus cooperation is needed between the Viet Nam Disaster Management Authority in MARD and MONRE's Department for Climate Change.

The NCCC and CCFSC were both set up within the CPV-state structures. Priorities in CPV (2013) include preventing and mitigating disasters related to sea-level rise in coastal areas and reducing greenhouse gas emissions. National policies and local action plans reflect these priorities. Scientific data on climate change are needed for good planning and are reasonably accessible for authorities, often summarized by the media, and discussed in many forums. The creation of the "Viet Nam Panel on Climate Change" under the NCCC has played a role in analysing and sharing climate change data and informing policies (Trần Thục et al. 2015).

Engagement with non-state stakeholders

Political discussions on climate change remain mainly within CPV-state structures. But in the context of the NTP-RCC and SPRCC, the NCCC meetings have occasionally engaged provincial leaders and the international community, offering them an opportunity for policy dialogue. Besides, MARD organized a REDD+ partnership with UN support, involving national and provincial departments, international agencies, and NGOs. MOIT is leading the Viet Nam Energy Partnership Group (VEPG), with EU support, in which energy efficiency and renewable energy policies take center stage. The focus of these partnership groups is on aid coordination, information exchange, and policy and technical dialogue, with different effects on climate change governance. International NGOs and UN agencies have worked with MARD on disaster management in working groups for decades, supporting coordination. International and Vietnamese NGO networks have jointly signed a memorandum of understanding with MONRE on regular exchange of views and information sharing, focusing on community-based climate change responses and lessons for national and international climate policy (CCWG 2018, 2019). NGOs are raising awareness, and they support local-level planning and capacity building on climate change. MONRE and MARD have thus heard opinions on the importance of community participation in adaptation, gender equality, and support to ethnic minority groups.

The ministries are leading national climate change governance. Local authorities and their climate change coordination offices are important too. These offices and some sectoral departments have led the formulation of local Climate Change Action Plans, Green Growth Action Plans, and other climate-related regulations, which have guided the allocation of provincial investment budgets on, e.g., water and disaster management (MPI-WB-UNDP 2015; UNDP-Viet Nam 2021).

Governance of climate change in the Mekong Delta region

A vulnerable region

Climate change adaptation in the Mekong Delta region is a national and even global top priority. Climate change and sea-level rise-related stresses and shocks are increasing, which has been highlighted in international publications (UNDP 2007; IPCC 2007; Dasgupta et al. 2007). Some risks to the region are enhanced by changes in the river flow and sediment load because of upstream dam building. Groundwater extraction in the region is causing land subsidence, and sand mining is increasing riverbank erosion. Major disasters that have affected the region over the past decades included Typhoon Linda in 1997, with 3,000 dead or missing people and 100,000 damaged homes; river floods in 2000, 2001, and 2011 causing hundreds of fatalities each, including many children; and severe drought and salinity intrusion in 1998, 2016, and 2020, affecting household water supply, agriculture, and aquaculture (Trần Thục et al. 2015; UNDP-Viet Nam 2016b). The Mekong Delta region is highly productive and the main source of Viet Nam's exports of rice, seafood and fruit, but it is lagging behind national average economic growth, per capita income, industrialization, and health and education standards.

Planning for resilience

There is no regional authority, but the 12 Mekong Delta provinces plus Cần Thơ as a 'centrally managed city' must cooperate for success in dealing with sea-level rise, land subsidence, saline water intrusion, drought, river floods, as well as riverbank and coastal erosion. Measures to increase resilience are reliant on policies and investments by the central government, international finance, and technical cooperation. The Netherlands mobilized experts to work with MONRE and MARD on the Mekong Delta Plan (MDP) (Strategic Partnership 2013), which follows principles that had been applied in the Netherlands (Delta Committee 2008). The MDP was debated at the national and province levels, its strategic outlook was widely appreciated, and the World Bank agreed a loan to put things in practice (World Bank 2016a). However, the MDP did not fit the Vietnamese planning hierarchy, style and required contents, and additional policy was needed. Cooperation in water management, transport, and agriculture were prioritized in a pilot on "linking" provinces "for socio-economic development in the Mekong Delta region" (SR Viet Nam 2016c). This included prioritization criteria for regional investment projects in these sectors, but there were no suitable regional project proposals and funds were distributed over the provinces. MPI and the CPV-related Southwest Steering Committee coordinated this pilot, but the latter was dissolved in 2018. There was thus a lack of success in enhancing cooperation and coordination in the context of climate change adaptation.

However, a Mekong Delta Conference in September 2017 was chaired by the Prime Minister and included several deputy prime ministers, ministers, and provincial leaders, showing unprecedented political commitment. The conference addressed issues raised in the MDP, including the need for strengthened governance. Its conclusions were reflected in Government Resolution 120 on Sustainable and Climate-Resilient Development of the Mekong Delta, which has climate change responses at its core (SR Viet Nam 2017b). Resolution 120 also states that a Mekong Delta masterplan had to be formulated, which the World Bank agreed to fund: the Mekong Delta Integrated Regional Planning (MDIRP) for the period 2021–2030. This has been drafted in consultation with ministries, provincial authorities, and other stakeholders through 2019 and 2020, and (at the time of writing in 2021) government approval is imminent (MPI 2020). It shows, e.g., that agriculture must transition from freshwater crops to commodities that thrive in brackish water, and in another zone towards freshwater floating crops and fish. Also, the Mekong Delta Regional Coordination Council was created in 2020, with the aim "to renew the regional coordination mechanism and promote sustainable development and climate change adaptation of the Mekong Delta" (SR Viet Nam 2020e). It is headed by a Deputy Prime Minister, the Minister of MPI is Standing Vice Chairman, and the ministers of MONRE, MARD and the Ministry of Transport (MOT) are Vice Chairmen, so its focus is similar to the above-mentioned linking mechanism. This 'advises and assists' the Prime

Minister, and it must propose regional investment programs and projects. Deciding on funding of regional programs and projects is where the comparative influence of different stakeholders matters, as the MDIRP is expected to be financed by central government funds and a new World Bank loan.

Throughout the MDIRP planning process, provincial stakeholders prioritized investments in roads, including a coastal road that will open areas strongly affected by sea-level rise, and a deep-sea port. It is, however, not clear how these would be climate proofed. Furthermore, expert analysis suggested that resolving inland waterway transport bottlenecks should be given high priority, whereas the deep-sea port capacity near Ho Chi Minh City is underused and could serve the Mekong Delta region's export needs until 2030. But, for various reasons, the roads and deep-sea port were prioritized. In the context of administrative decentralization and increased influence of provincial authorities on investment decisions, these local priorities may prevail, although it will take several years to be able to tell. However, the provincial stakeholders' interests may lead to decisions that are sub-optimal in terms of regional climate change adaptation, whereas emissions from road transport tend to be higher than those of water transport per unit transported.

Electricity planning

The MDIRP also includes proposals for future power generation and transmission in the region. Viet Nam's power demand is increasing rapidly, shortages are feared in the industrialized Southeast (Ho Chi Minh City), and the Mekong Delta would export power. The baseline for the MDIRP was PDP7-revised (SR Viet Nam 2016a), which included approvals of 14 coal-thermal power units for the Mekong Delta region by 2030; and the first of these started operation in 2016 (Neefjes & Hoang Anh 2017). Energy SoEs represented a strong voice in favor of coal-thermal power expansion. Reform roadmaps for energy transition had been proposed by UNDP, the World Bank, representatives of the business community, as well as NGOs (UNDP-Viet Nam 2014; World Bank 2016b; VBF 2016). This may have contributed to a gradual change in views, but until about 2016 resistance remained strong on the grounds that renewable energy is too expensive, unstable because it depends on variable sun and wind and cannot provide 'base load', and the power grid cannot cope. There were also fears of unpopular price increases. These points were countered and proposals for renewable support policies were made (GreenID 2016; UNDP-Viet Nam 2016a, 2017). There were estimates that solar and wind power costs per unit electricity would reach parity with coal and LNG power soon, especially if externalities were included in costs (VEPG 2018), but this advice was also not accepted. The resistance against renewable energy and voices in favor of more coal-thermal power were explained as 'vested interests' (Neefjes & Dang Thi Thu Hoai 2017), who would be winners from coal-power expansion. An analysis of interests and influence of different stakeholders in energy transition in Viet Nam showed that this included the energy SoEs, groups connected to MOIT, and professional organizations – all outside the climate change circles. Some of their interests were easy to understand, as coal-power plants require large transport infrastructure to supply coal and manage waste, transport takes place for the economic lifetime of the plants, and coal involves lucrative supply contracts. Furthermore, large power plants produce significant public revenue for the hosting provinces (Neefjes & Hoang Anh 2017).

However, the views regarding GHG emissions were changing, as was the case for renewable energy. There were public expressions that could not be ignored, including protests over local pollution around a coal-power plant (2015) and pollution from a steel-cum-coal-power plant in 2016, both in central Viet Nam (Neefjes & Dang Thi Thu Hoai 2017). There were more and more calls for deployment of solar and wind power, as NGOs and international aid agencies were arguing for the phase-out of indirect subsidies on coal power and promoted policy instruments to bring the cost of renewable energy down (UNDP-Viet Nam 2016a). Cautiously, the first small wind power plants were built by private investors with some participation of subsidiaries of energy SoEs (i.e., the vested interests in coal power participated early on in renewable energy). There were reports about looming power shortages affecting the Southeast as well as difficulties in (international) financing of coal-power plants, whereas solar and wind could be deployed relatively fast if the projects were bankable. Bac Lieu province in the Mekong Delta region hosted the first wind park in Viet Nam, and provincial leaders were concerned with potentially major impacts of coal power on coastal mangrove and aquaculture. They proposed to expand offshore wind power instead of building a coal-power plant as per PDP7-revised, which was accepted by the Prime Minister (Neefjes & Dang Thi Thu Hoai 2017). Agreement was reached on a large thermal power plant with imported LNG in Bac Lieu too, and in 2018–2019 other plans for coal power were considered for conversion to LNG as well (MOIT 2019; SR Viet Nam 2019b). Government Resolution 120 expressed caution on coal power and encouraged renewable energy (SR Viet Nam 2017b). Subsequently, FiTs led to spurts of solar PV deployment in 2019 and 2020, mainly in central Viet Nam but in the Mekong Delta region too. At the time of writing, additional wind power is under development in the Mekong Delta east coast provinces, aiming to complete before 31 October 2021.

At the start of the MDIRP planning process in 2019, most provinces in the region already had action plans on wind power or solar PV for the period to 2020 and beyond. This means that spatial priorities had already been identified, but achieving such plans depends on national regulation and private investment. The CPV issued Resolution 55 on Orientations for the Viet Nam National Energy Development Strategy to 2030 and Outlook to 2045 (CPV 2020). This looks favorably upon renewable energy and high-efficiency coal-power technology, which makes coal power more expensive and solar and wind power more competitive. All subsequent energy policy must follow this political guidance, including PDP8 that has been drafted in parallel with the MDIRP through 2019 and 2020. NGOs, energy professionals, and international agencies have argued that PDP8 should be a major departure from PDP7-revised including a moratorium on new coal-power plants (Neefjes & Dang Thi Thu Hoai 2017; GreenID 2017; VEPG 2018), and these stakeholders welcomed Resolution 55.

The draft MDIRP is indeed proposing no more new coal-power plants, substantial solar PV and wind power, and LNG power expansion (MPI 2020). The total power generation capacity proposed would maintain the Mekong Delta as a power exporting region. This requires increased power transmission capacity and power storage capacity in the national system. Provincial stakeholder feedback on the MDIRP proposals was supportive and provincial influence can be substantial in directing private investment. However, the draft PDP8 suggests limited solar and wind power deployment, and new coal and gas-thermal power plants in the region. The MDIRP and PDP8 are yet to be approved and only actual investment will prove the actual shift, which will take some years.

Concluding remarks

Both electricity production and climate change adaptation policy and politics, nationally and in the Mekong Delta, are about contested environments, including access to land, clean water and air, and competition for public and private investment capital. Actual investments in climate change actions are a function of the political economy, with interests of some stakeholder groups prevailing over others in regards to controlling capital, land, water, and other natural resources.

International stakeholders have had some influence on the emergence of climate change policies as well as governance structures, through ODA with subtle conditionalities (e.g., World Bank, EU, Japan) or through research and advice (e.g., UN, international NGOs). But international finance was captured by leading ministries such as MONRE, MARD, and MPI who formulated climate change and green growth policies that reflect their views and interests. MONRE co-directed investment to nationally or locally agreed priorities and reinforced their position vis-à-vis other stakeholders. MONRE and MARD have established partnerships with NGOs on climate change, disaster risk management and REDD+, and they hear opinions on the importance of community participation in adaptation, gender equality, and support to ethnic minority groups and areas. The VEPG is a partnership for policy dialogue between MOIT and (mainly) the international community. Provincial authorities, empowered by administrative decentralization and with increased influence over public investment, have voiced their views on adaptation and they co-determine private investment in power generation in their territories.

As climate change-relevant policies have been issued from 2008 onwards, governance structures have been created. CPV Resolutions 24 and 55 have consolidated shifts in societal views on the importance of climate change adaptation and energy (CPV 2013, 2020). These outcomes of processes in Viet Nam's constricted political space carry authority and are expected to be translated into polices such as the forthcoming MDIRP and PDP8. With MONRE leadership, Viet Nam has updated the NDC with increased GHG mitigation ambition. The recent renewable energy deployment suggests that the modest emissions mitigation targets might be overachieved, and the energy transition is supported by local authorities and stakeholders such as NGOs, who prefer development of renewables over coal power. But some vested interests may prevail, as the draft PDP8 suggests high and rising long-term emissions.

Strengthened governance structures such as the Mekong Delta Regional Coordination Council are needed and could both add to the influence of local authorities and improve cooperation between provinces, e.g., to enable agricultural transitions. But whether that leads to optimal adaptation and mitigation investments is not certain. Despite changes in public discourse in support of climate change adaptation and renewable energy, vested interests proposing a coal and gas-dominated electricity future of the Mekong Delta, or voices in favor of climate-vulnerable transport infrastructure might prevail.

Note

1 REDD+ = "reducing emissions from deforestation and forest degradation in developing countries; and the role of conservation, sustainable management of forests and enhancement of forest carbon stocks in developing countries" (UNFCCC document FCCC/CP/2007/6/Add.1, 14 March 2018).

References

Adger, N. & Kelly, M. 2000, 'Theory and Practice in Assessing Vulnerability to Climate Change and Facilitating Adaptation', *Climatic Change*, vol. 47, pp. 325–352.

CCWG 2018, *Greenhouse Gas Mitigation Activities in Viet Nam Including Inventory of Advocacy Materials on Energy and Other Climate Change Mitigation Activities*, Climate Change Working Group (CCWG), Hanoi.

CCWG 2019, *Community-based Climate Change Initiatives in Vietnam - Experiences of the Members of the Climate Change Working Group (CCWG)*, CARE, Malteser International, SNV, Oxfam, WWF, Hanoi [updated from the 2015 version].

Chaudhry, P. & Ruysschaert, G. 2007, *Climate Change & Human Development in Viet Nam: A case study for the Human Development Report 2007/2008*, UNDP-Viet Nam, Hanoi.

CPV 2013, 'Resolution No. 24-NQ/TW on Pro-actively Responding to Climate Change, Enhancing Natural Resource Management and Environmental Protection', Central Executive Committee of the Communist Party of Viet Nam (CPV), 3 June 2013.

CPV 2020, 'Politburo Resolution nr. 55-NQ/TW of February 11, 2020, on Orientations for the Viet Nam National Energy Development Strategy to 2030 and Outlook to 2045', Central Committee of the Communist Party of Viet Nam (CPV).

Dasgupta, S., Laplante, B., Meisner, C., Wheeler, D. & Yan, J. 2007, *The Impact of Sea Level Rise on Developing Countries: A Comparative Analysis*, World Bank Policy Research Working Paper 4136, Washington D.C.

Delta Committee 2008, 'Working Together with Water – A Living Land Builds for its Future. Findings of the Deltacommissie 2008', Delta Committee of the Government of the Netherlands.

EVN 2021, 'Circular 6/ĐĐQG-TTĐ of 4/1/2021 on Coordinated Operation to Avoid Overloading the Regional Grid on January 5, 2021', National Electrical Load Dispatch Centre of Electricity Viet Nam (EVN).

GreenID 2016, 'Debunking Renewable Energy Myths in Viet Nam' Green Innovation and Development Centre (GreenID).

GreenID 2017, 'Analysis of Future Generation Capacity Scenarios for Viet Nam', Report by Nguyen Quoc Khanh for the Green Innovation and Development Centre.

IMHEN and NCAP. 2008, 'Climate Change Impacts in Huong River Basin and Adaptation in its Coastal District Phu Vang, Thua Thien Hue Province. Final Report', Institute of Meteorology, Hydrology and Environment (IMHEN) & the Netherlands Climate Assistance Program (NCAP), Hanoi.

IPCC 2007, *Fourth Assessment Report of the Intergovernmental Panel on Climate Change*, Cambridge University Press, Cambridge, UK and New York, NY.

McGrath, T. 2018, 'Decentralization in Vietnam: Resolving Central–Provincial Relations', in D. Goldston (ed.), *Engaging Asia: Essays on Laos and Beyond in Honour of Martin Stuart-Fox. Nordic Institute of Asian Studies – Studies in Asian Topics no. 67*, NIAS Press, Copenhagen.

MOIT 2019, 'Report Nr 58/BC-BCT (04/06/2019) on the Situation of Implementation of Electricity Projects in Power Development Plan VII-Revised'.

MPI 2020, 'Report on Mekong Delta Region Masterplan for the Period 2021–2030 with Outlook to 2050', Draft for consultation of 28 December 2020, Ministry of Planning and Investment (MPI), Hanoi [in Vietnamese only].

MPI-WB-UNDP 2015, 'Financing Vietnam's Response to Climate Change: Smart Investment for a Sustainable Future. Laying The Foundation for Resilient Low-Carbon Development Through the Climate Public Expenditure and Investment Review', Ministry of Planning and Investment (MPI), World Bank (WB) and the United Nations Development Programme (UNDP), Hanoi.

Neefjes, K. 2000, *Environments and Livelihoods: Strategies for Sustainability*. Oxfam, Oxford.

Neefjes, K. & Dang Thi Thu Hoai. 2017, *Towards a Socially Just Energy Transition in Viet Nam: Challenges and Opportunities*, Friedrich-Ebert-Stiftung Viet Nam Office, Hanoi.

Neefjes, K. & Nguyen Trinh Hoang Anh. 2017, 'The Influence of Provincial Stakeholders in Determining Regional and National Vietnamese Energy Policy', Assessment for Agora-Energiewende and European Climate Foundation.

Nguyen Huu Ninh 2007, 'Flooding in Mekong River Delta, Viet Nam. A Case Study for the Human Development Report 2007/2008', UNDP, New York.

SR Viet Nam 2003, 'Viet Nam Initial National Communication Under the United Nations Framework Convention on Climate Change', Socialist Republic of Viet Nam, Ministry of Natural Resources and Environment, Hanoi.

SR Viet Nam 2007, 'Statement by the Head of Delegation at the UNFCCC-COP 13 and KP-CMP 3, Bali - Indonesia, December 03–14, 2007, HE Dr. Pham Khoi Nguyen, Minister of Natural Resources and Environment, SR Viet Nam'.

SR Viet Nam 2008a, 'Prime Minister Decision no. 158/2008/QD-TTg of December 2, 2008 on Approving the National Target Programme to Respond to Climate Change'.

SR Viet Nam 2008b, 'Speech by HE Dr. Nguyen Thien Nhan, Deputy Prime Minister of the Socialist Republic of Viet Nam at the High-level Segment of the Fourteenth United Nations Climate Change Conference (UNFCCC-COP14 and KP-CMP4)'.

SR Viet Nam 2011, 'Prime Minister Decision 2139/QĐ-TTg of December 5, 2011 on the Approval of the National Climate Change Strategy'.

SR Viet Nam 2012a, 'Prime Minister Decision no. 43/QD-TTg of January 9, 2012 on the Establishment of the of National Committee on Climate Change'.

SR Viet Nam 2012b, 'Prime Minister Decision no. 799/QD-TTg of June 27, 2012 on the Approval of the National REDD+ Action Plan (NRAP)'.

SR Viet Nam 2012c, 'Prime Minister Decision 1393/QĐ-TTg National of 25/09/2012 on Approval of the National Green Growth Strategy 2011–2020 with Outlook to 2050'.

SR Viet Nam 2013, 'Constitution of the Socialist Republic of Viet Nam. Adopted on November 28, 2013 by the Thirteenth National Assembly'.

SR Viet Nam 2014, 'The Law on Environmental Protection, no. 55/2014/QH13', The National Assembly, Socialist Republic of Viet Nam, 23 June 2014.

SR Viet Nam 2015, 'Intended Nationally Determined Contribution of Viet Nam'.

SR Viet Nam 2016a, 'Prime Minister Decision 428/QD-TTg, 18/03/2016, on Approval of Adjustments of the National Power Development Plan for the 2011–2020 Period with Outlook to 2030'.

SR Viet Nam 2016b, 'Prime Minister Decision no 2053/QĐ-TTg of 28/10/2016 on Promulgating the Plan for Implementing the Paris Agreement on Climate Change'.

SR Viet Nam 2016, 'Prime Minister Decision No. 593/QD-TTg of April 6, 2016 on the Pilot on Linking for Socio-economic Development in the Mekong Delta Region, 2016–2020'.

SR Viet Nam 2017a, 'Prime Minister Decision no 11/2017/QD-TTg of April 11, 2017 on the Support Mechanisms for the Development of Solar Power Projects in Viet Nam'.

SR Viet Nam 2017b, 'Government Resolution 120/NQ-CP on Sustainable and Climate-Resilient Development of the Mekong Delta, November 17, 2017'.

SR Viet Nam 2018, 'Prime Minister Decision no 39/2017/QD-TTg of September 10, 2018 on Revisions and Additions of Prime Minister Decision 37/2011/QD-TTg of June 29, 2011 on the Support Mechanisms for the Development of Wind Power Projects in Viet Nam'.

SR Viet Nam 2019a, 'Prime Minister Decision No. 280/2019/QĐ-TTg (2019) on the Approval of the National Program on Energy Efficiency and Conservation Phase 2019–2030 (VNEEP3)'.

SR Viet Nam 2019b, 'Prime Minister Order 1725/TTg-CN 19/12/2019 on Addition of the Bac Lieu LNG Thermal Power Centre to the National Power Development Plan'.

SR Viet Nam 2020a, 'Updated Nationally Determined Contribution (NDC) of Viet Nam'.

SR Viet Nam 2020b, 'Law on Environmental Protection, no. 72/2020/QH14. The National Assembly, Socialist Republic of Viet Nam, 17/11/2020'.

SR Viet Nam 2020c, 'Prime Minister Decision No. 1055/QĐ-TTg (2020) on Issuing the National Plan for Adaptation to Climate Change 2021–2030, with Outlook to 2020'.

SR Viet Nam 2020d, 'Prime Minister Decision no. 13/2020/QĐ-TTg of April 6, 2020 on the Mechanism to Encourage the Development of Solar Power in Viet Nam'.

SR Viet Nam 2020e, 'Prime Minister Decision no. 825/QĐ-TTg of 12/06/2020 on the Establishment and Issuing the Operation Regulations of the Coordination Council for the Mekong River Delta for the Period 2020–2025'.

Strategic Partnership 2013, 'Mekong Delta Plan: Long-Term Vision and Strategy for a Safe, Prosperous and Sustainable Delta', Strategic Partnership Between the Governments of Viet Nam and the Netherlands.

Stern, N. 2007, *The Economics of Climate Change - The Stern Review. Report for HM Treasury*. Cambridge University Press, Cambridge, UK.

Trần Thục, KoosNeefjes, Tạ ThịThanh Hương, Nguyễn Văn Thắng, Mai Trọng Nhuận, Lê Quang Trí, Lê Đình Thành, Huỳnh Thị Lan Hương, Võ Thanh Sơn, Nguyễn Thị Hiền Thuận, Lê Nguyên Tường 2015, *Viet Nam Special Report on Managing the Risks of Extreme Events and Disasters to Advance Climate Change Adaptation*, Viet Nam Natural Resources, Environment and Cartographic Publishing House, Hanoi.

UNDP 2007, *Human Development Report 2007/2008 - Fighting Climate Change: Human Solidarity in a Divided World*, Palgrave Macmillan, New York.

UNDP-Viet Nam 2014, *Green Growth and Fossil Fuel Fiscal Policies in Viet Nam - Recommendations on a Roadmap for Policy Reform*, United Nations Development Programme, Hanoi.

UNDP-Viet Nam 2016a, *Greening the Power Mix: Policies for Expanding Solar Photovoltaic Electricity in Viet Nam*, United Nations Development Programme, Hanoi.

UNDP-Viet Nam 2016b, *Viet Nam Drought and Saltwater Intrusion: Transitioning from Emergency to Recovery Analysis Report and Policy Implications*, United Nations Development Programme, Hanoi.

UNDP-Viet Nam 2017, *Ensuring Social Equity in Viet Nam's Power Sector Reforms. Policy Discussion Paper*, United Nations Development Programme, Hanoi.

UNDP-Viet Nam 2018, *Long-term Greenhouse Gas Emission Mitigation Opportunities and Drivers in Viet Nam*, United Nations Development Programme, Hanoi.

UNDP-Viet Nam 2021, *Climate Public Expenditure and Investment Review of Viet Nam*, (draft of February 2021).

Vasavakul, T. 2019, *Vietnam, A Pathway from State Socialism - Elements in Politics and Society in Southeast Asia*. Cambridge University Press, Cambridge, UK.

VBF 2016, 'Made in Viet Nam Energy Plan', prepared by Economic Consulting Associates for Viet Nam Business Forum (VBF).

VEPG. 2018, 'Proceedings EAG-EU Seminar June 26, 2018', EU Technical Assistance Facility for the Sustainable Energy for All Initiative (SE4ALL): Support the Operation of the Viet Nam Energy Partnership Group (VEPG).

VNRC & NRC 2006, 'Preparedness for Disasters Related to Climate Change', a project implemented by the Viet Nam Red Cross Society and supported by the Netherlands Red Cross/ Red Cross/Red Crescent Climate Centre / Dutch government, May 2003–December 2005.

World Bank 2016a, 'Project Appraisal Document on A Proposed Credit in the Amount of SDR 218.8 Million (US$310 Million Equivalent) to the Socialist Republic of Vietnam for a Mekong Delta Integrated Climate Resilience and Sustainable Livelihoods Project', Report No. PAD1610, International Development Association.

World Bank 2016b, 'A Financial Recovery Plan for Viet Nam Electricity (EVN), with Implications for Viet Nam's Power Sector', Washington D.C.

Part 3

Regional perspectives

13 Fossil capitalism the ASEAN way

Oliver Pye

Introduction

Southeast Asia is one of the world's regions most affected by climate change. The escalating destruction and deaths wreaked by typhoons in the Philippines, changing weather patterns, impending threats to low-lying cities, or the possibility of large-scale dieback or even extinction of the reefs of the coral triangle – all these are alarming and require an immediate and forceful response. The historical responsibility for climate change and, therefore, for the threats to the peoples and ecologies of Southeast Asia, lies squarely in the Global North and in the way the development of capitalism became synonymous with extracting and burning fossil fuels (Angus 2016; Malm 2016; Roberts & Parks 2009).

This chapter argues that this responsibility goes beyond the accumulated emissions of the Global North, but extends to the way colonialism and post-colonial structures hardwired fossil extractivism into the development trajectory of Southeast Asia. Centuries of colonial rule established path dependencies in mining, forestry, and agribusiness extractivism, as well as in fossil-fuel-based energy production and transportation systems. The armed character of national liberation struggles then led to the predominant role of the military in postcolonial development, a role that increased as the armed forces were propped up and funded in the Cold War and one that is often ignored in the political economy of climate change. The new ruling classes of the liberated Southeast Asia became, and still are, dominated by 'fossil capital' (Malm 2016).

Despite professed concern and climate mitigation targets that Southeast Asian governments regularly declare at the United Nations Framework Convention on Climate Change (UNFCCC) summits, the region's political economy is addicted to fossil capitalism. Key climate-relevant sectors are in the grip of corporate-political networks that are ultimately rooted in Southeast Asia's colonial past.[1] Looking at the key sectors of extractivist industries, energy production, and mobility systems, I show that the development model pursued by all the governments in the region is the conventional postcolonial one of increasing economic growth – of fossil-fuel based sectors. Rather than

DOI: 10.4324/9780429324680-16

a comprehensive plan for transforming key emissions-intensive sectors such as energy, industry, transport, agriculture or forestry, the region seems to be hell-bent on a path towards climate catastrophe.

Today, governments use the rhetoric of anti-colonialism and of climate justice to justify their own inaction on climate change. The article ends with a discussion of the challenges that this poses the climate justice movements across the region.

Southeast Asia and the global climate change regime

At first glance, Southeast Asian governments seem committed to effective action against climate change and the international climate architecture of the UNFCCC. The Association of Southeast Asian Nations (ASEAN), in particular, has made frequent statements expressing their concern over the heightened risk the region faces due to climate change and affirming their support for the UNFCCC process (ASEAN 2007, 2011, 2014, 2017, 2019).[2] A closer reading of these declarations, however, reveals a particular pattern. Firstly, while the commitment to the UNFCCC is reiterated, the contribution by ASEAN member countries to mitigation remains typically vague (Eucker 2014). For example, the declaration given at the 17th Conference of Parties of the UNFCCC "recalls" that "the purpose of ASEAN [is] to promote sustainable development for a clean and green ASEAN as envisioned in ASEAN Vision 2020" (ASEAN 2011). The declarations never include concrete targets, budgets, or plans of action. Secondly, the ASEAN consistently (and quite rightly) calls for mandatory action by 'developed countries' and funding by the Global North of mitigation and adaptation measures in the Global South. In contrast, governments of ASEAN member countries only commit to voluntary Nationally Appropriate Mitigation Actions (NAMAs). This is consistent with the Group of 77 (i.e., the group of 134 'developing countries' at the UN) negotiating position at the UNFCCC. Thirdly, ASEAN often makes specific reference to the emissions-trading program, Reducing Emissions from Deforestation and Forest Degradation (REDD+).

A look at the Intended Nationally Determined Commitments (INDCs) that individual Southeast Asian governments submitted to the UNFCCC in the context of the Paris Agreement (see Table 13.1) reveals a similar pattern. The statements are filled with a general affirmation of the need to combat climate change and at first glance, the commitments seem ambitious, particularly for 'developing countries'. Thailand promises a reduction of 20 percent, Indonesia of 29 percent, Malaysia and Singapore 35 percent and 36 percent by 2030. If financial and technological help from the 'developed countries', as agreed upon in the Paris Agreement, is forthcoming, then the reductions would be even more significant, with Viet Nam's commitment rising from 8 percent to 25 percent and the Philippines declaring a 70 percent reduction by 2030, for example.

Table 13.1 Intended Nationally Determined Contributions (INDC) of selected Southeast Asian Countries as part of the Paris Agreement

Country	INDC pledged reduction	Conditional on financial transfer	Qualification	Real pledged increase[3] in INDC (MTCO$_2$e/a)
Indonesia	29%	41%	of BAU	200
Malaysia	35%	45%	of emissions intensity	470
Thailand	20%	25%	of BAU	219
Viet Nam	8%	25%	of BAU	250
Philippines		70%	of BAU	29
Singapore	36%		of emissions intensity	25
Cambodia	27%		of BAU	9

Source: INDCs submitted to UNFCCC

However, a second reading of the INDCs uncovers their Orwellian nature. The qualification needed is the question: reduction in relation to what? In most cases, the reference point is the business-as-usual (BAU) scenario for 2030. For example, the 29 percent reduction promised by Indonesia is a reduction of the planned *increase* of emissions by 2030 to 2.9 *billion* tonnes carbon dioxide equivalent per year (CO$_2$e/a) in the BAU scenario. This 29 percent reduction would mean that Indonesia plans to increase its annual GHG emissions by 200 million tons of carbon dioxide equivalent per year (MtCO$_2$e/a) compared to the 2005 baseline of 1.8 billion tonnes. Viet Nam's reduction of 8 percent is in relation to the expected 787 MtCO$_2$e/a in 2030. An 8 percent reduction would be 724 MtCO$_2$e/a; in other words, this works out as an increase by 293 percent compared to the 2010 baseline of 247 MtCO$_2$e/a (Government of Viet Nam 2015).

Malaysia and Singapore, in contrast, do not promise to reduce emissions at all, neither in relation to a BAU scenario nor in real terms. Instead, they plan to reduce their 'emissions intensity of GDP' by 45 percent (Government of Malaysia 2015) or by 35 percent (Government of Singapore 2015) by 2030. This means that Malaysia has officially promised to increase its emissions by 759 MtCO2e per year compared to the reference year 2005, within 15 years. Similarly, Singapore pledges to increase its emissions every year up to 2030, in which year the emissions are expected to "peak" and "stabilise at around 65 MtCO$_2$e". This means that Singapore will be producing 25 MtCO$_2$e *more* GHG than in 2005 – an increase of over 60 percent.

The military in the room

The military is the elephant in the room when it comes to climate change. One strand of environmental sociology has started to look into the "treadmill of destruction" caused by the military (Clark & Jorgenson 2012). Various studies have begun to calculate the military's direct and indirect contribution to GHG emissions (Crawford 2019). Even when wars are not included in the calculation, the impact is considerable (Belcher et al. 2020). Although no one has specifically looked at the GHG emissions of the military in Southeast Asia, the general patterns are similar. Emissions directly related to the military include those embedded within the military infrastructure, weapons and fleets of vehicles, including warships and warplanes – in addition to the fuel and the workforce used in everyday operations, manoeuvres, or military crackdowns.

The influence of the military in most Southeast Asian countries has its roots in colonialism and imperialism. Armies evolved out of national liberation struggles against colonial powers (e.g., Viet Nam, Indonesia), or were formed by colonial powers themselves before gaining independence (Malaysia, Myanmar, the Philippines, Singapore). During the Cold War, they were bolstered, financed, and weaponized by their relevant superpower. For different reasons, the military took on an active role in postcolonial Southeast Asia's political and economic development. Most countries were under direct and indirect military dictatorships for long stretches of time (Thailand, Viet Nam, Laos, Myanmar, Cambodia, the Philippines, Indonesia).

Fossil fuels are materially inscribed into military operations and infrastructure. The large fleets of vehicles run on petrol or diesel, and armies need roads, ports, and airports for their operations. Producing the vehicles and weapons is dependent on extractive industries, while the necessary infrastructure consumes large amounts of concrete. Both industries are major contributors to global emissions. There is no military based on renewables nor any military that could operate in a carbon-neutral way. The military-industrial complex, in which the military apparatus and key industrial sectors are co-dependent and dominate much of the state apparatus as a whole, needs to be taken into account when cross-examining climate policies across Southeast Asia. Climate change commitments given by governments dominated by this complex cannot be taken seriously, while any country with a serious strategy towards a zero-carbon world would have to get rid of its military. Therefore, military spending should be characterized as emissions and climate change-inducing investment that could and should be diverted to investment into transforming key sectors of the economy to a carbon-neutral production. These funds are substantial, as outlined in table 13.2.

Extractivist development

The military has played an important role in developing the strategy of industrial development in Southeast Asia. This strategy has been, and still is,

Table 13.2 Largest military expenditures in Southeast Asian countries

Country	Military Spending 2018 ($USD Million)
Singapore	10,835
Indonesia	7,557
Thailand	6,876
Viet Nam	5,451
Malaysia	3,470
Myanmar	3,155
The Philippines	2,843
Cambodia	543

Source: https://www.sipri.org/databases/milex;https://tradingeconomics.com/myanmar/military-expendi ture, https://tradingeconomics.com/Viet Nam/military-expenditure

one of extractivism, i.e., "a development model focused on the extraction of and dependence on raw materials for export" (Brad et al. 2015, p.101). This is also rooted in the region's colonial history, which established the pathways that are still trodden today.

For the United States, their former colony, the Philippines, were literally a gold mine. By the 1930s, hundreds of mining companies and thousands of mines were operating across the country. After independence, the nationalist government focused on nationalizing and regulating (not replacing) the lucrative industry. The Marcos military regime developed an 'active, state-led' mineral regime in which state capital, military generals, and Marcos himself held major shareholdings in large mining corporations with minority stakes by foreign mining companies. After the dictatorship was toppled, the new, more democratic regime was still closely aligned to the extractivist development model. It liberalized the industry, ushering in a new wave of foreign direct investment in mining, in which neoliberal governance techniques dovetail with active repression by the military (Camba 2015, 2016). In the words of Delina (2021, p.2), this led to "a development trajectory predicated upon neoliberalism and punctuated by extractivism" in which coal had primacy. Accordingly, the Department of Energy's Coal Roadmap 2017–2050 plans to expand domestic coal production rather than phase it out (ibid.).

A similar pattern can be observed across the region. Colonial regimes established the extractive industries, which were then taken over and expanded by state capitalist and military regimes. In Indonesia, the intimate relationship between the mining giant Freeport and the Suharto regime is notorious (Leith 2002). In Myanmar, the military junta "ruled over the country's extractive industries [...] for over half a century" (Hatcher 2020, p.330), while lucrative oil and gas deals with corporations such as Unocal and Total (for example in the notorious Yadana Gas Pipeline) have lubricated the military-industrial complex (Larsen 1998). In both countries, liberalized investment regimes (for example with the infamous Omnibus law passed this year in Indonesia) are characterized by the continued influence of state capital and state institutions in close partnership with global mining and fossil fuel capital.

Today, Southeast Asia is a major producer of minerals such as gold, nickel, tin, copper, bauxite, and bismuth (Hatcher 2020). Mining for metals creates substantial emissions in its own right (aside from the other disastrous environmental impacts) and is then used in industries that are key planks of the fossil-fuel-industrial complex, such as the automobile and the weapons industries. Another climate-relevant industry that is reliant on minerals is cement, which, because of the chemical process and high energy requirements involved in its production, accounts for 7 percent of all emissions worldwide (Ali, Saidur & Hossain 2011). Cement producers are often large-scale conglomerates with significant political clout, such as HeidelbergCement in Indonesia or the Siam Cement Group in Thailand, whose biggest shareholder is the Crown Property Bureau (Porphant 2008).

The collusion between industry and politics is particularly shameless in Indonesia's coal industry. A report by the anti-mining coalition #Bersihkan Indonesia (2018) revealed extensive political patronage networkers between the big coal corporations and leading generals and politicians. Aside from countless local politicians and district heads who have received bribes from the coal industry in exchange for permits or sanitized EIAs, coal sits in the highest echelons of power in the Indonesian state. For example, Aburizal Bakrie, the former Golkar Party (i.e., the political party of the Suharto regime) chairman and former 'Minister for People's Welfare' (sic) owns the Bakrie Group, whose company, Bumi Resources, is the largest producer of thermal coal in the country. The ex-general and business partner of President Joko Widodo, Luhut B. Pandjaitan, who controls the mining company Kutai Energi, is Coordinating Minister for Maritime Affairs, and, in that role – how convenient – oversees the Ministry of Energy and Mineral Resources. To add insult to injury, the ex-commander of the 'Strategic Reserve Command' and Suharto's ex-son-in-law Prabowo Subianto, accused of severe human rights abuses, has now been appointed Minister of 'Defense' by Widodo. Prabowo is the leader of the fascist Great Indonesia Movement (Gerindra) Party and a major shareholder of the Nusantara Group, which now controls significant coal reserves in East Kalimantan.

It would be no exaggeration to claim that the coal industry 'owns' the three main political parties in Indonesia. Any claims of plans to reduce emissions by the Indonesian government should be taken with a pinch of salt. The coal industry has every intention of continuing with its incredibly profitable and ecologically destructive business model. President Joko Widodo has publicly stated his concern that "without proper management, Indonesia's coal reserves will only last for the next 83 years" (#Bersihkan Indonesia 2018, p.346). This statement implies that his government intends to carry on digging up and burning Indonesia's vast coal reserves in the foreseeable future and preferably well into the next century. Meanwhile, Singapore's largest financial institutions are bankrolling the extractivist boom, including coal (Brown & Spiegel 2017).

Another significant contributor to climate change is industrial, mono-culture agriculture. This can also be understood as a form of extractivism, by which biomass is extracted and exported, draining the soil of nutrients and carbon rather than preserving the nutrient cycle. The colonial roots of this form of agriculture is clear in the tobacco, sugar, and rubber plantations across the region. The obvious example is the palm oil industry (Brad et al. 2015). Because of its high-input usage of crude-oil-derived fertilizers and pesticides, and because of the methane emitted in the milling process, palm oil is already a net emitter of GHG (Pye 2018). Once the direct and indirect land-use changes associated with its continued expansion into forest areas are factored in, the sector becomes one of the most significant sources of emissions in the region. This is particularly the case for the expansion on peat-lands and the associated forest fires and haze that regularly plagues Southeast Asia (Alisjahbana & Busch 2017). The 'Palm Oil Industrial Complex' (Pye 2009, p.86) is embedded within the Malaysian developmental state, while in Indonesia, the industry wields significant political influence via extensive patronage networks (Varkkey 2016). Authoritarian regimes are busy promoting palm oil plantations in Myanmar, Thailand, Cambodia, and the Philippines. Other agribusiness activities, such as the contract farming systems of maize and broiler production run by the multi-billion dollar corporation Charoen Pokphand (Pananond 2001), are creating similar problems.

Burning coal

Extracting fossil fuels is closely related to energy production. In Southeast Asia, the energy sector gives ample evidence that governments are not just dragging their feet on climate change but are actively promoting and investing in fossil fuel-based energy production. Overall, Southeast Asia is "dependent on three main carriers of energy, namely, oil, gas, and coal" (Bakhtyar et al. 2013, p.513). In Thailand, only 4 percent of energy was produced from renewable sources in 2013, the rest from fossil fuels. In the Philippines, seen as the leading proponent of renewable energy in the region, coal, oil, and gas accounted for 68 percent of the total installed capacity of power generation in 2003 (Das & Ahlgren 2010). Similarly, in Viet Nam, power is generated mainly by coal and gas (62 percent). While the contribution of hydropower is substantial at 38 percent of capacity, this is problematic for other reasons (Das & Ahlgren 2010).

The energy sector as a whole is not only moving too slowly in the direction of renewables and in reducing emissions; instead, the emissions from this sector are expanding, and the growth rate of that expansion is also growing. For example, Thailand's energy-related emissions rose at an annual rate of nearly 8 percent between 1980 and 2010, from 33.56 $MtCO_2$ to 278.49

MtCO$_2$, and are expected to double again by 2030 (Aumnad & Teeradej 2013). Viet Nam is also highly reliant on coal, oil, and gas, which is predicted to rise in tandem with economic growth. It has no plan for any significant role for renewables; instead, the government is planning to reduce their role still further to 10 percent by 2050. Billions are being invested in expanding electricity generated by oil and gas (Do & Sharma 2011). Across ASEAN, energy-related emissions are expected to rise by 60 percent by 2025 (International Renewable Energy Agency & ASEAN Center for Energy 2016).

The region is addicted to coal. Coal-generated electricity went from 53 billion kWh in 2000 to 120 billion in 2007 in the Philippines, i.e., it trebled within seven years. In Thailand, the figures were 75 and 140 billion respectively, in Malaysia 40 and 90, in Indonesia 35 and 60 billion (Bakhtyar et al. 2013, p.510). Is this what a transition to renewables looks like? If the Paris Agreement target is to be met, energy production from coal would have to be significantly reduced across ASEAN by 2030 and nearly phased out by 2040. Instead, as Table 13.3 shows, nearly all Southeast Asian governments plan to *expand* the capacity of coal-fired power plants.[4]

Indonesia, whose thermal coal reserves are more than one hundred billion tonnes (Gunningham 2013, p.186), seems to be intent on digging them all up to burn them. Coal corporations produce a staggering 500 million tonnes of coal every year, supplying one third of the world's coal exports (Dutu 2016, p.515). Since 2008, the government has installed 20,000 MW new coal-fired power plants. Could the billions of dollars used to pay for coal plants not have been invested instead into solar, geothermal, and wave energy?

There is no evidence that the region, as it develops, will become any more environmentally sustainable, as suggested by Ecological Modernization Theory (Mol, Spaargaren & Sonnenfeld 2014). Generally, the more developed the country, the more emissions per capita it produces. There is also an inverse correlation between prosperity measured as per capita GDP and the contribution of renewables to energy production (Bakhtyar et al. 2013, p.513). In Singapore, hardly a developing nation, renewable energy's contribution to the total produced was "about 0%" in 2013, with 80 percent coming from gas, i.e., fossil fuels (Bakhtyar et al. 2013, p.512). Since then, solar power capacity has risen to 290 MW, a paltry 2 percent of total power production (Energy Market Authority 2020).

Table 13.3 Planned expansion of coal power plant capacity in selected Southeast Asian countries

Country	Thailand	Indonesia	Philippines	Viet Nam	Cambodia	Myanmar
Planned expansion of coal power plant capacity	57%	62%	78%	183%	471%	3188%

Source: Climate Analytics 2019a, p.37.

The addiction to fossil fuels is another path dependency rooted in the political economy of power production in the region and in the "evolution of relationships among the government, state-owned utilities, the private sector, and civil society" (Chuenchom & Greacen 2004, p.517). In most Southeast Asian countries, we see a combination of a state energy generating company within a partly liberalized energy market, both of which are entangled in the history of military regimes (Simpson & Smits 2019; Morrow 1975). As the example of the Energy Generating Authority of Thailand (EGAT) shows, the path dependency of these relations, arising out of militarized developmental states, has led to an unhealthy dependence on fossil fuels. EGAT was created in the 1960s by military governments and with the support of the World Bank. As a bureaucratized state corporation, its primary purpose was to supply enough energy for industrial development. To this end, it copied the 'modern' development model of the former colonial powers, with large-scale fossil fuel power plants and a national grid. Thailand received technology, knowledge, and loans from the West to do so. EGAT's bureaucrats, engineers, and managers have been following this path for over 60 years. Across Southeast Asia, similar state energy corporations pursued the same direction.

The liberalization of energy markets in the 1980s and 1990s has now created a peculiar combination of a state-run bureaucratic energy corporation dominating an energy market which includes independent power producers (IPPs) (Smith 2003). EGAT managers and former managers bought up shares of the partly privatized EGAT and of the new IPPs. The IPPs, often privatized coal or gas power plants, then sell their energy to EGAT via power purchase agreements (PPAs). They are awarded guaranteed profits for long contract periods of ten or 25 years (Chuenchom & Greacen 2004). For the privately owned power plants, this is a gravy train. If they have guaranteed profits to burn coal for the next 25 years, they will not magically switch to renewables. Conversely, for the upper-level management of EGAT, the control over PPAs is not only an easy way to ensure power supply but concentrates power and capital in their hands. They will also resist plans to decentralize or democratize power production or any fundamental shift towards renewables.

Untapped potential

These vested interests are preventing an energy transition that could be commenced immediately. Southeast Asia has vast and untapped renewable energy sources. Wind and solar power alone could cover today's energy consumption and future demand several times over. Besides, the region has great potential in geothermal and ocean energy. Switching to renewables would be not only able to provide sufficient and clean energy for the working classes on and off-grid, but also new jobs in greener industries.

Solar energy potential is high across the region, much higher than current and projected energy consumption. Solar radiation rates are between 4 and 5 kWh/m2/day in all Southeast Asian countries, making decentralized and off-grid or 'smart-grid' electricity production via solar panels a viable option. If the goal was to provide electricity to the population, of which many households live off-grid, then a comprehensive solar panelling roll-out program would be the way to go. Thailand, for example, could produce 1557 TWh from solar panelling on 1.5 percent of its total area – i.e., eight times its current electricity consumption. Despite the potential, in Thailand, electricity produced from solar energy was virtually non-existent in 2009 and a modest 2.6 GW in 2017 (Climate Analytics 2019b; Nares & Somchai 2016). Similarly, in the Philippines, 1.5 percent of land covered with solar PV could produce 792 TWh of electricity, i.e., ten times consumption in 2016 (Climate Analytics 2019c). In Indonesia, solar photovoltaic panels could provide over 500 GW capacity. The current installed capacity stands at just 58 MW (Climate Analytics 2019d).

Wind energy potential is also high, the technology to tap it is tried and tested and relatively straightforward to produce. In combination with solar energy, wind can offer another off-grid renewable energy solution. The Philippines, in particular, has the technical potential to produce 7,404 MW from wind power but has only realized 427 MW, or 6 percent (Bakhtyar et al. 2013, p.510, Climate Analytics 2019a). Thailand has an overall potential of 150 GW large-scale wind turbines on land and further potential in the Gulf of Thailand (Sakkarin & Somchai 2014). Offshore wind alone could provide a fifth of the current overall capacity, but, so far, Thailand has only installed 628 MV onshore and nothing offshore (Climate Analytics 2019a).

In addition to 'traditional' renewables, the Philippines and particularly Indonesia have enormous largely untapped geothermal resources. Indonesia "boasts an estimated 40 percent of the world's geothermal energy reserves" (Dutu 2016, p.519) but uses less than 5 percent of its potential. This is despite the fact that geothermal power plants have lower levelized costs of electricity than fossil fuel plants (EIA 2020, p.7). The technically feasible potential is estimated at 30 GW (IRENA 2017), or half of the current capacity of total power production in Indonesia.

Largely unexplored are the possibilities of Ocean Thermal Energy Conversion (OTEC, which makes use of the temperature difference between the warm upper layers of the ocean and the colder deeper layers), tidal energy, and wave energy. These are particularly interesting for insular Southeast Asia. Indonesia has a theoretical potential of 57 GW for OTEC, 160 GW for tidal current, and 510 GW for wave energy. In practice, this could translate as 43 GW OTEC, 4.8 GW tidal, and 1.2 GW wave energy – or most of the current capacity. The Philippines' potential ocean energy is estimated at 170 GW (Quirapas et al. 2015), significantly more than overall capacity predictions.

Despite the huge potential for renewables in the region, investment in transitioning towards a carbon-neutral power production is exceedingly modest. This is not because of lack of funds, nor because Southeast Asian countries lack trained engineers. In fact, wind, solar, and OTEC are all technologically simpler than coal plants to build, and, as decentralized power sources, require less capital to initiate.[5] There are many examples for decentralized off-grid renewable energy solutions that already exist in the region.[6] But because governments, in the name of development, are committed to increasing overall energy production without significantly investing in renewables, their role will remain marginal in the following years and decades. The share of renewables in electricity production actually *decreased* between 2000 and 2016, down to 16 percent in Indonesia, 36 percent in Viet Nam, and 24 percent in the Philippines (Climate Analytics 2019a, p.61). Instead, Southeast Asian governments are promoting and investing in power generation based on extracting and burning fossil fuels.

Traffic jam

While energy is the flagship sector of emissions, the transport sector is also a significant source. Key climate-relevant challenges are the unchecked growth of the Southeast Asian air network (rather than developing a network of fast, transnational train connections) or the problem of developing carbon-neutral shipping. However, the most urgent transition across the region is shifting to carbon-neutral mobility systems in urban areas. Most emissions are caused in or by cities (Sovacool & Brown 2010). Road transport is responsible for nearly one quarter of global emissions, and road transport accounts for over 90 percent of transport sector emissions in Indonesia, Malaysia, Thailand, and Viet Nam (Andong & Sajor 2017). Across the region, transportation is a major consumer of primary energy, and because this is predominantly fossil fuel, a significant contributor to emissions (Ahanchian & Biona 2014, Aumnad & Teeradej 2013).

The refusal of the ruling classes in Southeast Asia to pursue a development path that does not just copy the former colonial powers can be seen in the transport sector. Across the region, transport is predominantly organized as a road-car-based system, with no plans to change this in any significant way. One reason is that roads, trucks, and tanks are hard-wired into the infrastructure of the military-industrial complex. More generally, society's motorization and the development of automotive industries are seen as national goals to be aspired to (Han 2010). Motorization is linked to wealth, and wealthy people – those running society – want to own cars.

Industrial development in Southeast Asia has been synonymous with the motorization of society. Rapid urbanization was paralleled by considerable investments in car and road infrastructure (Han 2010) instead of public transport systems. The urban sprawl of industrializing megacities and the factory belts around Bangkok, Manila, Ho Chi Minh City, Phnom Penh, or in JABODETABEK in Indonesia has led to huge problems of traffic congestion and air pollution. City planners reacted to traffic jams by building more roads. The more affluent urbanites responded by buying more cars (Sodri & Garniwa 2016; Wismadi et al. 2013; Sovacool & Brown 2010).

In Bangkok, the transport sector accounted for the most significant share of energy consumption and for more than half of total emissions; both are expected to rise significantly in the foreseeable future (Aumnad 2010). The number of vehicles has been continuously growing, from 6 to 10 million between 2007 and 2015, with emissions rising in tandem. The huge number of cars creates traffic jams, air pollution, and serious human health risks (Penwadee et al. 2017). In Jakarta, the number of motor vehicles increased from 11 to 15 million between 2010 and 2017 (Both et al. 2013). Fuel consumption rose from under 5 billion litres in 2001 to over 20 billion litres in 2014, with emissions from transport accelerating from under 10 billion to 45 billion tonnes CO_2e/a over the same time frame (Sodri & Garniwa 2016, p.732). The number of registered vehicles in Metro Manila increased from 5 million in the 1980s to nearly 16 million in 2006. Emissions in the same period rose from 5 to 18 $MtCO_2e/a$ (Andong & Sajor 2017) and are expected to almost treble to 28 $MtCO_2e/a$ by 2040 (Ahanchian & Biona 2014). Public transport, as it currently stands, is not necessarily a solution. In Manila, for example (and the situation is similar in many Southeast Asian cities), 'public' transport consists of privately owned and largely uncoordinated jeepneys and buses using inefficient and polluting second-hand diesel engines. The latter contributed to seven times the amount of emissions than private cars (Andong & Sajor 2017).

The business, military, and bureaucratic elites that live in major cities drive cars and want to drive cars. However, from a working-class perspective, a transport system in cities based on private vehicles and roads is the worst possible scenario. Most urbanites in Southeast Asia do not own a car and are dependent on public transport (Andong & Sajor 2017; Wismadi et al. 2013). As the urban sprawl continues and megacities expand, and as rents increase in the inner cities, workers have to commute longer distances (Andong & Sajor 2017). They not only have to spend a substantial portion of their day traveling to and from work and in traffic jams, the journey to work is also a major health hazard. Southeast Asian cities such as Manila (Ahanchian & Biona 2014), Jakarta (Both et al. 2013, p.965), and Bangkok (Fold et al. 2020) are amongst the most polluted in the world, with high concentrations of particulate matter (PM2.5) that is linked to severe illnesses including "allergic disorders, asthma, cognitive deficits, brain abnormalities, decreased lung function, cardiovascular disease, cardiopulmonary disease, and death" (Both et al. 2013, p.965). PM2.5 is a killer (Fold et al. 2020) – and, as its cause, road transport is too (Khan et al. 2016).

More recent investments in expensive metro and skytrain options are but auxiliary measures that are subordinated to the interests of private vehicle owners and to city planning based on road networks. The problem is evident in the most advanced city of the region, Singapore, which operates a first-rate public transport system and has developed a renowned management of car population and use with a vehicle quota system and road pricing. Singapore is therefore often promoted as a "sustainable, safe, smart" city (Haque, Chin & Debnath 2013, p.20) and as "a model for Asian countries and cities" (Han 2010, p.320).

However, if Singapore indeed became the model, then this would be bad news for climate change. Singapore has the highest per capita carbon emissions globally, and 17 percent of these emissions come from the transport sector (Sovacool & Brown 2010). Singapore does not have a strategy to *overcome* a mobility system based on cars but has tried to "balance motorization with a public transport system" by encouraging "parallel growth in motorization and transport" (Han 2010). While expanding public transport, between 1996–2005, the city also invested heavily in *more* roads, and car ownership increased by 70 percent over the same period. Transport emissions increased between 2000 and 2010 (Su, Ang & Li 2017). More efficient engines have achieved a slight decrease in emissions in recent years. Singapore intends to further green its vehicle fleet by gradually expanding the number of electric vehicles. But replacing fuel-run motor vehicles with electric vehicles is not the way forward (García-Olivares, Solé & Osychenko 2018). They use significantly more energy per capita than public transport, and their production fuels the extractive industries. If electricity is produced with fossil fuels (see above, and cities such as Singapore have no intention of shifting to renewables), they are no cleaner than fuel-based transport. Although modern and attractive as a 'global city', Singapore's transport policy is not part of a transition to a carbon-neutral city or a "sustainable transport city" (Newman 2010, p.67). A similar pseudo 'greening' of transport can be seen across the region in the promotion of 'biofuels', which reinforce the emission-high agribusiness development (i.e., palm oil) described above.

Today, it is possible to build or adapt cities to be carbon neutral and based 100 percent on renewables (Newman 2010). Car-free cities based mainly on trains, metro systems, and trams, in combination with light electric vehicles (such as electric rickshaws and electric scooters) would not only ensure carbon-neutral transportation but would also be largely pollution-free and would put an end to traffic jams and pollution (García-Olivares, Solé & Osychenko 2018). This would be in the interest of the large majority of (working class) urban dwellers. But no regime in Southeast Asia is currently moving in this direction.

The climate justice movement

Given the many problems connected to the expansion of fossil fuel-based development, it is no surprise the climate politics have become highly contested in the region. The Climate Justice Movement (CJM) in Southeast Asia is made up of a significant number of diverse organizations and movements. Larger climate justice movements exist in the Philippines, Indonesia, and Thailand, represented by coalitions such as the Philippine Movement for Climate Justice, the Indonesian Civil Society Forum on Climate Justice, or the Thai Working Group for Climate Justice. A certain amount of regional coordination occurs within the ASEAN Peoples' Forum or for larger events such as the Climate Summit in Bali in 2002. Many Southeast Asian groups are also involved in the global CJM, with two NGOs, the Malaysian-based Third World Network, and the Bangkok-based Focus on the Global South, playing a prominent role, particularly in the Climate Justice Now! (CJN) network.

These larger coalitions are embedded within networks of local struggles that emerge out of some of the contradictions of the extractivist development model. As fossil fuel, mining, and plantation companies expand into the "resource frontier ... accumulating by dispossessing" (Harvey 2003), they are met by territorial struggles over resources, pollution, and land (Martinez-Alier et al. 2016, Chatterton, Featherstone & Routledge 2013). Anti-coal protests in Indonesia, for example, have usually taken the form of local opposition by villagers to land grabs and pollution in concession areas, which are coordinated by national organizations such as the anti-coal network JATAM or the environmental justice network WALHI (Brown & Spiegel 2019). Because a lot of these conflicts take place 'at the margins', they often impact indigenous peoples, who are then at the frontline of struggles against extractivism, including fossil fuels. These struggles are coordinated by indigenous federations such as the Cordillera Peoples Alliance (CPA) from the Philippines and Aliansi Masyarakat Adat Nusantara (Indigenous Peoples' Alliance of the Archipelago, AMAN) from Indonesia (Manaysay 2020). Therefore, the social base of the climate justice movement in Southeast Asia can be characterized as predominantly "small farmers or fisherfolk, indigenous peoples or forest dwellers" (Bullard & Müller 2012, p.59, Tramel 2018, Darlington 2014).

In the context of the politicization of climate change, those activist networks engaged with local struggles against extractivism reframed their political messaging as one of climate justice (Manaysay 2020). Globally, the CJM can be seen as the fusion between environmental justice struggles and the anti-globalization movement (Martinez-Alier et al. 2016, Bullard & Müller 2012). By converging different movements against land grabbing, mining, or monoculture agriculture, the CJM could create a powerful narrative that could unite disparate movements around a common cause (Tramel 2018). This narrative started with the concept of climate debt, i.e., that historically accumulated and current per capita emissions were and are mainly produced in the Global North, whilst climate vulnerabilities are strongest in Global South (Saraswat & Kumar 2016). Southeast Asia is no exception. Climate Justice Now! called first and foremost for the repayment of this climate debt historically caused by the former colonial powers, and for a meaningful reduction of consumption by the Global North and 'Southern elites'.

At the same time, those most vulnerable to the impact of climate change, seen as indigenous peoples, small-scale farmers, and fisherfolk, were given center-stage in the alternatives to the current development model. The way out of the climate crisis, in addition to renewable energy, was via indigenous rights, "peoples' sovereignty over energy, forests, land and water" (Bullard & Müller 2012, p.56), and small-scale farming, aka food sovereignty. It is in these rural areas, particularly regarding agriculture, where the CJM alternatives are most convincing. The concept of food sovereignty, for example, is based on ecological agriculture and on improving the humus content of the soil so that more carbon is stored than emitted (Claeys & Delgado Pugley 2017).

Whilst the CJM has had some success, as can be seen with some local victories and the coal moratorium in the Philippines (Delina 2021), the defeat of the planned coal power plant in Sabah, Malaysia, or the moratorium on peatland conversion in Indonesia, it faces many challenges, of which I would highlight three. Firstly, climate justice activists face severe repression in many Southeast Asian countries. When vested interests are threatened by local opposition, the importance of the military becomes evident when leading activists are threatened or murdered (Delina 2020). Tragically, the number of 'environmental defenders' murdered by military or paramilitary bodies is on the rise. In 2016, for example, environmental justice activists were killed in Myanmar, Malaysia, Thailand, and the Philippines (Global Witness 2017).

Secondly, the framing of climate justice "as primarily a struggle between global Northern and Southern states within UNFCCC process" (Chatterton, Featherstone & Routledge 2013, p.606) inadvertently strengthened the ruling classes in the region. Within the negotiations, climate debt and climate justice became the negotiating position of the G-77 group in which the demand for decisive action and binding commitments by the industrialized countries was coupled with the 'right to develop' in the sense of a 'fair share' of polluting industries (Roberts & Parks 2009). By adopting the *"nation-state* logics" (Roosvall & Tegelberg 2015, p.40) of the negotiations, governments in Southeast Asia could portray themselves as pro-climate whilst misusing the issue of climate debt to continue their disastrous expansion of fossil-fuel-based development in extractive industries, energy, and transport at home. So, for example, whilst painting itself in a progressive light on the international stage, the Filipino government was (directly or indirectly) responsible for the murder of 28 environmental justice activists in 2016 (Global Witness 2017) and became "the world's deadliest country for environmental defenders" in 2019 (Delina 2021, p.11).

Thirdly, the post-development critique of extractivism and land grabbing, offering small-scale agriculture and fisheries as a (subsistence-oriented) alternative, largely ignores the urban majority in Southeast Asia. It leaves the question of industrial planning (i.e., how to produce everything apart from food) of cities, energy, and transport to the current regimes. Whilst there are urban-focused NGOs working on these and other issues, such as resilience against floods for slum dwellers, there remains a lot to be done to appeal to the working-class majorities in the cities. In particular, road-based transport systems in the city, coal mining, and (because of forest fires and haze) the palm oil industry have serious health impacts on millions of people across the region. More could also be done by the trade unions and labour movements of Southeast Asia, for example by developing concrete labor-led 'Just Transition' perspectives for the social-ecological transformation of extractive industries, power production, and urban mobility systems.

Conclusion: Class and climate in Southeast Asia

Climate change in Southeast Asia is not a mistake or an unfortunate side-effect of poor countries desperately struggling to develop. Instead, it is a process intentionally organized by a very, very rich minority of capitalists, military generals, and leading bureaucrats of key state institutions across Southeast Asia. These groups have no intention of phasing out their extractivist development model, switching to renewables, creating car-free cities, or supporting carbon-neutral and ecological farming systems.

Across Southeast Asia, fossil fuel companies and corporations deeply committed to a high-emission trajectory control the heights of industry. As sectors that are 'too big to fail', they exert a great deal of political power. Their influence can be traced back to path dependencies created by the way capitalism emerged as an economic system wedded to fossil fuels (Malm 2016) and how this was imposed on the region by colonialism. In a system of political-economic co-dependencies, extractivist industrial development, capital-oriented urbanization, road-based transportation systems, and fossil fuel energy production all reinforce each other (e.g., more cars need more metals, fuel and 'biofuels', and cement, which require more fossil-fuel power production, etc.).

Overall, despite huge and in part heroic efforts, the CJM in Southeast Asia has yet to significantly shift the fossil-fuel-based development strategy that dominates the region.[7] No country has seen the emergence of a political force that offers a viable programmatic alternative to the military-industrial 'carbon complexes' (Tramel 2018) currently in power. A carbon-neutral development future is possible for Southeast Asia. To achieve it, the CJM needs to dismantle the military-industrial carbon complexes currently in power. This can only happen if it expands its base and appeal to include the urban working classes in the region, who would stand to gain from green industrialization, pollution-free cities, and renewable energy. The democracy movements in Thailand and Myanmar that are currently challenging their military regimes are also confronting one of the key actors in the 'carbon complexes' of those countries. In that sense, their victory would also be one of the drivers for climate justice.

Notes

1 This also applies to Thailand, which was never formally colonized. The Bowring Treaty of 1855, imposed by military force, opened up forestry and mining sectors to European capital. In response to the threat from colonial Britain and France, the Kingdom of Siam modernized, leading ultimately to the Revolution of 1932 which established the dominant role of the military until the present day, see Chaiyan 1994.
2 Each new declaration typically lists the previous declarations too, e.g., "RECALLING our commitments made in the ASEAN Leaders' Statement on Joint Response to Climate Change (2010), the ASEAN Joint Statement on Climate Change to the 15th Session of the Conference of the Parties to the United Nations

Framework Convention on Climate Change (UNFCCC) and the 5th Session of the Conference of Parties serving as the Meeting of Parties to the Kyoto Protocol (2009); the ASEAN Declaration on the 13th session of COP to the UNFCCC and the 3rd session of the CMP to the Kyoto Protocol (2007), and the ASEAN Declaration on Environmental Sustainability (2007)." (ASEAN 2011).

3 Difference between reduction of BAU 2030 and 2005 emissions.

4 In the Philippines, despite the announcement of a temporary moratorium on some new coal plants at the end of 2020, coal-generated power capacity is expected to expand by a further 9,550 MW in the next two decades, https://mb.com.ph/2021/01/12/doe-excludes-expansion-projects-in-coal-moratorium.

5 The engineering challenges of geothermal energy are sometimes given as a reason why Indonesia or the Philippines have not developed this energy source to a greater extent. But the technology involved is much less sophisticated than nuclear power, for example, and less dangerous. Despite the huge cost and technological challenges, countries across the region have plans to develop nuclear power. Indonesia had started to build a nuclear power plant on Mount Muria – a (currently) dormant volcano (Sovacool 2010).

6 For example, the Green Island project on Palawan. This consists of a hybrid power plant combining a 20 kW biomass gasifier (using coconut husks), a 5 kW wind turbine, and a 2.5 kW solar panel array replacing diesel generators for 40 families and producing fresh water and ice (for cooling fish) into the bargain (Roxas & Santiago 2016, p.1403).

7 This applies to the CJM in general, which, in the words of two of its key protagonists, "failed to substantially reconfigure the political field around climate change [and] to do anything to significantly advance the fight for climate justice" (Bullard and Müller 2012, p.57).

References

#Bersihkan Indonesia, 2018, 'COALRUPTION: Shedding Light on Political Corruption in Indonesia's Coal Mining Sector', Jakarta, viewed at, <https://auriga.or.id/resources/reports/24/coalruption-shedding-light-on-political-corruption-in-indonesia-s-coal-mining-sector>.

Ahanchian, M. & Biona, J. B. M. 2014, 'Energy Demand, Emissions Forecasts and Mitigation Strategies Modeled Over a Medium-Range Horizon: The Case of the Land Transportation Sector in Metro Manila', *Energy Policy*, vol. 66, pp. 615–629, doi:10.1016/j.enpol.2013.11.026.

Ali, M. B., Saidur, R. & Hossain, M. S. 2011, 'A Review on Emission Analysis in Cement Industries', *Renewable and Sustainable Energy Reviews*, vol. 15, no. 5, pp. 2252–2261, doi:10.1016/j.rser.2011.02.014..

Alisjahbana, A. S. & Busch, J. M. 2017, 'Forestry, Forest Fires, and Climate Change in Indonesia', *Bulletin of Indonesian Economic Studies*, vol. 53, no. 2, pp. 111–136, doi:10.1080/00074918.2017.1365404.

Andong, R. F. & Sajor, E. 2017, 'Urban Sprawl, Public Transport, and Increasing CO2 Emissions: The Case of Metro Manila, Philippines', *Environment, Development and Sustainability*, vol. 19, no. 1, pp. 99–123, doi:10.1007/s10668-015-9729-8.

Angus, I. 2016, *Facing the Anthropocene: Fossil Capitalism and the Crisis of the Earth System*, Monthly Review Press, New York.

ASEAN 2007, 'ASEAN Declaration on Environmental Sustainability', ASEAN, Singapore.

ASEAN 2011, 'ASEAN Leaders' Statement on Climate Change to the 17th Session of the Conference of the Parties to the United Nations Framework Convention on Climate Change (UNFCCC) and the 17th Session of the Conference of Parties Serving as the Meeting of Parties to the Kyoto Protocol', ASEAN, Bali.

ASEAN 2014, 'ASEAN Joint Statement on Climate Change 2014', ASEAN, Nay Pyi Taw.

ASEAN 2017, 'ASEAN Joint Statement on Climate Change to the 23rd Session of the Conference of the Parties to the United Nations Framework Convention on Climate Change (COP23)', ASEAN, Manila.

ASEAN 2019, 'ASEAN Joint Statement on Climate Change to the 23rd Session of the Conference of the Parties to the United Nations Framework Convention on Climate Change (UNFCCC COP25)', ASEAN, Bangkok.

Aumnad, P. 2010, 'Integrated Energy and Carbon Modeling with a Decision Support System: Policy Scenarios For Low-Carbon City Development in Bangkok', *Energy Policy*, vol. 38, no. 9, pp. 4808–4817, doi:10.1016/j.enpol.2009.10.026.

Aumnad, P. & Teeradej, W. 2013, 'Analyses of the Decarbonizing of Thailand's Energy System Toward Low-Carbon Futures', *Renewable and Sustainable Energy Reviews*, vol. 24, pp. 187–197, doi:10.1016/j.rser.2013.03.050.

Bakhtyar, B., Sopian, K., Sulaiman, M.Y. & Ahmad, S. H. 2013, 'Renewable Energy in Five South East Asian Countries: Review on Electricity Consumption and Economic Growth', *Renewable and Sustainable Energy Reviews*, vol. 26, pp. 506–514, doi:10.1016/j.rser.2013.05.058.

Belcher, O., Bigger, P., Neimark, B. & Kennelly, C. 2020, 'Hidden Carbon Costs of the "Everywhere War": Logistics, Geopolitical Ecology, and the Carbon Boot-Print of the US Military', *Transactions of the Institute of British Geographers*, vol. 45, no. 1, pp. 65–80, doi:10.1111/tran.12319.

Both, A. F., Westerdahl, D., Fruin, S., Haryanto, B. & Marshall, J. D. 2013, 'Exposure to Carbon Monoxide, Fine Particle Mass, and Ultrafine Particle Number In Jakarta, Indonesia: Effect Of Commute Mode', *The Science of the Total Environment* 443, pp. 965–972, doi:10.1016/j.scitotenv.2012.10.082.

Brad, A., Schaffartzik, A., Pichler, M. & Plank, C. 2015, 'Contested Territorialization and Biophysical Expansion of Oil Palm Plantations in Indonesia', *Geoforum*, vol. 64, pp. 100–111, doi:10.1016/j.geoforum.2015.06.007.

Brown, B. & Spiegel, S. J. 2017, 'Resisting Coal: Hydrocarbon Politics And Assemblages Of Protest In The UK And Indonesia', *Geoforum* 85, pp. 101–111, doi:10.1016/j.geoforum.2017.07.015.

Brown, B. & Spiegel, S. J. 2019, 'Coal, Climate Justice, and the Cultural Politics of Energy Transition', *Global Environmental Politics*, vol. 19, no. 2, pp. 149–168, doi:10.1162/glep_a_00501.

Bullard, N. & Müller, T. 2012, Beyond the 'Green Economy': System Change, Not Climate Change?', *Development*, vol. 55, no. 1, pp. 54–62, doi:10.1057/dev.2011.100.

Camba, A. 2015, 'From Colonialism to Neoliberalism: Critical Reflections on Philippine Mining In The "Long Twentieth Century"', *The Extractive Industries and Society*, vol. 2, no. 2, pp. 287–301.

Camba, A. 2016, 'Philippine Mining Capitalism: The Changing Terrains of Struggle in the Neoliberal Mining Regime', *Austrian Journal of South-East Asian Studies*, vol. 9, no. 1, pp. 69–86, doi:10.14764/10.aseas-2016.1-5.

Chaiyan, R. 1994, *The Rise and Fall of the Thai Absolute Monarchy*, White Lotus, Bangkok.

Chatterton, P., Featherstone, D. & Routledge, P. 2013, 'Articulating Climate Justice in Copenhagen: Antagonism, the Commons, and Solidarity', *Antipode*, vol. 45, no. 3, pp. 602–620, doi:10.1111/j.1467-8330.2012.01025.x.

Cheewaphongphan, P., Junpen, A., Garivait, S. & Chatani, S. 2017, 'Emission Inventory of On-Road Transport in Bangkok Metropolitan Region (BMR) Development during 2007 to 2015 Using the GAINS Model', *Atmosphere*, vol. 8, no. 12, p.167, doi:10.3390/atmos8090167.

Chuenchom S. G. & Greacen, C. 2004, 'Thailand's Electricity Reforms: Privatization of Benefits and Socialization of Costs and Risks', *Pacific Affairs*, vol. 77, no. 3, pp. 517–541.

Claeys, P. & Delgado Pugley, D. 2017, 'Peasant and Indigenous Transnational Social Movements Engaging With Climate Justice', *Canadian Journal of Development Studies/Revue canadienne d'études du développement*, vol. 38, no. 3, pp. 325–340.

Clark, B. & Jorgenson, A. K. 2012, 'The Treadmill of Destruction and the Environmental Impacts of Militaries 1', *Sociology Compass*, vol. 6, no. 7, pp. 557–569, doi:10.1111/j.1751-9020.2012.00474.x.

Climate Analytics 2019a, 'Decarbonising South & South East Asia', viewed at, <https://climateanalytics.org/publications/2019/decarbonising-south-and-south-east-asia/>.

Climate Analytics 2019b, 'Decarbonising South & South East Asia. Country Profile – Thailand', viewed at, <https://climateanalytics.org/media/decarbonisingasia2019-profile-thailand-climateanalytics.pdf>.

Climate Analytics 2019c, 'Decarbonising South & South East Asia. Country Profile – Philippines', viewed at, <https://climateanalytics.org/media/decarbonisingasia2019-profile-philippines-climateanalytics.pdf>.

Climate Analytics 2019d, 'Decarbonising South & South East Asia. Country Profile – Indonesia', viewed at, <https://climateanalytics.org/media/decarbonisingasia2019-profile-indonesia-climateanalytics.pdf>.

Crawford, N.C. 2019, 'Pentagon Fuel Use, Climate Change, and the Costs of War', viewed at, <https://watson.brown.edu/costsofwar/files/cow/imce/papers/2019/Pentagon%20Fuel%20Use,%20Climate%20Change%20and%20the%20Costs%20of%20War%20Final.pdf/>.

Das, A. & Ahlgren, E. 2010, 'Implications of Using Clean Technologies to Power Selected ASEAN Countries', *Energy Policy*, vol. 38, pp. 851–1871, doi:10.1016/j.enpol.2009.11.062.

Darlington, S. 2014, 'Environmental Justice in Thailand in the Age of Climate Change', in B. Schuler (ed.), *Environmental and Climate Change in South and Southeast Asia*, BRILL, Leiden, pp. 211–230.

Delina, L. L. 2021, 'Topographies of Coal Mining Dissent: Power, Politics, and Protests in Southern Philippines', *World Development*, vol. 137, no. 105194.

Delina, L. L. 2020, 'Indigenous Environmental Defenders and the Legacy of Macli-Ing Dulag: Anti-Dam Dissent, Assassinations, and Protests in the Making of Philippine Energyscape', *Energy Research & Social Science*, vol. 65, no. 101463, doi:10.1016/j.erss.2020.101463.

Do, T. M. & Sharma, D. 2011, 'Vietnam's Energy Sector: A Review of Current Energy Policies and Strategies', *Energy Policy*, vol. 39, pp. 5770–5777. doi:10.1016/j.enpol.2011.08.010.

Dutu, R. 2016, 'Challenges and Policies in Indonesia's Energy Sector', *Energy Policy*, vol. 98, pp. 513–519, doi:10.1016/j.enpol.2016.09.009.

EIA 2020, 'Levelized Cost and Levelized Avoided Cost of New Generation Resources', viewed at, <https://www.eia.gov/outlooks/aeo/pdf/electricity_generation.pdf>.

Energy Market Authority 2020, 'Singapore Energy Statistics 2020', viewed at, <https://www.ema.gov.sg/Singapore_Energy_Statistics.aspx>.

Eucker, D. 2014, 'Institutional Dynamics of Climate Change Adaptation in Southeast Asia: The Role of Asean', in B. Schuler (ed.), *Environmental and Climate Change in South and Southeast Asia*, BRILL, Leiden, pp. 254–279.

Fold, N., Allison, M., Wood, B., Thao, P. T. B., Bonnet, S. & Garivait, S. *et al.* 2020, 'An Assessment of Annual Mortality Attributable to Ambient PM2.5 in Bangkok, Thailand', *International Journal of Environmental Research and Public Health*, vol. 17, no. 19, doi:10.3390/ijerph17197298.

García-Olivares, A., Solé, J. & Osychenko, O. 2018, 'Transportation in a 100% Renewable Energy System', *Energy Conversion and Management*, vol. 158, pp. 266–285, doi:10.1016/j.enconman.2017.12.053.

Global Witness 2017, 'Defenders of the Earth', viewed at, <https://www.globalwitness.org/en/campaigns/ environmental-activists/defenders-earth/>.

Government of Malaysia 2015, 'Intended Nationally Determined Contribution of Malaysia', viewed at, <https://www4.unfccc.int/sites/ndcstaging/PublishedDocuments/Malaysia%20First/INDC%20Malaysia%20Final%2027%20November%202015%20Revised%20Final%20UNFCCC.pdf/>.

Government of Singapore 2015, 'Intended Nationally Determined Contribution of Singapore', viewed at, <https://www4.unfccc.int/sites/ndcstaging/PublishedDocuments/Singapore%20First/Singapore%20INDC.pdf/>.

Government of Viet Nam 2015, 'Intended Nationally Determined Contribution of Viet Nam', viewed at, <https://www4.unfccc.int/sites/ndcstaging/PublishedDocuments/Viet%20Nam%20First/VIETNAM%27S%20INDC.pdf/>.

Gunningham, N. 2013, 'Managing the Energy Trilemma: The Case of Indonesia', *Energy Policy*, vol. 54, pp. 184–193, doi:10.1016/j.enpol.2012.11.018.

Han, S. S. 2010, 'Managing Motorization in Sustainable Transport Planning: The Singapore Experience', *Journal of Transport Geography*, vol. 18, no. 2, pp. 314–321.

Haque, M. M., Chin, H. C. & Debnath, A. K. 2013, 'Sustainable, Safe, Smart—Three Key Elements of Singapore's Evolving Transport Policies', *Transport Policy*, vol. 27, pp. 20–31, doi:10.1016/j.tranpol.2012.11.017.

Harvey, D. 2003, *The New Imperialism*, Oxford University Press, Oxford.

Hatcher, P. 2020, 'The Political Economy of Southeast Asia's Extractive Industries: Governance, Power Struggles and Development Outcomes', in T. Carroll, S. Hameiri & L. Jones (eds.), *The Political Economy of Southeast Asia*, Springer International Publishing, Cham.

International Renewable Energy Agency (IRENA) 2017, 'Renewable Energy Prospects: Indonesia', viewed at, <https://www.irena.org/publications/2017/Mar/Renewable-Energy-Prospects-Indonesia>.

International Renewable Energy Agency (IRENA) & ASEAN Centre for Energy (ACE) 2016, 'Renewable Energy Outlook for ASEAN: A REmap Analysis', viewed at, < https://www.irena.org/media/Files/IRENA/Agency/Publication/2016/IRENA_REmap_ASEAN_2016_report.ashx?la=en&hash=A2C8C02BFD443584645401A6A8931733733C0FC4>.

Khan, M. F., Latif, M. T., Saw, W. H., Amil, N., Nadzir, M. S. M. & Sahani, M. *et al.* 2016, 'Fine Particulate Matter in the Tropical Environment: Monsoonal Effects,

Source Apportionment, and Health Risk Assessment', *Atmospheric Chemistry and Physics*, vol. 16, no. 2, pp. 597–617, doi:10.5194/acp-16-597-2016.

Larsen, J. 1998, 'Crude Investment: The Case of the Yadana Pipeline in Burma', *Bulletin of Concerned Asian Scholars*, vol. 30, no. 3, pp. 3–13.

Leith, D. 2002, *The Politics of Power: Freeport in Suharto's Indonesia*, University of Hawaii Press, Honolulu.

Malm, A. 2016, *Fossil Capital. The Rise of Steam Power and the Roots of Global Warming*, Verso, London.

Manaysay, F. V. 2020, 'Norms from Above, Movements from Below: Climate Change and Global-Local Dynamics of Indigenous Resistance in the Philippines and Indonesia', *Journal of Southeast Asian Human Rights*, vol. 4, no. 1, pp. 226–252.

Martinez-Alier, J., Temper, L., Del Bene, D. & Scheidel, A. 2016, 'Is There a Global Environmental Justice Movement?', *The Journal of Peasant Studies*, vol. 43, no. 3, pp. 731–755, doi:10.1080/03066150.2016.1141198.

Mol, A. P. J., Spaargaren, G. & Sonnenfeld D. A. 2014, 'Ecological Modernization Theory: Taking Stock, Moving Forward', in S. Lockie & D. A. Sonnenfeld (eds), *Routledge International Handbook of Social and Environmental Change*, Routledge, London, pp. 15–30.

Morrow, M. 1975, 'The Politics of Southeast Asian Oil', *Bulletin of Concerned Asian Scholars*, vol. 7, no. 2), pp. 34–43, doi:10.1080/14672715.1975.10406371.

Nares, C.& Somchai, W. 2016, 'Critical Review of the Current Status of Solar Energy in Thailand', *Renewable and Sustainable Energy Reviews*, vol. 58, pp. 198–207, doi:10.1016/j.rser.2015.11.005.

Newman, P. 2010, 'Green Urbanism and its Application to Singapore', *Environment and Urbanization ASIA*, vol. 1, no. 2, pp. 149–170, doi:10.1177/097542531000100204.

Pananond, P. 2001, 'The Making of Thai Multinationals. A Comparative Study of the Growth and Internationalization Process of Thailand's Charoen Pokphand and Siam Cement Groups', *Journal of Asian Business*, vol. 17, no. 3, pp. 41–70.

Penwadee, C., Agapol, J., Garivait, S.& Chatani, S.,2017, 'Emission Inventory of On-Road Transport in Bangkok Metropolitan Region (BMR) Development during 2007 to 2015 Using the GAINS Model', *Atmosphere*, vol. 8, no. 167, pp. 1–34, doi:10.3390/atmos8090167.

Porphant, O. 2008, 'The Crown Property Bureau in Thailand and the Crisis of 1997', *Journal of Contemporary Asia*, vol. 38, no. 1, pp. 166–189, doi:10.1080/00472330701652018.

Pye, O. 2009, 'Palm Oil as a Transnational Crisis in Southeast Asia', *ASEAS - Austrian Journal of South-East Asian Studies*, vol. 2, no. 2, pp. 81–101.

Pye, O. 2018, 'Commodifying Sustainability: Development, Nature and Politics in the Palm Oil Industry', *World Development*, doi:10.1016/j.worlddev.2018.02.014.

Quirapas, M. A., Lin, H., Lochinvar, M., Abundo, S., Brahim, S., Santos, D., 2015, 'Ocean Renewable Energy in Southeast Asia: A Review', *Renewable and Sustainable Energy Reviews*, vol. 41, pp. 799–817.

Roberts, J. T. & Parks, B. C. 2009, 'Ecologically Unequal Exchange, Ecological Debt, and Climate Justice', *International Journal of Comparative Sociology*, vol. 50, no. 3–4, pp. 385–409, doi:10.1177/0020715209105147.

Roosvall, A. & Tegelberg, M. 2015, 'Media and the Geographies of Climate Justice: Indigenous Peoples, Nature and the Geopolitics of Climate Change', *tripleC*, vol. 13, no. 1, doi:10.31269/triplec.v13i1.654.

Roxas, F. & Santiago, A. 2016, 'Alternative Framework for Renewable Energy Planning in The Philippines', *Renewable and Sustainable Energy Reviews*, vol. 59, pp. 1396–1404, doi:10.1016/j.rser.2016.01.084.

Sakkarin, C. & Somchai, W. 2014, 'Critical Review of the Current Status of Wind Energy in Thailand', *Renewable and Sustainable Energy Reviews*, vol. 31, pp. 312–318.

Saraswat, C. & Kumar, P. 2016, 'Climate Justice in Lieu of Climate Change: A Sustainable Approach to Respond to the Climate Change Injustice and an Awakening of the Environmental Movement', *Energy, Ecology and Environment*, vol. 1, no. 2, pp. 67–74, doi:10.1007/s40974-015-0001-8.

Simpson, A. & Smits, M. 2019, 'Illiberalism and Energy Transitions in Myanmar and Thailand', *Georgetown Journal of Asian Affairs*, vol. 4, no. 2, pp. 45–57.

SIPRI 2020, 'Military Expenditure by Country, In Constant (2018) US$ M., 1988–2019', viewed at, <https://www.sipri.org/sites/default/files/Data%20for%20all%20countries%20from%201988%E2%80%932019%20in%20constant%20%282018%29%20USD.pdf>.

Smith, T. B. 2003, 'Privatising Electric Power in Malaysia and Thailand: Politics and Infrastructure Development Policy', *Public Administration and Development*, vol. 23, no. 3, pp. 273–283, doi:10.1002/pad.267.

Sodri, A. & Garniwa, I. 2016, 'The Effect of Urbanization on Road Energy Consumption and CO2 Emissions in Emerging Megacity of Jakarta, Indonesia', *Procedia - Social and Behavioral Sciences*, vol. 227, pp. 728–737, doi:10.1016/j.sbspro.2016.06.139.

Sovacool, B. K. 2010, 'A Critical Evaluation of Nuclear Power and Renewable Electricity in Asia', *Journal of Contemporary Asia*, vol. 40, no. 3, pp. 369–400, doi:10.1080/00472331003798350.

Sovacool, B. K. & Brown, M. 2010, 'Twelve Metropolitan Carbon Footprints: A Preliminary Comparative Global Assessment', *Energy Policy*, vol. 38, no. 9, pp. 4856–4869, doi:10.1016/j.enpol.2009.10.001.

Su, B., Ang, B. W. & Li, Y. 2017, 'Input-Output and Structural Decomposition Analysis of Singapore's Carbon Emissions', *Energy Policy*, vol. 105, pp. 484–492, doi:10.1016/j.enpol.2017.03.027.

Tramel, S. 2018, 'Convergence as Political Strategy: Social Justice Movements, Natural Resources and Climate Change', *Third World Quarterly*, vol. 39, no. 7, pp. 1290–1307, doi:10.1080/01436597.2018.1460196.

Varkkey, H. 2016, *The Haze Problem in Southeast Asia: Palm Oil and Patronage*, Routledge, London and New York.

Wismadi, A., Soemardjito, J. & Sutomo, H. 2013, 'Transport Situation in Jakarta', in I. Kunardi (ed.), *Study on Energy Efficiency Improvement in the Transport Sector Through Transport Improvement and Smart Community Development in the Urban Area* (Research Project Report, 2012–2029), Jakarta, pp. 29–58.

14 Climate change governance in Southeast Asia

Commonalities, complexities and contestations

Jens Marquardt, Laurence L. Delina and Mattijs Smits

Dealing with complexity: Multiple contexts, multiple tensions

Debates among academics and practitioners about how to govern the climate crisis have mushroomed over the last decades. The complexity of climate change governance, its characterization as a "super wicked problem" (Lazarus 2009), and the growing number of relevant actors involved have nurtured a field that is interested not only in technological advancements and supportive policies but also in the competing discourses and visions, questions of legitimacy and justice, power dynamics, and issues of contestation (Kuyper, Linnér & Schroeder 2018). Classical governance perspectives that focus on relevant actors, institutions, and policies were gradually supplemented by a wide range of other social science concepts dealing with knowledge-making (Beck & Forsyth 2015), climate justice (Okereke & Coventry 2016) and the role of power (Marquardt 2017). Critical voices from social sciences, like political science, human geography, or science and technology studies, enriched these debates with perspectives on the politics of space (Rigg 2003), the contested nature of temporality (Marquardt & Delina 2021), and the co-production of technological advancements and social order (Jasanoff 2010), among others.

In Southeast Asia, the richness of these conceptual debates meets a heterogenous reality that refuses simple textbook governance solutions to tackle climate change. As a result, the region's striking social, political, economic, and environmental diversity became this book's starting point. The authors have brought together multiple ways of dealing with the climate crisis in such a heterogenous world region, where highly developed, wealthy, and emissions-intensive countries like Singapore and Brunei Darussalam are located side by side with least developed countries, such as Myanmar, Laos, and Cambodia (UNDP 2020). International institutions like the World Bank (2018) describe Southeast Asia as one of the world's most dynamic regions while presenting it as a positive example of a growth-oriented development model that should be emulated elsewhere in the Global South. At the same

DOI: 10.4324/9780429324680-17

time, rapid economic growth, demographic changes, and steadily increasing demand for natural resources led to severe environmental degradation across the region (Rigg 2003; Hirsch 2016).

The previous chapters provide in-depth perspectives into all 11 Southeast Asian countries. While they bring in different insights, they should not be considered as comprehensive accounts since they focus only on particular varieties of climate change governance. Arguably, these diverse cases and approaches make them hard to compare; still, we can carefully draw key lessons from them. The most obvious takeaway from these chapters is to acknowledge the countries' heterogeneous contexts and recognize the situated nature of climate action. At the same time, we also recognize some general patterns, commonalities, and tensions arising from what we found as a contested field of politics. In drawing these lessons, we can summarize three significant governance challenges and reflect these contributions against concepts from various disciplines such as global governance research, sustainability transitions research, development studies, anthropology, and science and technology studies.

Climate change governance – as this collection shows – reflects a complex endeavor revolving around societal conflicts that involve not only structural elements such as policies, norms, and organizational frameworks (Bernstein & Hoffmann 2018; Jordan et al. 2018) but also questions of agency and leadership when formulating and implementing climate action (Tosun & Schoenefeld 2017). Acting upon these conflicts are not only government officials but also a variety of non-state actors. In addition, climate change politics is irreducible to national contexts. Instead, it involves relations and interactions across jurisdictional boundaries – both vertically and horizontally – from the local to the international level. These connections are crucial for spreading global norms, formulating national goals, and implementing subnational climate action (Hooghe & Marks 2001, 2003).

In Southeast Asia, climate change is an issue of growing public concern (Lee et al. 2015). Awareness of the severe effects of rapidly changing and extreme weather patterns is generally high. As this volume shows, governments have been creating institutions to address the climate crisis, and climate change policies have been formally institutionalized. Yet this volume also reveals that these processes and outcomes are fragmented at best. Actors, both internal and external to these countries, have made several attempts to strengthen climate action, but evidence about their impacts remains scant. National governments have been struggling not only to formulate but also to implement and enforce climate change policies across their jurisdictions; yet, we also saw how non-state actors and networks are fostering climate action. Initiatives for climate-resilient development (Gallagher 2018), urban climate planning (Daniere & Garschagen 2019), or coordination facilitated by the Association of Southeast Asian Nations (ASEAN) (Elliott 2012) are three typical examples.

The Paris Agreement has called upon parties for their individual and context-specific Nationally Determined Contributions (NDCs), which has become a unique opportunity to re-contextualize climate action through more bottom-up approaches, while still considering specific contexts in which these efforts are embedded. National governments around Southeast Asia have made their NDCs central reference points for their climate change commitments. These pledges reflect national ambitions to reduce greenhouse gas (GHG) emissions – or at least to limit the increase of emissions – over the next decades. For example, Southeast Asia's biggest country in terms of land area and population, Indonesia, has pledged to reduce 29 percent of its emissions compared to a business-as-usual scenario by 2030. The target would be increased to 41 percent depending on additional support from the Global North. Similarly, Singapore, Thailand, and Malaysia have proposed more ambitious conditional emissions reduction targets if they receive international climate finance, technology transfer, and capacity building. The NDCs also outline each country's plan to adapt to climate change impacts, which is particularly relevant in Southeast Asia. In 2020, the "Long-Term Climate Risk Index" (Eckstein et al. 2020, p.9) had ranked Myanmar, the Philippines, Viet Nam, and Thailand among the world's top ten countries most affected by extreme weather events, underlining the region's exceptional exposure to severe effects of a changing climate.

Southeast Asian governments offer different responses to the climate crisis. The Climate Action Tracker – a consortium evaluating NDCs worldwide – has rated Viet Nam's NDC as "critically insufficient" to limit global warming to 1.5 °C. In contrast, the Philippines' NDC is described as "2 °C compatible" (Climate Analytics & Next Climate 2021). While the Climate Action Tracker attempts to operationalize and compare climate change commitments across countries, comparing NDCs is a limiting exercise in terms of understanding these countries' climate action plans. Opening up the black box of the nation-state and revealing the deep societal struggles, political conflicts, and inequalities within these countries – as presented in the previous chapters – is more expansive analytically.

The contributions in this book vividly demonstrate that governing climate change is a highly context-specific endeavor in terms of relevant actors, political institutions, and societal environments. In her chapter on Indonesia, **Monica Di Gregorio** unpacks how the lack of recognition and awareness among sectoral ministries hampered effective climate change policy integration in the country's land use sector. Sectoral governing systems are essential to innovate, introduce, and enforce climate change policies even in countries with powerful central governments like in Brunei Darussalam, as **Romeo Pacudan** explains. The situation looks similarly dysfunctional in the Philippines despite the country's long track-record of climate change policy development and the early creation of climate-responsible institutions, as shown by **Antonio Gabriel La Viña** and **Jameela Joy M. Reyes**. Multi-level governance complexity also presents a challenge in Malaysia, where non-state actors such as private entities and universities jumped in to support state-driven climate action according to **Irina Safitri Zen** and **Zeeda Fatimah Binti Mohamad**. In addition, **Adam Simpson** and **Mattijs Smits** point at the conflictual relations between top-down climate change governance and grassroots activism in Thailand. In contrast, the chapter by **Adam Simpson** and **Ashley South** brings to light how authoritarian and military-dominated governance in Myanmar dramatically limits the room for non-state actor participation or even public debate about climate change politics. Insights from centrally organized countries like Singapore also show how national governments often prioritize growth over ambitious climate change mitigation targets as outlined by **Natasha Hamilton-Hart**. With insights from Viet Nam, **Koos Neefjes** demonstrates that climate change politics is inseparable from the country's contested environments such as access to land or water. Along these lines, **Miles Kenney-Lazar** shows in his contribution on Laos how power-laden interactions between relevant actors have shaped climate change politics in that country. Echoing a relational approach, **Tim Frewer** unpacks Cambodia's climate change assemblage as a loose web of relations among donors, government departments, non-governmental organizations, and recipients. For Timor-Leste, **Alexander Cullen** reveals how epistemological and ontological differences between the national and the local level have created conflicts. Finally, **Oliver Pye**'s regional perspective brings to light how current power structures and vested interests, particularly related to fossil fuels, prevent a sustainable transformation in Southeast Asia.

All these contributions underline the primacy of acknowledging the multilayered socio-political environments in which climate action flourishes or fails. Governing climate change in Southeast Asia should therefore not be reduced to standardized GHG emissions reduction targets or the provision of technical solutions. Instead, climate change politics are deeply intertwined within broader societal issues, such as development (Janardhanan & Mitra 2018), equality (Lee & Zusman 2018), and the socio-economic impacts of climate change on the most vulnerable people (Islam & Khan 2018).

The chapters in this book have addressed many of these issues by exploring the role of institutional structures, different actors, and their interactions. At the structural level, the chapters on Brunei Darussalam and the Philippines unpack these countries' climate change governance architectures and policy developments. The contributions on Myanmar and Viet Nam emphasize the role of key actors, such as the armed forces, or the ruling Communist Party in climate change governance. Efforts to coordinate climate change politics from the national to the local level are at the heart of the chapters on Laos and Indonesia, with the latter illustrating the challenge of policy integration in a highly decentralized political environment. These snapshots teach us about the need for contextualization, which stands in sharp contrast to the global climate change governance discourse that is too often focused on global solutions, technological fixes, and carbon emissions reduction potentials (e.g., Bäckstrand and Lövbrand 2019; Dryzek 2013).

Governing climate change: Three challenges

In addition to the broad commonalities reflected in the NDCs of Southeast Asian countries, the chapters also led us to identify at least three tensions or themes of contestation. These themes point at the fundamental conflicts emerging from the complex challenges in climate change governance in Southeast Asia, and include: i) the tensions between authoritarian and more democratic modes of governance, ii) the tensions between state-driven climate change politics and the role of non-state interventions, and iii) the tensions between state regulations and market forces. While these three dimensions resemble the classic analytical distinction between the role of the state, civil society, and the market in political decision-making, unpacking these three against Southeast Asian contexts helps us situate the different meanings attached to climate change governance in this world region. These dimensions also allow us to point at future research directions, as discussed below.

Climate change governance from above: Authoritarian versus democratic politics

The chapters in this book cover a wide range of forms of government ranging from absolute (Brunei Darussalam) and parliamentarian monarchies (Cambodia, Malaysia, Thailand) to parliamentarian (Laos, Viet Nam), semi-presidential (Singapore, Timor-Leste), and presidential republics (Indonesia, Philippines). In addition, various changes have occurred over the last decades, where military regimes regained power as in the recent case of Myanmar. Aurel Croissant and Philip Lorenz (2018) once described Southeast Asia as a "natural laboratory" for political scientists since it provides a variety of partly unstable forms of governments and regime types.

Not surprisingly, governments across Southeast Asia employ various political approaches to address climate change. While some – like Singapore and Viet Nam – promote an idea of environmental authoritarianism (Beeson 2010; Lederer et al. 2020) by announcing top-down measures with little room for critical debates or open confrontation, others – like Indonesia and Malaysia – are confronted with increased pressures to publicly justify their climate action plans and commitments in more democratic and decentralized fashion. Powerful subnational entities in those states offer examples as to how public debates on climate change can enhance inclusiveness and open up political topics for deliberation (as suggested by Dryzek and Niemeyer 2019). Yet, they also leave questions in terms of practical implementation. The Philippine government under President Rodrigo Duterte illustrates the tensions between a democratic government with a robust institutional framework for tackling climate change and an authoritarian style of leadership. There, the government's oppression of critical voices and resistance to critique have severely hampered an open and democratic debate on climate action (Curato 2017).

Countries with strong central governments and absent open opposition, such as Viet Nam and Cambodia, demonstrate how central state authorities can streamline climate discourses and action with little disruption from within their jurisdictions. Even an absolute monarchy like Brunei Darussalam is committed to stakeholder cooperation to implement its climate change policy framework. However, as the Viet Nam case shows, authoritarian settings do not necessarily lead to more legitimate climate action and effective GHG emissions reduction. The lack of state capacity to implement and enforce climate action is present in democratic and authoritarian systems alike, which is arguably most evident in Southeast Asia's youngest nation, Timor-Leste. Here, climate action is hampered by missing government capacities and the different priorities of a post-conflict state.

Governing climate change from below: The role of non-state actors

Questions of climate change governance and leadership are closely related to issues around the inclusion and participation of non-state actors. The chapters reveal that national options for climate action reflect the respective country's political constitution and the status by which fundamental human rights and freedoms of expression are upheld (Schapper & Lederer 2014). In addition to the chapters in this volume, extant research also suggests that environmental non-governmental organizations (NGOs) in Southeast Asia are mainly engaged in project implementation, but less so in policy formulation. How NGOs engage with climate change policymaking depends largely on the concerned government's openness to critical debates (Haris, Mustafa, & Raja Ariffin 2020). In this collection, the influence of various non-state actors has been equally diverse, which had led to multiple degrees of success and failure when it comes to governing climate change from below.

In Thailand, activists and grassroots movements promote renewable energy development; in the Philippines, a growing number of activists mobilize against coal power; and, in Myanmar, environmental civil society organizations fight for a climate-friendly energy transition, at least until the military coup in 2021 (see also Delina 2021; Marquardt & Delina 2019; Simpson & Smits 2018). These examples of resistance and organized protests against fossil fuel-dominated economies demonstrate the important role civil society actors play when demanding climate action from their governments. In addition, we also observe a growing vibrant climate justice movement throughout the region; yet, these activists are often confronted with powerful public and private advocates of a fossil fuel-dominated economy. Private companies play an essential role here since they control energy markets in (partly) liberalized countries like the Philippines or Singapore. At the same time, disruptive renewable energy businesses and entrepreneurs can challenge the dominance of fossil fuels and develop climate-friendly solutions, as shown in Indonesia and Malaysia. Yet, the dominance of big companies in these two countries and their preference for coal makes them critical barriers in expanding renewable energy and in substituting carbon-intensive modes of production.

Insights from the chapters also suggest that civil society is irreducible to formally registered NGOs alone; instead, they include informal networks and practices as well as groups of affected people who struggle with the everyday impacts of climate change. In Timor-Leste, the government needs to recognize rural customary systems, knowledge practices, and deep relations to nature to avoid conflicts with or the refusal of state-driven climate action. In Laos, peasant farmers deal with climate change on a daily basis; yet, climate adaptation programs are often disconnected from their needs and experiences. We can draw similar lessons from Cambodia, where the state and professional non-state actors accumulated funding for climate change capacity building, while village people remained neglected. As for other areas in Southeast Asian environmental politics, the literature suggests that both formal and informal rules, practices, and institutions are needed when dealing with climate change governance (e.g., Miller et al. 2020).

These insights speak to gaps in terms of extant knowledge about informal practices on climate change governance. To address these, we encourage sustainability transitions studies and environmental governance scholars to recognize the roles of informal political networks and institutions and to critically reflect upon knowledge claims and potential alternatives to science-driven interventions, particularly at the local level as many of our authors observed throughout Southeast Asia.

Market-driven climate change governance: Incentives for solutions

Market conditions and actors shape climate change governance in Southeast Asia. Their importance is strongly connected to the political economy of climate action and the role of market mechanisms in fostering change. Financial incentives for climate-friendly technologies, such as renewable energy, have been introduced across Southeast Asia, most notably in Thailand, Indonesia, and the Philippines (see also Erdiwansyah et al. 2019). While external actors and foreign donors often support these market-based mechanisms and financial incentives, policies rarely trigger significant decarbonization or an energy transition. Instead, policymakers and businesses would integrate climate-friendly solutions into the established fossil fuel-dominated systems. This has, among other things, to do with vested interests in control of the market, the political economy of decarbonization, and the colonial legacy of many Southeast Asian countries (as discussed in the regional chapter by Oliver Pye). These historical path dependencies matter when it comes to climate change governance (Abayao 2020; Vaddhanaphuti 2020). For example, the Philippine energy sector is liberalized, but still dominated by a few influential business families who are also well-connected to the country's political elites. Incumbent networks continue to prefer and support large-scale coal power facilities over decentralized and locally owned renewable energy sources. Less liberalized economies, such as Brunei Darussalam, also struggle to balance the need to diversify their economies with climate action.

Coordination between governments and private actors is a particular challenge for climate change policy implementation, as shown in the Malaysia and Singapore chapters. Both countries prioritized efficiency measures and fostered a pro-economic growth agenda to avoid the trade-offs between climate change mitigation and economic growth. Not surprisingly, international organizations would often prioritize market-driven mechanisms to drive climate action. The Cambodian chapter suggests that the climate change governance assemblage in that country is a product of Western forms of finance and managerialism rather than a response to the Cambodian context. In Laos, donor-driven climate action is also considered apolitical, technical, and managerial, which led to the neglect of the socio-economic contexts that constitute climate change politics. The importance of market- and competition-oriented rationales is also reflected in the conditional commitments outlined in the NDCs across Southeast Asia. To avoid economic disadvantages, countries such as Indonesia and Thailand demanded financial support and access to new technologies, allowing for them to formulate more ambitious climate commitments.

Outlook and cautionary trends

Addressing climate change in Southeast Asia represents a challenging and highly context-specific endeavor. The examples offered in this book provide some in-depth perspectives into these specificities, but as discussed, these examples can hardly be comprehensive. Nonetheless, they point at the pressing tasks on expanding climate action, which relates to numerous challenges, including policy implementation, political stability, and climate justice.

All Southeast Asian countries have adopted the Paris Agreement and developed their NDCs to outline their future climate adaptation and mitigation efforts. Even more so, many countries have long formulated solid and comprehensive climate change legislation, most notably in the Philippines. However, the biggest challenge remaining is the effective implementation of these policies and the institutionalization of climate action beyond the framework provided by the international climate change regime. Challenges arising from complex, multi-level governance arrangements and the contestations between national and subnational authorities are most evident in decentralized countries like Indonesia. Still, also countries with central governments, like Laos, often lack the institutional capacity to implement, monitor, and evaluate their own climate plans and policies. As observed in Myanmar, Thailand, and the Philippines, these institutional challenges are multiplied by forms of political disruption by populist leaders, or even military interventions.

Southeast Asia has seen dramatic political shifts over the last decades, reflecting the region's history of political instability. Many of these changes have been violent. Yet, there are also signs of hope with peaceful and democratic transitions of power, as experienced in Indonesia, Malaysia, and, at one time, in the Philippines. Political stabilization processes are essential in developing legitimate and plannable climate action. At the same time, consolidating democracies with vibrant civil societies stand side by side with more authoritarian yet seemingly stable governments, as can be observed in Singapore, Brunei Darussalam, Laos, and Viet Nam. Ironically, political stability can also hamper change towards more progressive climate action. The strong connections between political elites and established fossil fuel-dominated industries can be problematic for inclusive and just transitions to less carbon-intensive economies. Mobilizing for climate justice, thus, will be vital in generating acceptable climate action in a region with growing inequalities and disadvantaged groups disproportionally vulnerable to climate change. In addition, transnational climate initiatives (Kuriyama & Tamura 2018) and regional climate change governance fostered by ASEAN could also facilitate a regional response to climate change (Elliott 2012).

The challenges outlined above leave us with three cautionary notes: First, we observe the trend towards reducing climate change governance to purely technical processes and apolitical arguments, as evidenced by the preponderance of measurable commitments and GHG emissions reduction targets in climate change policymaking, overshadowing the socio-political conflicts they entail. This understanding runs the risk of neglecting the social and political dimensions of climate change governance and thereby avoids discussions about the winners and losers of any envisioned decarbonization pathway. While a democratic setting might not provide more effective climate change policies in the short run, it does offer room for public debates and participation that could increase legitimacy and, at the same time, the pressure on powerful elites and vested interests. Second, there is a clear trend towards stakeholder engagement, yet these modes of participation tend to be symbolic and grant limited power to non-state actors. Informal networks can serve as forums for political debates where narratives are allowed to form around climate action, but this should not be construed to mean the state not leading in its responsibility. Third, we observe a trend towards the use of market-based solutions to tackle climate change, including carbon pricing and financial incentive structures like feed-in tariffs for renewables. These attempts at one-size-fits-all solutions neglect historical path dependencies and established market structures that are upheld by vested interests in favor of fossil fuel economies. While market solutions are important, stronger regulations and public funding towards climate action should also contribute to climate change governance.

This volume provides fresh insights from Southeast Asia relevant not only for scholars and policymakers interested in the region but also the international climate change governance community. Current debates on environmental governance or sustainability transitions often embody a Western bias regarding what institutions, policies, and discourses ought to define these transformations. Alternative solutions from non-Western countries, like the ones presented in this volume, can also be helpful in scrutinizing these established assumptions. As shown in the preceding chapters, these examples are also helpful in terms of how climate action can be implemented and how more context-specific theoretical ideas can be developed from these experiences. The range of climate action presented here is instructive: from resilience and adaptation plans for improved livelihoods, to mitigation plans that prioritize efficiency gains, to measures that are consistent with a pro-economic growth agenda. They provide mirrors on the various ways climate action is done across Southeast Asia.

With its growing economic power, political influence, and contribution to global GHG emissions, Southeast Asia inevitably becomes a more and more important region in international climate change governance. A better understanding of the domestic contexts and circumstances will help us identify the challenges and opportunities for climate action in this region, while providing us ways to critically reflect upon how the globalized climate change narrative pans out regionally, nationally, and subnationally. To foster these debates, we hope that more scholars, especially from Southeast Asia, will shape and advance our understanding of climate change governance in this vibrant part of the world.

References

Abayao, L. 2020, 'Disaster Risk Governance in Northern Philippine Communities', in K.-T. Chou, K. Hasegawa, D. Ku, & S.-F. Kao (eds), *Climate Change Governance in Asia*, Routledge, London & New York.

Bäckstrand, K. & Lövbrand, E. 2019, 'The Road to Paris: Contending Climate Change Governance Discourses in the Post-Copenhagen Era', *Journal of Environmental Policy & Planning*, vol. 21, no. 5, pp. 519–532.

Beck, S. & Forsyth, T. 2015, 'Co-Production and Democratizing Global Environmental Expertise: The IPCC and Adaptation to Climate Change', in S. Hilgartner, C. Miller, & R. Hagendijk (eds), *Science and Democracy: Making Knowledge and Making Power in the Biosciences and Beyond*, Routledge, London, pp. 113–132.

Beeson, M. 2010, 'The Coming of Environmental Authoritarianism', *Environmental Politics*, vol. 19, no. 2, pp. 276–294.

Bernstein, S. & Hoffmann, M. 2018, 'The Politics of Decarbonization and the Catalytic Impact of Subnational Climate Experiments', *Policy Sciences*, vol. 51, no. 2, pp. 189–211.

Climate Analytics & Next Climate 2021, 'Countries', *Climate Action Tracker*, viewed at, <https://climateactiontracker.org/countries>.

Croissant, A. & Lorenz, P. 2018, 'Government and Political Regimes in Southeast Asia: An Introduction', *Comparative Politics of Southeast Asia*, pp. 1–14.

Curato, N. 2017, 'Politics of Anxiety, Politics of Hope: Penal Populism and Duterte's Rise to Power', *Journal of Current Southeast Asian Affairs*, vol. 35, no. 3, pp. 91–109.

Daniere, A. G. & Garschagen, M. (eds) 2019, *Urban Climate Resilience in Southeast Asia*, Springer, Wiesbaden.

Delina, L. L. 2021, 'Topographies of Coal Mining Dissent: Power, Politics, and Protests in Southern Philippines', *World Development*, vol. 137, p. 105194.

Dryzek, J. S. 2013, *The Politics of the Earth. Environmental Discourses*, Oxford University Press, Oxford.

Dryzek, J. S. & Niemeyer, S. 2019, 'Deliberative Democracy and Climate Change Governance', *Nature Human Behaviour*, vol. 3, no. 5, pp. 411–413.

Eckstein, D., Künzel, V., Schäfer, L., & Winges, M. 2020, *Global Climate Risk Index 2020: Who Suffers Most from Extreme Weather Events?*, Germanwatch, Bonn.

Elliott, L. 2012, 'ASEAN and Environmental Governance: Strategies of Regionalism in Southeast Asia', *Global Environmental Politics*, vol. 12, no. 3, pp. 38–57.

Erdiwansyah, Mamat, R., Sani, M. S. M. & Sudhakar, K. 2019, 'Renewable Energy in Southeast Asia: Policies and Recommendations', *Science of The Total Environment*, vol. 670, pp. 1095–1102.

Gallagher, E. 2018, *Building Climate Resilience in Southeast Asia: A Framework for Private Sector Action*, BSR, New York.

Haris, S. M., Mustafa, F. B., & Raja Ariffin, R. N. 2020, 'Systematic Literature Review of Climate Change Governance Activities of Environmental Nongovernmental Organizations in Southeast Asia', *Environmental Management*, vol. 66, no. 5, pp. 816–825.

Hirsch, P. (ed.) 2016, *Routledge Handbook of the Environment in Southeast Asia*, Routledge, London.

Hooghe, L. & Marks, G. 2001, 'Types of Multi-Level Governance', *European Integration Online Papers*, viewed 16 August 2016 at, <http://eiop.or.at/eiop/texte/2001-011a.htm>.

Hooghe, L. & Marks, G. 2003, 'Unraveling the Central State, but How? Types of Multi-Level Governance', *American Political Science Review*, vol. 97, no. 02, pp. 233–243.

Islam, M. R. & Khan, N. A. 2018, 'Threats, Vulnerability, Resilience and Displacement Among the Climate Change and Natural Disaster-Affected People in South-East Asia: An Overview', *Journal of the Asia Pacific Economy*, vol. 23, no. 2, pp. 297–323.

Janardhanan, N. & Mitra, B. K. 2018, 'Developing Asia's Response to Climate Change', in S. Bhattacharyya (ed.), *Routledge Handbook of Energy in Asia*, Routledge, London & New York, pp. 296–310.

Jasanoff, S. 2010, 'A New Climate for Society', *Theory, Culture and Society*, vol. 27, no. 2, pp. 233–253.

Jordan, A., Huitema, D., Schoenefeld, J., van Asselt, H., & Forster, J. 2018, 'Governing Climate Change Polycentrically: Setting the Scene', in A. Jordan, D. Huitema, H. van Asselt & J. Forster, *Governing Climate Change: Polycentricity in Action?*, Cambridge University Press, Cambridge, pp. 3–26.

Kuriyama, A. & Tamura, K. 2018, 'Importance of Regional Climate Policy Instruments Towards the Decarbonisation of Electricity System in the Great Mekong Sub-region', in S. Bhattacharyya (ed.), *Routledge Handbook of Energy in Asia*, Routledge, London & New York.

Kuyper, J. W., Linnér, B. O., & Schroeder, H. 2018, 'Non-State Actors in Hybrid Global Climate Change Governance: Justice, Legitimacy, and Effectiveness in a Post-Paris Era', *Wiley Interdisciplinary Reviews: Climate Change*, vol. 9, no. 1, pp. 1–18.

Lazarus, R. J. 2009, 'Super Wicked Problems and Climate Change: Restraining the Present to Liberate the Future', *Cornell Law Review*, vol. 94, no. 5, pp. 1153–1233.

Lederer, M., Wallbott, L., Urban, F., Siciliano, G., & Dang Nguyen, A. 2020, 'Implementing Climate Change Top-Down: Climate, Energy and Industrial Politics in Vietnam', *Sociology*, vol. 8, no. 2, pp. 12–29.

Lee, S.-Y. & Zusman, E. 2018, 'Participatory Climate Change Governance in Southeast Asia', in T. Jafry, M. Mikulewicz, & K. Helwig (eds), *Routledge Handbook of Climate Justice*, Routledge, London & New York.

Lee, T. M., Markowitz, E. M., Howe, P. D., Ko, C. Y., & Leiserowitz, A. A. 2015, 'Predictors of Public Climate Change Awareness and Risk Perception Around the World', *Nature Climate Change*, vol. 5, no. 11, pp. 1014–1020.

Marquardt, J. 2017, 'Conceptualizing Power in Multi-level Climate Change Governance', *Journal of Cleaner Production*, vol. 154, pp. 167–175.

Marquardt, J. & Delina, L. L. 2019, 'Reimagining Energy Futures: Contributions from Community Sustainable Energy Transitions in Thailand and the Philippines', *Energy Research and Social Science*, vol. 49, October 2018, pp. 91–102.

Marquardt, J. & Delina, L. L. 2021, 'Making Time, Making Politics: Problematizing Temporality in Energy and Climate Studies', *Energy Research and Social Science*, vol. 76, March, p.102073.

Miller, M.A., Middleton, C., Rigg, J., & Taylor, D. 2020, 'Hybrid Governance of Transboundary Commons: Insights from Southeast Asia', *Annals of the American Association of Geographers*, vol. 110, no. 1, pp. 297–313.

Okereke, C. & Coventry, P. 2016, 'Climate Justice and the International Regime: Before, During, and After Paris', *Wiley Interdisciplinary Reviews: Climate Change*, vol. 7, no. 6, pp. 834–851.

Rigg, J. 2003, *The Human Landscape of Modernization and Development*, Routledge, London.

Schapper, A. & Lederer, M. 2014, 'Introduction: Human Rights and Climate Change: Mapping Institutional Inter-Linkages', *Cambridge Review of International Affairs*, vol. 27, no. 4, pp. 666–679.

Simpson, A. & Smits, M. 2018, 'Transitions to Energy and Climate Security in Southeast Asia? Civil Society Encounters with Illiberalism in Thailand and Myanmar', *Society and Natural Resources*, vol. 31, no. 5, pp. 580–598.

Tosun, J. & Schoenefeld, J. J. 2017, 'Collective Climate Action and Networked Climate Change Governance', *Wiley Interdisciplinary Reviews: Climate Change*, vol. 8, no. 1.

UNDP 2020, *Human Development Report 2020. The Next Frontier: Human Development and the Anthropocene*, United Nations Development Programme, New York.

Vaddhanaphuti, C. 2020, 'Governing Climate Knowledge', in K.-T. Chou, K. Hasegawa, D. Ku, & S.-F. Kao (eds), *Climate Change Governance in Asia*, Routledge, London & New York.

World Bank 2018, *Riding the Wave: An East Asian Miracle for the 21st Century*, Washington D.C.

Index

Printed in the United States
by Baker & Taylor Publisher Services